Lecture Notes in Physics

Springer-Verlag Berlin Heidelberg GmbH

The Editorial Policy for Proceedings

The series Lecture Notes in Physics reports new developments in physical research and teaching – quickly, informally, and at a high level. The proceedings to be considered for publication in this series should be limited to only a few areas of research, and these should be closely related to each other. The contributions should be of a high standard and should avoid lengthy redraftings of papers already published or about to be published elsewhere. As a whole, the proceedings should aim for a balanced presentation of the theme of the conference including a description of the techniques used and enough motivation for a broad readership. It should not be assumed that the published proceedings must reflect the conference in its entirety. (A listing or abstracts of papers presented at the meeting but not included in the proceedings could be added as an appendix.)
When applying for publication in the series Lecture Notes in Physics the volume's editor(s) should submit sufficient material to enable the series editors and their referees to make a fairly accurate evaluation (e.g. a complete list of speakers and titles of papers to be presented and abstracts). If, based on this information, the proceedings are (tentatively) accepted, the volume's editor(s), whose name(s) will appear on the title pages, should select the papers suitable for publication and have them refereed (as for a journal) when appropriate. As a rule discussions will not be accepted. The series editors and Springer-Verlag will normally not interfere with the detailed editing except in fairly obvious cases or on technical matters.
Final acceptance is expressed by the series editor in charge, in consultation with Springer-Verlag only after receiving the complete manuscript. It might help to send a copy of the authors' manuscripts in advance to the editor in charge to discuss possible revisions with him. As a general rule, the series editor will confirm his tentative acceptance if the final manuscript corresponds to the original concept discussed, if the quality of the contribution meets the requirements of the series, and if the final size of the manuscript does not greatly exceed the number of pages originally agreed upon. The manuscript should be forwarded to Springer-Verlag shortly after the meeting. In cases of extreme delay (more than six months after the conference) the series editors will check once more the timeliness of the papers. Therefore, the volume's editor(s) should establish strict deadlines, or collect the articles during the conference and have them revised on the spot. If a delay is unavoidable, one should encourage the authors to update their contributions if appropriate. The editors of proceedings are strongly advised to inform contributors about these points at an early stage.
The final manuscript should contain a table of contents and an informative introduction accessible also to readers not particularly familiar with the topic of the conference. The contributions should be in English. The volume's editor(s) should check the contributions for the correct use of language. At Springer-Verlag only the prefaces will be checked by a copy-editor for language and style. Grave linguistic or technical shortcomings may lead to the rejection of contributions by the series editors. A conference report should not exceed a total of 500 pages. Keeping the size within this bound should be achieved by a stricter selection of articles and not by imposing an upper limit to the length of the individual papers. Editors receive jointly 30 complimentary copies of their book. They are entitled to purchase further copies of their book at a reduced rate. As a rule no reprints of individual contributions can be supplied. No royalty is paid on Lecture Notes in Physics volumes. Commitment to publish is made by letter of interest rather than by signing a formal contract. Springer-Verlag secures the copyright for each volume.

The Production Process

The books are hardbound, and the publisher will select quality paper appropriate to the needs of the author(s). Publication time is about ten weeks. More than twenty years of experience guarantee authors the best possible service. To reach the goal of rapid publication at a low price the technique of photographic reproduction from a camera-ready manuscript was chosen. This process shifts the main responsibility for the technical quality considerably from the publisher to the authors. We therefore urge all authors and editors of proceedings to observe very carefully the essentials for the preparation of camera-ready manuscripts, which we will supply on request. This applies especially to the quality of figures and halftones submitted for publication. In addition, it might be useful to look at some of the volumes already published. As a special service, we offer free of charge LaTeX and TeX macro packages to format the text according to Springer-Verlag's quality requirements. We strongly recommend that you make use of this offer, since the result will be a book of considerably improved technical quality. To avoid mistakes and time-consuming correspondence during the production period the conference editors should request special instructions from the publisher well before the beginning of the conference. Manuscripts not meeting the technical standard of the series will have to be returned for improvement.

For further information please contact Springer-Verlag, Physics Editorial Department II, Tiergartenstrasse 17, D-69121 Heidelberg, Germany

Michael F. Shlesinger   George M. Zaslavsky
Uriel Frisch (Eds.)

Lévy Flights and Related Topics in Physics

Proceedings of the International Workshop
Held at Nice, France, 27–30 June 1994

 Springer

Editors

Michael F. Shlesinger
Physics Division,
Office of Naval Research
Arlington, VA 22217-5660, USA

George M. Zaslavsky
Courant Institute of Mathematical Sciences
Magneto-Fluid Dynamics Division
251 Mercer Street
New York, NY 10012, USA

Uriel Frisch
Observatoire de Nice, BP 229
F-06304 Nice Cedex 4, France

ISBN 978-3-662-14048-2

Library of Congress Cataloging-in-Publication Data
Lévy Flights and related topics in physics: proceedings of the international workshop
held at Nice, France, 27–30 June 1994 / M. F. Shlesinger, G. M. Zaslavsky, U. Frisch
(eds.). p.cm. – (Lecture notes in physics; 450)
  ISBN 978-3-662-14048-2   ISBN 978-3-540-49225-2 (eBook)
  DOI 10.1007/978-3-540-49225-2

1. Statistical mechanics–Congresses. 2. Probabilities–Congresses. 3. Fractals–Congresses. 4. Lévy, Paul, 1886- . I. Shlesinger, Michael F.   II. Zaslavskiĭ, G. M. (Georgiĭ
Moiseevich) III. Frisch, U. (Uriel), 1940- . IV. Series QC174.7L48 1995 530.1'59282–
dc20 95-17873

Typesetting: Camera-ready by the editors
SPIN: 10481127   55/3142-543210 - Printed on acid-free paper

Preface

Newtonian physics began with the attempt to make precise predictions about natural phenomena which could be accurately checked through observation and experiment. The goal was to understand nature as a deterministic "clockwork" universe. The application of probability distributions to physics developed much more slowly. The Gaussian distribution has been used widely since the early 1800s to express errors in measurement. It has a well-defined mean and variance. The mean traditionally represents the most probable value for the actual size, and the variance is connected to the error bars of the measurement. For, say, the Maxwell-Boltzmann distribution or the Planck distribution, the whole distribution has physical meaning so being away from the mean is not an error. In fact, the whole distribution is a prediction. This is a major advance in thought, but it was developed for distributions which are well characterized by their first two moments.

Other distributions have appeared in physics where the mean and variance do not well represent the process. For example, all the moments of both the lognormal distribution and the stretched exponential distribution are finite, but

$$\lim_{n \to \infty} \left[\langle x^n \rangle / n! \right]^{1/n} = \infty$$

where $\langle x^n \rangle$ is the nth moment of the distribution. This implies a lot of weight is in the tail of the distribution where rare, but extreme, events occur.

One can now take this idea to its logical limit of a distribution where all of the moments, starting with the mean, are infinite. In the study of critical phenomena one actually does encounter distributions with infinite moments. For example, at the critical point one finds clusters of all sizes, and the mean of the distribution of cluster sizes diverges. Concurrent with the development of critical phenomena, the fields of fractals and chaos emerged. Both of these new fields were characterized by objects with features on all scales with no characteristic scale existing. Analysis shifted away from the earlier intuition of moments to newer notions involving calculations of exponents (e.g. Lyapunov, fractal, spectral) , new techniques (e.g. renormalization groups, multifractals, embedology) and topics such as strange attractors and strange kinetics.

There were (as always) predecessors of the modern work on chaos and fractals. First, in the early 1850s Cauchy introduced his distribution

$$p(x) = \frac{1}{\pi} \left(\frac{1}{1 + x^2} \right)$$

which has by symmetry a zero first moment, and whose second moment is infinite. In the context of atomic lineshapes this is also called a Lorentzian. Its scale invariant properties are usually ignored and practically one characterizes this distribution by the value of its full width at half maximum, as if it were a Gaussian and one just needed its variance to determine the distribution. The first in-depth grappling with probability distribution with infinite moments is due to Paul Lévy. His work, and the work it inspired, is the topic of this book.

From the physicist's view point, Lévy was concerned with a random walk whose probability distribution for each jump has infinite moments. He wanted to investigate under what circumstances the probability distribution for the position of the random walker after N steps looked the same as after one step, except for scale factors. This is the paradigm of fractals of when does the part (the distribution for a single step) look like the whole (the distribution for N steps). His answers, the Lévy distributions, are scale invariant and turn out to be most simply expressed in Fourier space. For the simplest case of a symmetric one dimensional random walk, Lévy found in Fourier space the following form for the probability

$$f(k) \equiv \int_{-\infty}^{\infty} \exp(ikx)p(x)dx = \exp(-D|k|^{\beta}).$$

The values of β lie between zero and two, with $\beta = 1$ being the Cauchy distribution and with $\beta = 2$ being the Gaussian and the only situation where the second moment is finite. Lévy's infinite moment random walk trajectories are now called Lévy flights. Today, Lévy flights have been expanded into areas such as turbulent diffusion, polymer transport and Hamiltonian chaos. Lévy's ideas and algebra of random variables with infinite moments seemed out of place in the 1920s and 1930s when they were developed. It is only now in the 1990s that the greatness of Lévy's work is more fully appreciated as a foundation for probabilistic aspects of fractals and chaos along with his role as the early pioneer in the mathematics of scale-invariant processes.

In the 1980s, systems with chaotic dynamics became a vast area for the application of Lévy's ideas on processes with fractal and multifractal self-similarity. The phenomenon of dynamical chaos promises to become a real laboratory for developing generalizations of the Lévy process and building new tools to study scaling properties of nonlinear dynamics and kinetics. Those who are interested in the foundations of statistical mechanics can find inspiration in analyzing the role of dynamical chaos in statistical physics. On this path one will unavoidably encounter Lévy-type processes and their influence on long time statistical asymptotics, which may bring a revision of the Poincaré recurrences distribution and Maxwell's Demon design.

This is the first book for physical scientists devoted to Lévy processes. It is mostly based on a series of lectures given at the Observatoire de Nice at a workshop sponsored by the US Office of Naval Research from June 28 to July 1, 1994. Lévy himself, in fact, enjoyed support from ONR in the 1950s. It is the hope of the Editors that these essays will make Lévy's work known to a wider audience and inspire further developments. One student who took

Lévy's courses and put them to good use was Benoit Mandelbrot as a component for founding the field of fractals.

April 6, 1995

Arlington, VA, USA   Michael F. Shlesinger

New York, NY, USA   George M. Zaslavsky

Nice, France   Uriel Frish

Paul Lévy *(1886 - 1971)*

(Reproduced with the help of Studio Harcourt, Paris)

THE PAUL LÉVY I KNEW

Benoit B. Mandelbrot[1,2]
[1]Mathematics Department, Yale University
New Haven, CT 06520-8283
[2]IBM T.J. Watson Research Center, Yorktown Heights, NY 10598-0218

Paul Lévy (1886-1971) stated on many occasions that he never had a student, and I had stated that I never had a teacher. Yet, in a real though indirect fashion, Paul Lévy was, after all, the teacher of several members of his family, and also mine. Besides, it may well be that his name will survive most conspicuously and least ambiguously through his work's influence on fractal geometry. Indeed, the best known of my fractional Brownian reliefs generalize Lévy's own generalization of the Brownian random process to fields. And no one here has to be reminded of the Lévy stable processes, which I used in science and named "Lévy flights". It is a delight to see that this terminology has taken root, and that interest in Lévy flights has grown and warranted this workshop, leading to many new variants on that theme.

What a contrast with the period around 1960! Then, Lévy stability was viewed as a very special and uninteresting example deserving one page in textbooks. One exception was the book on "Limit Distributions" by B. V. Gnedenko and A. N. Kolmogorov. The English translation of 1954 expresses on p. 11 the hope that Lévy stable limits "will also receive diverse applications in time,... in, say, the field of statistical physics." But no actual application was given or referred to, and the preceding words may simply reflect the Russian equivalent of the style of many O.N.R. grant proposals!

Once again, a few of Lévy's creations bear his name and that sub-branch of analysis one may call "mainstream probability theory" bears his mark all over, but seldom bears his name.

I seem to be the youngest survivor of those who knew this giant well, without belonging to his family (though it was a close call). Therefore, I was asked to reminisce about him as a person. I shall try not dwell on what is in my book "The Fractal Geometry of Nature", or what Lévy himself wrote in his autobiography, "Quelques aspects de la pensée d'un mathématicien." To me, his story is interesting precisely because so many people find it surprising. History as theater wants to simplify and it often seems that those who are remembered played one of two roles:

winners who lived in the palace or losers who lived in the garret. But life is far more complicated than that.

Chance makes me recall very well the day when I heard who Lévy was. In March 1945, after Paris was freed, my uncle resumed his chair of mathematics at the Collège de France. (The precise day is on the record somewhere). Later, those who heard his first lecture proceeded to the courtyard, mostly to tell each other who had or had not survived the whirlwind. I was the only young person present, and I can see in my mind's eye my uncle introducing me to the few others and commenting on my scandalous behavior: I had entered the École Normale but left after two days and was about to enter École Polytechnique. He could not understand why anyone would look for mathematics different from his and that of Bourbaki. One of those present was Michel Loève, who eventually moved to UC Berkeley. He tried to reassure everyone that Polytechnique was fine: I would study under Paul Lévy, who was a great man and a major figure in the exciting field of probability theory. No one could anticipate on that day the effect this brief encounter was to have on the rest of my life.

Loève deserves to be remembered for making Lévy's publications known worldwide, and (an even harder task) for making the Paris mathematics community look down at Lévy less harshly. In 1945, probability theory was a mainstream topic in Russia, but not in France. Lévy was already a tenured professor at Polytechnique in the twenties, when he moved to the study of probability, but he soon became regarded as a borderline mathematician. A few outright mistakes (big or small) made it to print (there was no refereeing) and were neither forgotten nor forgiven. Even the École Polytechnique eventually removed probability from his course of analysis and gave it to the naval engineer teaching applied mathematics. Lévy sought repeatedly to replace Polytechnique by Poincaré's old professorship in probability theory at the Sorbonne, but each time he failed. Remaining remarkably active until an "abnormally" late age, he was eager to find an audience for his newest results. But this audience could only be found at the Sorbonne, and non-credit research lectures were controlled by the man who had won that coveted professorship (and whose name few would recognize today). His approval had to be renegotiated each year, until Lévy had enough and gave up.

Let me retell a few other significant small indignities. When Paul Lévy was nearing age 70, I tried to organize a celebration for him and Fréchet. Maurice Fréchet, was born in 1878; he was not highly respected as a probabilist, but had been once upon a time an Establishment figure.

This seemed to be a good time to put an end to his many past clashes with Lévy. But the project degenerated into an unfocussed and toothless meeting. When Lévy died in 1971, I tried to convince the "Institute of Mathematical Statistics" to hold some kind of memorial, but few people cared. When a memorial meeting was held at Polytechnique sometime in 1972-3 few came. However, the centennial in 1986 told a different story. By then, Lévy's mistakes and idiosyncrasies were forgotten and forgiven, and a large meeting was organized by "pure" mathematicians. I was invited, late in the process, and discreetly informed that opposition to my attending had been very strong, and advised to avoid even seeing the loudest opponents. I wondered if Lévy would have been invited, or would have felt comfortable.

Back to his mini-courses. I attended several, and they marked my whole life. A thin, grey, distinguished and arcetic-looking man, he was not a charismatic lecturer. In some ways, he looked frail and withdrawn. The auditors were few and I recall (wrongly, I hope) having often been alone. Of course, everything was small then. In France during that period, the requirement in intermediate analysis was one big exam. At the University of Paris (by far the largest in France, and one of the largest in Europe), fewer than 20 candidates passed in 1947 (J.P. Kahane and I were in that group). Nevertheless, compared to other graduate courses, Lévy's lacked a constituency of auditors who felt professional or political pressure to attend.

Being imbued with Lévy's style since Polytechnique, I was among the few who read him easily, but these lectures were something else. They made us experience his thinking process in absolutely direct way, not only to understand his results but also his motivations and the errors that had to be overcome before reaching the truth.

I also watched Lévy closely at the weekly seminar on probability. One speaker began by describing a problem on the blackboard, then faced Lévy squarely and invited him to guess the answer. The guess was correct. This episode reminds me of something in the 1965 book, "Diffusion Processes and their Sample Paths", by K. Ito and H. McKean. This volume is "dedicated to P. Lévy, whose work has been our spur and admiration," and page 44 contains this comment: "The difficult point of this proof is the jump between 8a) and 8b); although the meaning is clear, the complete justification escapes us." Then the book proceeds through a long sequence needed to avoid the jump in question.

In those years, my life projects were already firmly focussed on science, for which Lévy had little interest. I did not want to become anything like his clone or shadow. Besides, once again, he was the least flashy person on earth. If so, how to explain the profound influence his work and manner had on me and many other scientists, in this room and elsewhere? Herein lies a familiar and always surprising story concerning the very nature of probability theory. One half of the story is part of the mystery Wigner called the "unreasonable effectiveness of mathematics in the sciences." A symmetric mystery, which should never be forgotten, deserves to be called the "unreasonable effectiveness of the sciences in mathematics." The only explanation I know for these mysteries acknowledges that human thinking is unified within itself (and even with feeling), not in a trendy "New Age" fashion, but very fundamentally. It is not true, contrary to the opinion of George Cantor, that the "essence of mathematics resides in their freedom." The best mathematics does "not" pick problems from thin air for the pleasure of solving them. To the contrary, the mark of "good taste" in any field, that is, of greatness, is the ability of identifying the most interesting problem in the framework of what is already known. And the highest level of the label "interesting" just does not come accompanied by a restrictive label, such as "in mathematics" or "in physics." My admiration for Lévy's "mathematical taste" increases each time his mark is revealed on yet another tool I need when tackling a problem in science that he could not conceivably have had on his mind.

Once again, Lévy's taste was not shared by his contemporaries in the Paris I knew in 1945-1958. Many saw the study of probability as permanently different from mathematics and viewed Lévy as hopelessly abstract. Other persons worked hard at creating a probability theory completely acceptable to mathematicians and were ashamed of Lévy. The latter group took over, but what did they win? When asked to define "probability theory," their answer is "a part of measure theory" or "a part of analysis." But such words imply that the field no longer exists as a clearly distinct subject. This may in fact be the case. Probability theory is like other big and important subjects, for example, "science" or "the theory of complex functions of a complex variable." The underlying motivation is very clear, but a picky definition proves elusive, hence is not worth pursuing. I am sure Lévy would not have given a durable definition, but, once again, a great definition for today is less important than great work, like Lévy's, that promises to never wear out.

Contents

Part 3: Lévy Flights in Dynamical Systems

Part 4: Lévy Flights and Statistical Mechanics

PART 1:

LÉVY FLIGHTS IN FLUIDS

VARIABILITY OF ANOMALOUS TRANSPORT EXPONENTS VERSUS DIFFERENT PHYSICAL SITUATIONS IN GEOPHYSICAL AND LABORATORY TURBULENCE

A.Tsinober

Department of Fluid Mechanics and Heat Transfer,Faculty of Engineering,
Tel Aviv University, Ramat Aviv 69978 Tel Aviv, Israel

Abstract.

Geophysical turbulent flows are characterized by large Reynolds numbers. Therefore, it has been a common expectation that universal relations (such as energy spectrum $E(k) \sim k^{-5/3}$, passive scalar spectrum $E_c(k) \sim k^{-5/3}$, diffusivity $\mathcal{K} \sim \ell^{\,4/3}$) should be valid in such flows as well as their "two-dimensional" analogs in quasi-two-dimensional situations.

We present an overview of results of observations in the atmosphere, ocean and laboratory (including those used by Richardson in his famous paper in 1926) which can be interpreted in terms of anomalous diffusion of passive scalar in turbulent flows, i.e. *not obeying the above universal relations.*

One of the natural candidates among the possible reasons for the deviations from the Richardson law is the phenomenon of spontaneous breaking of statistical isotropy (rotational and/or reflectional) symmetry, locally or globally.

An attempt is made to provide a quantitative explanation of anomalous diffusion in terms of this phenomenon.

Some of the results are of speculative nature and further analysis is necessary to validate or disprove the claims made, since *the correspondence with the experimental results may occur for the wrong reasons* as happens from time to time in the field of turbulence.

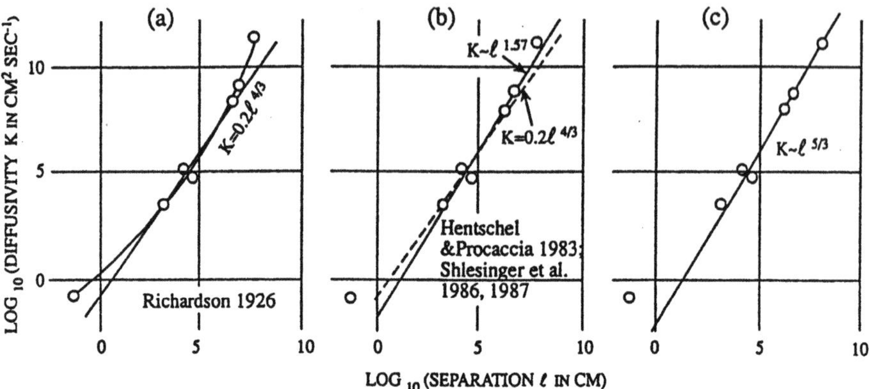

Fig. 1.1. Three interpretations of the data used by Richardson 1926 [3].

1 Introduction

Geophysical turbulent flows are characterized by large Reynolds numbers. Therefore, it has been a common expectation that universal relations (such as energy spectrum $E(k) \sim k^{-5/3}$, diffusivity $\mathcal{K} \sim \ell^{\,4/3}$) should be valid in such flows as well as their "two-dimensional" analogs in quasi-two-dimensional situations.

Richardson in his famous paper [3] initiated the modern approach to the subject of turbulent diffusion [4], stressing the importance of relative diffusion rather than single-particle diffusion. In particular, to find out how the coefficient of eddy diffusivity $\mathcal{K}$ varies with scale ℓ Richardson plotted $\mathcal{K}$ versus ℓ ranging from 0.05 to $10^8 cm$. His original plot is reproduced in figure $1a$. Discussing this result Taylor writes:

It will be seen that if the lowest point, which refers to molecular diffusion, and the highest point, which refers to transfer over distances of thousands of kilometers, are left out of consideration the straight line

$$\mathcal{K} = 0.2\ell^{4/3} \tag{2}$$

is a very good approximation to the curve between $\ell = 10^2$ and $\ell = 10^6 cm$. Since the curve shown here seems to contain all the observational data that Richardson had when he announced the remarkable law (2), it reveals a well-developed physical intuition that he chose as his index 4/3 instead of, say, 1.3 or 1.4 but he had the idea that the index was determined by something connected with the way energy was handed down from larger to smaller and smaller eddies. He perceived that this is a process which, because of its universality, must be subject to some simple universal rule. It is perhaps rather surprising that he did *not take the step which Kolmogoroff (1941) and Obukhov took fifteen years later, namely to express his equation non-dimensionally using only the two physical quantities which could be relevant to a universal rule regulating the handling down of energy, namely ϵ the rate of energy dissipation and ν the dynamical viscosity.*

The Richardson law was *claimed* to be confirmed in a large number of experiments [7], [8]. However, in spite of the common expectation there exist many examples of turbulent flows in the atmosphere, ocean and laboratory, in which the turbulent diffusivity $\mathcal{K}$ as a function of scale ℓ **does not follow** the Richardson law [3]

$$\mathcal{K} \sim \ell^{\,4/3} \tag{1.1}$$

as well as its "two-dimensional" analog[1] [5]. Examples of such behavior are given in the main text of the paper for various situations.[2]

Here we give three examples of different interpretation of the data of Richardson's original paper.[3] Namely, it is claimed in [9] (see also [10]) that, excluding the lowest point which pertains to molecular diffusivity, these data are best fitted by a relation

$$\mathcal{K} \sim \ell^{4/3+2\mu/3} \tag{1.2}$$

with non-zero intermittency exponent $\mu = 0.36$ and a slope of 1.57 in (1.2) (see figure 1*b* adapted from [9]). Similar results have been obtained in [11], [12] using Lévy walks. However, this interpretation (as well as the original one by Richardson) does not take into account that the upper three points in figure 1 correspond to strongly anisotropic (quasi-two-dimensional – QTD) conditions. It is argued below that in such a situation the relevant parameter is the rate of production of helicity $\zeta = \langle | \, dh/dt \, | \rangle$ rather than ϵ the rate of energy dissipation.[4] This results in the relation

$$\mathcal{K} \sim \ell^{\,5/3} \tag{1.3}$$

---

[1]We refer to such situations as possessing anomalous diffusion. Note, that in broader contexts the term *anomalous diffusion* is used when $\mathcal{K}$ is different from $\mathcal{K} \sim const$ (see, for example, [6]).

[2]In fact there exists solid observational evidence on strong variability not only in the *scaling exponents* but also in the *"universal" constants* in scaling laws such as for $\mathcal{K}, E(k)$, etc.

[3]It is appropriate to mention here the *variability in interpretations* of the same experimental data.

[4]It is possible that more appropriate is a related quantity $\hat{\zeta} = \langle | \, d(\hat{h})/dt \, | \rangle$, where $\hat{h}_u = \hat{u} \cdot \omega$ and $\hat{u} = u + \nabla \phi$. It was shown in [13] (see also [14]) that $\nabla \phi$ can be chosen in such a way that – in contrast to h – $\hat{h}$ is a lagrangian invariant, i.e. it is conserved along the paths (pointwise) and therefore for any fluid volume. In the absence of boundaries (or with some special boundary conditions) the integrals of h and $\hat{h}$ coincide.

with the exponent 5/3. The straight line with this slope is shown in figure 1c together with the data of Richardson's original paper.

One of the natural candidates among the possible reasons for the deviations from the Richardson law is the phenomenon of spontaneous breaking of statistical isotropy (rotational and/or reflectional) symmetry, locally or globally[5], [6]. In the sequel an attempt is made to provide a quantitative explanation of anomalous diffusion in terms of this phenomenon.

2 Atmospheric boundary layer

It is argued in [21] that regions with large fluctuations of turbulent energy are characterized by strong anisotropy and a local cascade of angular momentum (breaking of rotational symmetry), i.e. of a quantity of the type of Loytsianskii's invariant

$$\Lambda = \int_V \langle \mathbf{u}(\mathbf{x})\mathbf{u}(\mathbf{x}+\mathbf{r})\rangle r^2 dr^{D_\infty}, \tag{2.1}$$

where D_∞ characterizes the subregions Ω_r with large fluctuations of turbulent energy

$$large \int_{\Omega_r} \mathbf{u}^2\,(\mathbf{x})dx \sim r^{D_\infty}. \tag{2.2}$$

It is argued further in [21] that the governing parameter is the rate of transfer of angular momentum

$$\mathcal{L} =\mid \frac{d(\Lambda/V)}{dt} \mid, \tag{2.3}$$

which has the following dimensionality

$$[\mathcal{L}] = [L]^{1+D_\infty}[T]^{-3}. \tag{2.4}$$

It is straightforward to obtain the numerical value of D_∞ from dimensional arguments

$$large \int_{\Omega_r} \mathbf{u}^2(\mathbf{x})dx \sim \mathcal{L}^{2/3}\, r^{13/5}, \tag{2.5}$$

i.e. that for the field of turbulent energy $D_\infty = 13/5$. These arguments are supported by laboratory and numerical data on asymptotic values of intermittency exponent μ_q for large q of turbulent energy [21], [24], [25], which gave a value of $D_\infty = 2.6 \pm 0.05$.

---

[5] This should be distinguished from *imposed* reflectional symmetry breaking as in [15] - [18].

[6] The intermittent (multi)fractal behavior of turbulence could be considered as another possible reason [9]. However, while both reasons seem to be intimately related [20] for the velocity field, there are some indications that the impact of (multi)fractality and intermittency on dispersion may be small [19].

In particular the parameter $\mathcal{L}$ becomes relevant in case when the energy of turbulence is supplied at different scales. In such a situation one can expect that the Richardson-Kolmogorov cascade process will not be realized since there will be not enough time to allow for the process of isotropization owing to the action of long range forces due to pressure gradients. However, the remaining anisotropy in such a case allows to assume that a 'cascade' of angular momentum mentioned above can be realized in a considerable range of scales. For example, one can expect such a 'cascade' in turbulent flow over urban or rocky landscapes as well as over complex terrains.

A scaling relation for the effective diffusivity $\mathcal{K}$ as a function of scale ℓ follows from dimensional arguments assuming that the only relevant parameter is $\mathcal{L}$ (2.5)

$$\mathcal{K} \sim \mathcal{L}^{1/3}\, \ell^{4/5}, \tag{2.6}$$

which is different from the Richardson law (1.1) as well as from the relations describing correspondingly the initial ($\mathcal{K} \sim \ell$) and the final ($\mathcal{K} \sim const$) stages of diffusion [7], [26]. An estimate of the spread σ of a puff from a source of a passive scalar as a function of characteristic time of its motion t can be found in a similar way[7]

$$\sigma \sim \mathcal{L}^{5/18}\, t^{5/6}. \tag{2.7}$$

In case when the puff is advected horizontally σ is taken from the vertical spread, while t is estimated as X/V, where X is the distance from the source and V is the mean horizontal velocity ([7], p.365.) In such a case

$$\sigma \sim X^{5/6}. \tag{2.8}$$

In figure 2.1 adapted from [26], p.218 are shown results obtained by Högström from a tube at a height of 50 m at Agesta, Sweden. A straight line with the slope '5/6' is drawn in this figure in order to make a comparison with the relation (2.8) and also a straight line with the slope '1/2' corresponding to the long time limit in the statistical theory [7]. The broken lines have the slopes '3/2' (Richardson-Kolmogorov theory) and '1' (short time limit in statistical theory). It is seen from the figure 2.1 that at the *initial stage* of the evolution of the puff of passive scalar it follows the relation (2.8), i.e. the process of turbulent diffusion seems to be controlled by the 'cascade' of angular momentum.

A similar trend is seen clearly for a number of experimental results shown in figure 2.2 also adapted from [26], p. 195.

---

[7]The exponents λ_1 and λ_2 in the relations $\mathcal{K} \sim \ell^{\lambda_1}$ and $\ell \sim t^{\lambda_2}$ are related by a simple relation $\lambda_1 = (2\lambda_2 - 1)/\lambda_2$ or $\lambda_2 = 1/(2 - \lambda_1)$.

8

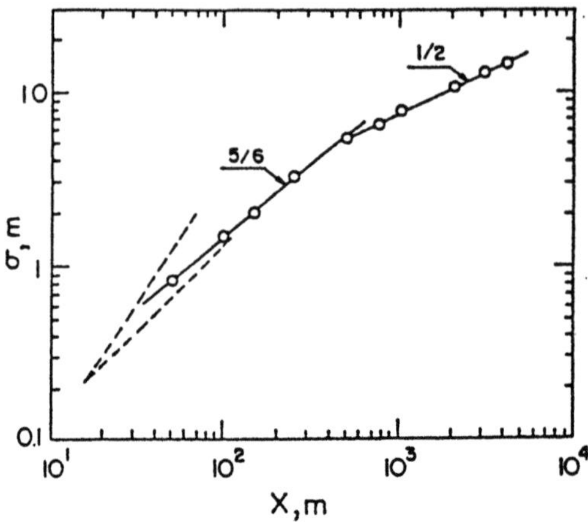

Fig. 2.1. Vertical spread from a source at a height 50 m at Agesta, Sweden measured by Högström 1964. Adapted from Pasquill and Smith 1983 [26]). The slope 5/6 corresponds to the relation $\mathcal{K} \sim \mathcal{L}^{1/3}\, \ell^{4/5}$.

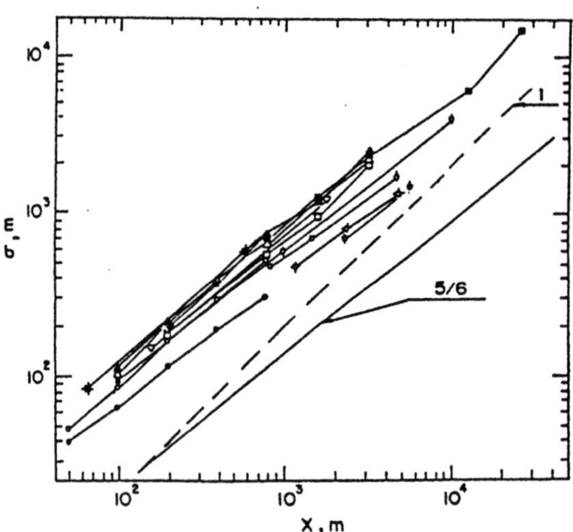

Fig. 2.2. Crosswind spread measured by different authors (adapted from Pasquill and Smith 1983 [26]). The slope 5/6 corresponds to the relation $\mathcal{K} \sim \mathcal{L}^{1/3}\, \ell^{4/5}$.

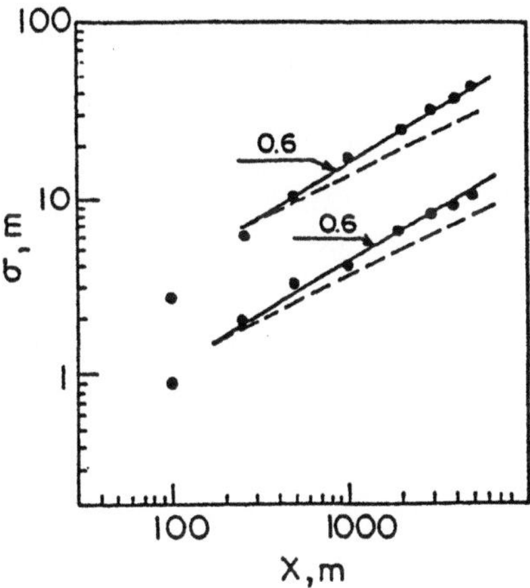

Fig. 2.3. Vertical spread of elongated smoke puffs in ABL. Adapted from Pasquill and Smith 1983 [26].

While the above considerations can be applied to the initial stage of diffusion, in case of the final stage one has to take into account the presence of organized structures, which can modify considerably the process of turbulent diffusion [27]. It has been shown in [23] that if in a turbulent flow there exist a finite number of large scale 'sinks' of turbulent energy (such as solitons, spontaneously formed large scale vortices, etc.) then at scales of the order of these objects it is more appropriate to use as *a governing parameter the 'dissipation' rate of energy per sink - G* and not the dissipation rate per volume unit $\langle \epsilon \rangle$ as in the Kolmogorov-Obukhov theory [7]. Since G and $\langle \epsilon \rangle$ are of different dimensionality ($[G] = [L]^5[T]^{-3}$, whereas $[\langle \epsilon \rangle] = [L]^2[T]^{-3}$), it follows from dimensional arguments that the scaling relation for diffusivity has the following form

$$K \sim G^{1/3} \ell^{1/3}. \tag{2.9}$$

Similarly

$$\sigma \sim G^{1/5} t^{3/5} \tag{2.10}$$

at the G-range of scales.

In fig. 2.3 (adapted form [26], p.225) we show the vertical spread of elongated smoke puffs observed by Högström [28] in Studswick (Sweden), source height 87 m for two values of stability category $\Lambda = 2.25$ (lower points) and $\Lambda = 1.5$ (upper points). We have drawn continuous lines with the slope 0.6 for comparison with the relation $\sigma \sim X^{3/5}$ (X is the horizontal distance from the source). The dotted lines correspond to the long time limit $\sigma \sim X^{1/2}$ ([26] p.194, [29]).

3 Diffusion in the troposphere and in the ocean

This question has been addressed in [22] by means of analysis of experimental data on helicity obtained in laboratory for turbulent grid, boundary layer and jet flows [1], [2]. It was shown that Kolmogorov (homogeneous) turbulence is unstable in respect to local states - fractons ([32]), which appear to be the *subregions with large helicity*. These self-organized states arise spontaneously in subregions of turbulent flow with essential *breaking of reflectional symmetry* with large helicity. The governing dimensional parameter for helical fractons is different from the Kolmogorov one. It is the so called renormalized dissipation rate $\tilde{\epsilon}$ [22] which has the dimensionality

$$\tilde{\epsilon} = [L^2][T]^{-1-D_f} \tag{3.1}$$

with fracton dimension $D_f = 4/3$ (for details see [22]).

In particular, the diffusivity $\mathcal{K}$ in fractons follows the relation

$$\mathcal{K} \sim \tilde{\epsilon}^{3/7} \ell^{8/7} \tag{3.2}$$

and not the law of Richardson (1.1) [3].

In case, when the number of helical fractons is large enough, the mean diffusivity (over the whole flow region) will follow the relation (3.2) too. Such a possibility is rooted in the properties of fractons enabling them to trap the passive scalar inside them for a very long time. Therefore, after some initial period most of the passive scalar will be located within the fractons. On the other hand, the interaction of fractons with their environment is controlled primarily also by the parameter $\tilde{\epsilon}$, i.e this parameter controls the statistical properties of the stochastic trajectories of fractons. In other words, the statistical properties of the fractons trajectories will be determined mainly by the properties of the fractons themselves and to a much lesser degree by the properties of their environment. This brings us to the conclusion that the relation (3.2) can be valid not only on the scales of the order of scales fractons, but also in a range of much larger scales. It is naturally to call this range the fracton range of scales. It is plausible that these properties of fractons form the basis of the extremely broad range of universal behavior of the dependence $\ell(t)$ in the troposphere (see figure 3.1).

Indeed, it follows from (3.1) and (3.2) that (see footnote 7)

$$\ell \sim t^{7/6}. \tag{3.3}$$

Looking at figure 3.1 - adapted from [30] and containing data on cloud width versus travel time of many authors in different conditions - one is amazed that

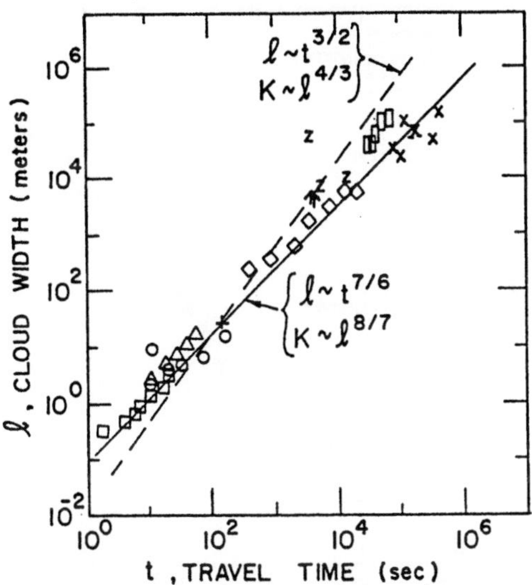

Fig. 3.1. Observations of widths (horizontal standard deviation) of diffusing tracer as a function of downwind travel time in troposphere. Different symbols correspond to the results of different authors (adapted from Gifford 1983 [30]). The slope 7/6 corresponds to the relation $K \sim \ell^{8/7}$.

all these results are well described by a *single universal* relation (3.3) in the range of scales (horizontal standard deviation) from *one meter* to *one hundred kilometers*. The straight line corresponds to the relation (3.2) and the dotted lines correspond to the relation (1.1) and $K \sim \ell$.

As seen from the figure 3.1 the universality of the relation (3.3) is manifested not only by the single exponent '7/6' but also by the *universal constant* in this relation. Apparently in all these experiments the fractons have been of the same type and the scales of cloud width fell into the fracton range.[8]

The relation of K versus ℓ shown in figure 3.2 is based on the results obtained in the ocean [31], where an empirical relation $\ell^2 \sim t^{2.34}$ was obtained. The relation (3.2) results in $\ell^2 \sim t^{7/3}$.

The relations (3.2, 3.3) are valid also in *some cases* for quasi-two-dimensional turbulence (large horizontal scales in the troposphere - figure 3.1, and in the ocean - figure 3.2), since fractons, which are *three*-dimensional formations of rather small scale, most probably can be effective in the transport of a passive scalar on much larger quasi-two-dimensional scales for the same reasons as argued above.

---

[8]Later data compilations seem to be in conformity with those shown in figure 3.1 (Gifford 1994, private communication).

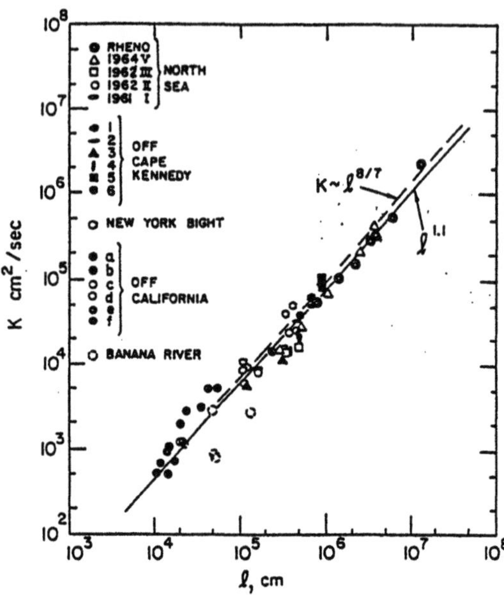

Fig. 3.2. Eddy diffusivity versus scale ℓ in the ocean. Adapted from Okubo 1971 [31].

Since the above results have been obtained in essentially different external conditions it is naturally to assume that the processes responsible for such universal behavior are realized on spatially *localized (and compact) carrier with universal dynamics* (we call this process - *fracton transfer of a passive scalar*).

It should be stressed that the above results are rather speculative since in the atmosphere and in the ocean there are observed relations of $\mathcal{K}(\ell)$ different from (3.2) and the geophysical conditions leading to the fracton transfer of a passive scalar are not clear yet. In particular, in the next section an example of a different behavior of turbulent diffusion in quasi-two-dimensional turbulence is given.

4   Diffusion in real quasi-two-dimensional turbulence - stratosphere

There exists a *qualitative* difference between strictly two-dimensional (2D) and real quasi-two-dimensional (Q2D) turbulence in spite of the "smallness" of the difference in their geometry. In fact, *this difference can be rather large primarily due to its topological nature.* In particular, helicity $\mathcal{H}_u = \int_V h_u dv$, helicity density $h_u = \mathbf{u} \cdot \omega$, superhelicity $\mathcal{H}_\omega = \int_V h_\omega dv$ and its density $h_\omega = \omega \cdot rot\, \omega$

and related quantities[9] *vanish identically in strictly two-dimensional turbulence,* *whereas in real Q2D turbulence $\mathcal{H}$ and h can be finite for whatever small rate of* *change of flow properties along the slow variation coordinate.*

Since purely two-dimensional turbulence is unstable to three-dimensional perturbations it cannot be realized in real 3-D space. However, the 3-D instabilities can be moderated or even totally suppressed by external factors and constraints such as stratification, rotation, magnetic field, rigid walls or strong velocity gradient in some direction. It is argued in [23] that in the presence of such factors the quasi-two-dimensional regime arises as a result of a spontaneous breaking of reflectional symmetry (parity breaking bifurcation), which in turn is a consequence of the instability of two-dimensional turbulence to three-dimensional helical traveling waves and solitons through super-and/or sub-critical bifurcations. Such instabilities can be realized on scales r_1 *much larger* than the the characteristic scale r_0 of energy input into the two-dimensional turbulent flow.[10]

The only source of energy for the 3-D disturbances is the basic two-dimensional turbulent flow with an energy input at the scale r_0. Since the characteristic scale of the traveling waves $r_1 \gg r_0$ there should occur an inverse (anisotropic) energy transfer to support their existence. This energy transfer cannot be of a cascade type, due to the scale separation $r_1 \gg r_0$. For this reason the mean rate of energy transfer $\langle \epsilon \rangle$ is not a governing parameter in this range of scales and its place it taken by the mean magnitude of rate of spontaneous helicity generation

$$\zeta = \langle |\, dh/dt \,| \rangle. \tag{4.1}$$

Then in the range $r_1 \gg r \gg r_0$ in analogy with the Kolmogorov theory it follows from dimensional arguments that the energy spectrum has the following form:

$$E_u(k) \sim \zeta^{2/3}\, k^{-7/3}, \tag{4.2}$$

where k is the modulus of the wave number in the plane of the primary two-dimensional turbulent flow.

The expression (4.2) was obtained in [35] for the case of *three*-dimensional isotropic turbulence.

However, since in the last case there seems to exist no natural mechanism of scale separation r_0 and r_1 (see above) the expression (4.2) appeared to be

---

[9]For a review on helicity in laminar and turbulent flows see [36].

[10]It is noteworthy that the situation is different in the case of 3-D instability of laminar flows. Here, short wave instability can play an essential role due to the absence of (2-D) turbulent diffusion and of a stabilizing factor [33], [34].

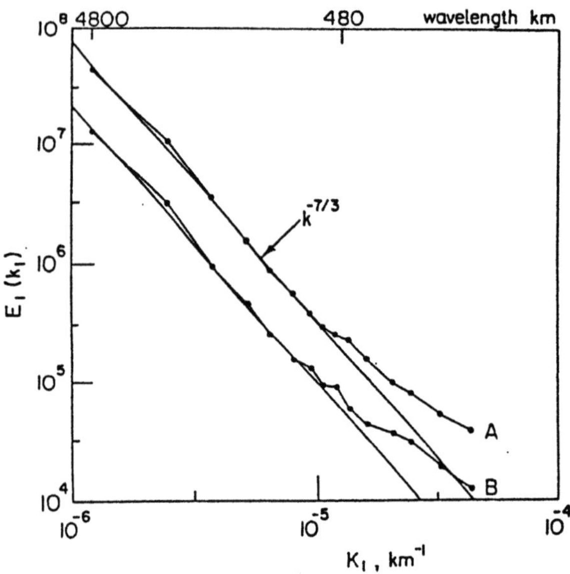

Fig. 4.1. Spectra from the GASP flights in stratosphere at least 4800 km long: A - kinetic energy; B - temperature. Adapted from Gage and Nastrom 1986 [38].

inadequate to the existing experimental data. By contrast in the case of quasi-two-dimensional turbulence there is a great variety of experimental and field observations of spectra with wide ranges in full agreement with (4.2). We will limit ourselves with examples in which the difference between 2D and Q2D turbulence is manifested in particular in dissimilar diffusive properties. Arguments similar to those used by Corrsin and Obukhov ([7], p.377) lead to a following expression for the spectrum of fluctuations of a passive scalar c

$$E_c(k) \sim \langle N \rangle \zeta^{-1/3} k^{-4/3}, \qquad (4.3)$$

where $N = | \, dc^2/dt \, |$.

An example of spectra of kinetic energy and temperature from the GASP flights in the stratosphere is shown in figure 4.1 [38]. While the energy spectra are seen to follow clearly the relation (4.2)[11], the temperature spectrum *does not follow* the power law (4.3) with the exponent '-4/3' (see also [39]) but rather the power law *with the same exponent '-7/3'*. Note, that in [23] it has been

---

[11]Other examples are given in [23]. In fact, velocity spectra with the slope close to '-7/3' were observed earlier [8], [40] - [43]. For example, in fig.13.1 Monin and Ozmidov [8] compiled data of different authors on one-dimensional spectra of large-scale meteorological fields. The slope '-7/3' is much closer to the data than the slope '-3' drawn in their figure. This is clearly seen from figure 4.2.

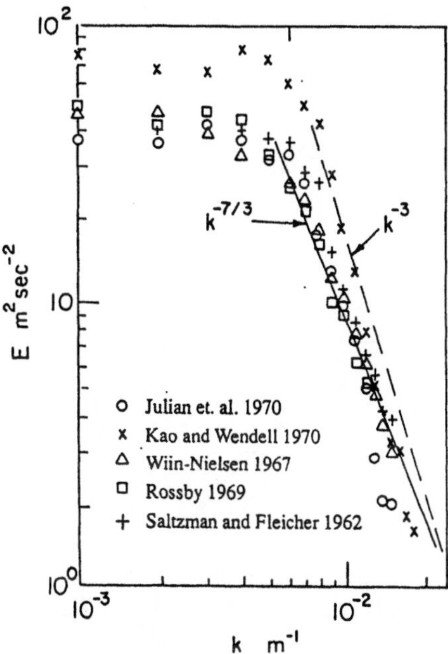

Fig. 4.2. One-dimensional spectra of large-scale meteorological fields compiled by Monin and Ozmidov 1985. Adapted from [8].

erroneously asserted that the expression of type (4.3) for E_c (!) can be obtained employing as a governing parameter ζ.

Before addressing this additional "anomaly" we show three results in which *a clear range with the exponent '-4/3' does appear.* The first result presented at figure 4.3 shows the low frequency part of fluctuations of temperature obtained from a month-long series of radiosonde soundings taken over Kharkov, USSR, in July 1966 [37]. The second result regarding the spectrum of ozone in the stratosphere in the GASP [44] is shown in figure 4.4 (an indication of similar behavior of spectra of carbon monoxide can be seen in [44] too). The third result has been obtained for temperature fluctuations in a totally different situation: in a laboratory flow past a circular cylinder at a distance about 100 diameters downstream of the cylinder on the wake centerline ($Re \sim 25000$ based on the cylinder diameter) [45] . This is shown in figure 4.5 and exhibits a slope '-4/3' over more than 1.5 decades (in [46] this slope was observed over more than two decades).

An important feature of this last result is that the '-4/3' *scaling at the low-wave number end extends to scales substantially larger than L* [45] (L - is the velocity correlation, or 'integral', length scale), i.e. this result is consistent both with the fact that the flow in the wake of a circular cylinder is dominated by

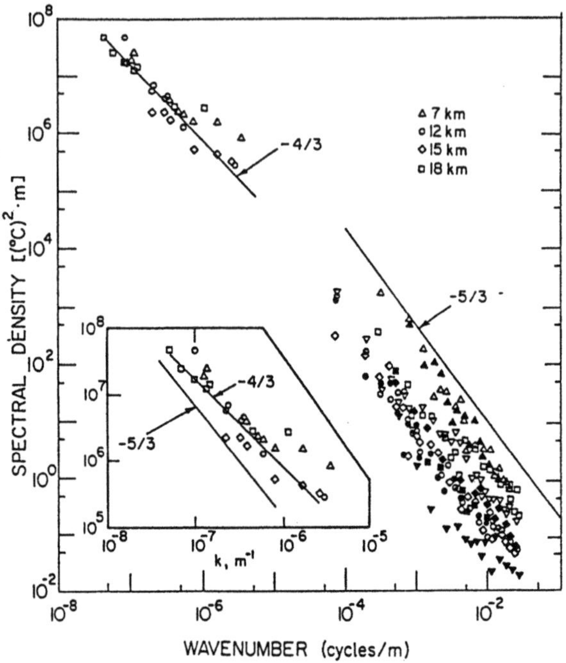

Fig. 4.3. Temperature spectra in the free atmosphere. The close up of the low frequency part is shown in the inset. Adapted from Vinichenko & Dutton 1969 [37].

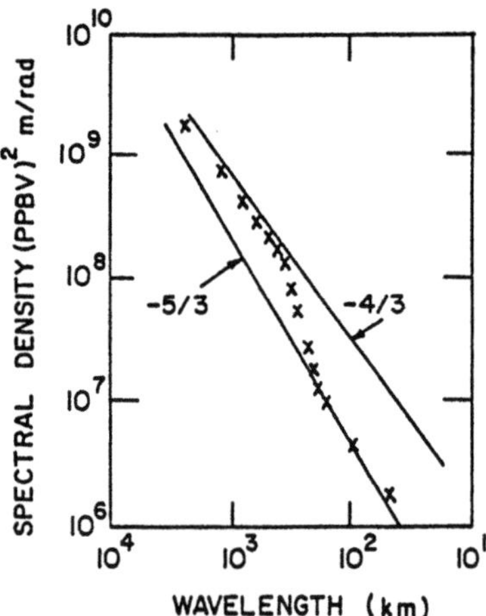

Fig. 4.4. Spectra of ozone in stratosphere. Adapted from Gage and Nastrom 1986 [44].

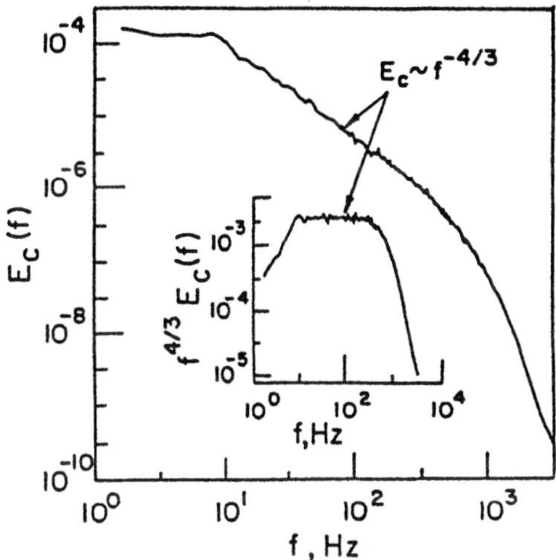

Fig. 4.5. Spectral density of temperature fluctuations in the wake of heated cylinder of circular cross section. Adapted from Sreenivasan 1991 [45].

large quasi-two-dimensional structures and with the use of ζ as a governing parameter as above.[12]

Let us return to the power law for $E_c(k)$ with the exponent '-7/3'. *It can be obtained via dimensional arguments too* taking instead of ζ the following governing parameter

$$\tilde{\zeta} = \langle | \frac{\partial \epsilon}{\partial z} | \rangle, \tag{4.4}$$

which has the meaning of average variation of "two-dimensional" dissipation in the direction z of slow variation of flow properties. It follows in this case that

$$E_c(k) \sim \langle | \frac{\partial N}{\partial z} | \rangle \, \tilde{\zeta}^{-1/3} \, k^{-7/3}. \tag{4.5}$$

The parameter $\tilde{\zeta}$ has the same dimensionality as ζ and therefore the energy spectrum in the form (4.2) can be obtained employing $\tilde{\zeta}$ as a governing parameter too [49]. The difference in spectra of $E_c(k)$ in (4.3) and (4.5) arises due to different contributions of $N = | dc^2/dt |$. One of the possible ways of resolving

---

[12] It should be noted that Sreenivasan [45] ascribes the '-4/3' exponent to insufficiently large Reynolds number. Howewer, the evidence given in [45] for larger Reynolds numbers is not of such quality as in fig. 4.5 and seems to be inconclusive. A slope very close to '-4/3' was recently obtained in [47] for temperature fluctuations in a grid flow turbulent flow when the temperature fluctuations were introduced by fine wires placed downstream from the grid in a *parallel array* (the spectra were different when the temperature fluctuations were introduced from a heated grid [48] or by a *toaster* [47]). In this last case the velocity spectrum had the *same* slope. Therefore it seems that the results of [47] cannot be explained using the argument based on the helicity invariant only and the question remains open.

the issue of the parameter $\tilde{\zeta}$ versus ζ is that in the GASP data the *potential temperature is not really passive*, since in this particular case *the magnitude and shape of potential temperature spectrum are determined by the same dynamics that govern the velocity spectra* [39], whereas ozone and carbon monoxide are passive [44]. It should be also emphasized, that in the laboratory experiments mentioned above the temperature [45] and the dye [46] *were passive* and obeyed (4.3). Therefore, it seems that the parameter *zeta* is the relevant one (see the last section for discussion).

Let us look now at other parameters related to turbulent diffusion.

The turbulent diffusion coefficient for quasi-two-dimensional turbulence can also be obtained via dimensional arguments in the following form

$$\mathcal{K} \sim d(\langle \ell^2 \rangle)/dt \sim \zeta^{1/3} \, \ell^{5/3}, \tag{4.6}$$

where ℓ is the characteristic scale of the cloud of the passive substance.

The expression (4.6) is different both from the case of 3-D turbulence with

$$\mathcal{K} \sim \ell^{4/3},$$

and from the case of purely two-dimensional turbulence with

$$\mathcal{K} \sim \ell^2$$

in the range of enstrophy transfer (i.e. $E(k) \sim k^{-3}$) (in the range of energy transfer in 2D case $\mathcal{K} \sim \ell^{4/3}$).

In a similar way an expression for the mean square relative velocity can be obtained.

$$\langle (d\ell/dt)^2 \rangle \sim \zeta^{2/3} \, \ell^{4/3}. \tag{4.7}$$

The relations (4.6) and (4.7) are in agreement with the results of experiments on diffusion of passive scalar in the lower stratosphere [50] as can be seen from figure 4.6 *a, b*. Similar behavior was observed in a laboratory experiment on turbulence in a rotating fluid [51] (figure 4.5 *c*). It is noteworthy that the data for $\mathcal{K}(\ell)$ complied in [52] are better fitted by(4.6) in the large scale range rather than by the '4/3' law (see figure 4.7).

The situation considered in this section (with *spontaneous* generation of helicity in Q2D turbulence) is different from the case of turbulence with *extrinsically* imposed mean helicity, which can have considerable influence on the transport properties (see [15] - [18] and references therein).

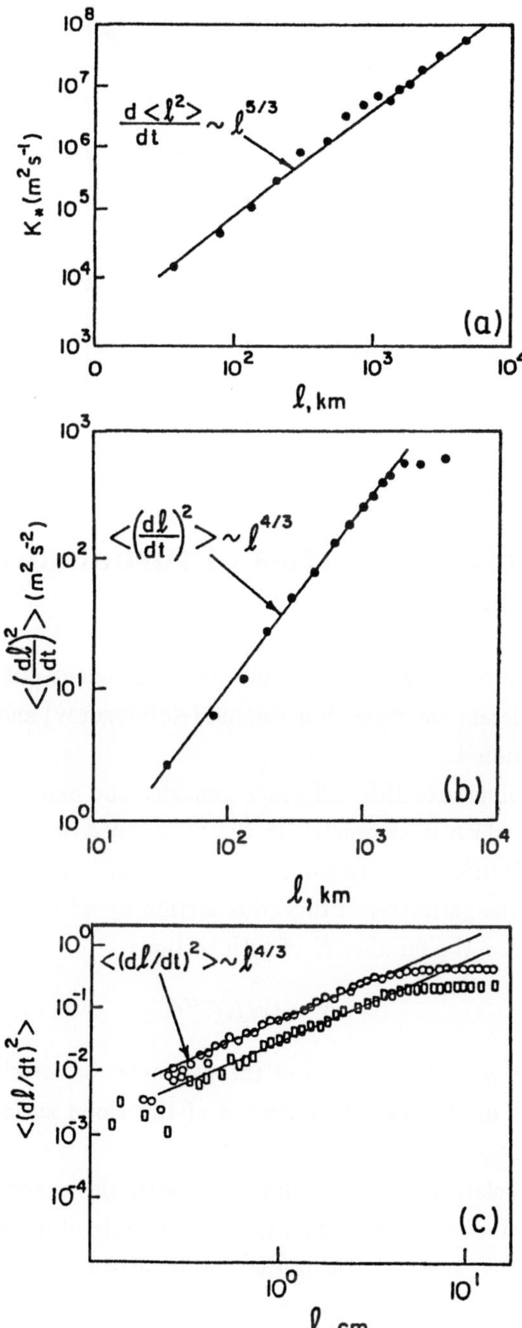

Fig. 4.6. a) Mean diffusivity. b) Mean square relative velocity of balloon pairs in stratosphere. Adapted from Morel & Larcheveque 1974 [50]. c) Mean square relative velocity of particle pairs in a rotating fluid. Adapted from Mory & Hopfinger 1986 [51].

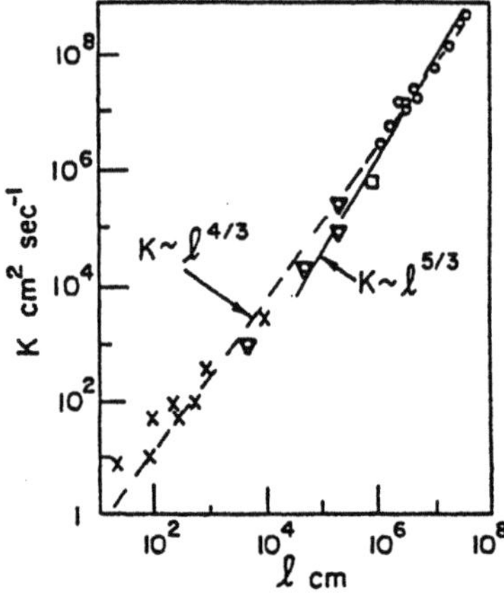

Fig. 4.7. Eddy diffusivity versus scale ℓ . Adapted from Olson and Ichie 1959 [52].

5 On fractal properties of turbulent diffusion of passive scalar

I. It can be expected that fractal properties of turbulent diffusion[13] should be qualitatively different for three-dimensional (Kolmogorov) and Q2D turbulence discussed in section 4.

In order to illustrate this difference consider the expansion of a cloud of turbulent fluid, which is symmetric in the mean an d its cross section passing through its geometric center (again in the mean) with an effective radius $R(t)$. Approximating the perimeter of this cross section by a broken line consisting of sections of length r the number N of such sections

$$N \sim (r/R)^{-D_p},\tag{5.1}$$

where D_p is the *fractal dimension* of the perimeter of the cloud cross section which is related to the fractal dimension of the cloud surface by the simple relation: $D_\sigma = D_p + 1$.

In order to relate the fractal dimension with the exponent in the power spectrum of passive scalar let us find the effective rate of increase of the area of the cloud cross section ds/dt. Using the simple relation

$$ds = r\,(\delta u_r dt)N,\tag{5.2}$$

where δu_r is the velocity of the section r normal to it [55], and (4) it follows

---

[13]Corrsin [53] gave a beautiful illustration of fractal nature of turbulent diffusion, see figure 9 in [53].

that

$$ds/dt \sim r^{1-D_P}\delta u_r R^{D_P}. \tag{5.3}$$

The velocity δu_r can be estimated via the well known relation [7]

$$\delta u_r \sim r^{\alpha}, \quad \text{where} \quad \alpha = (\gamma - 1)/2, \quad E(k) \sim k^{-\gamma}, \tag{5.4}$$

and the relation (5.3) becomes

$$ds/dt \sim r^{1-D_P+\alpha}. \tag{5.5}$$

Finally, since the rate of increase of area of the cloud cross section *is independent of r* (which has been used for its approximation) it follows from (5.4) that

$$D_p = 1 + \alpha = (1 + \gamma)/2. \tag{5.6}$$

Let us look at two important cases:

$\diamond$ – Kolmogorov turbulence (which is 3D). In this case $\gamma = 5/3$ and $D_p = 4/3$.

$\diamond$ – Quasi-two-dimensional turbulence. In this case (see 4.2) $\gamma = 7/3$ and $D_p = 5/3$.

These numbers are in good agreement with measurements of area $\mathcal{A}$ versus perimeter $\mathcal{P}$ of rain and cloud areas, determined from radar and satellite data [54], shown in figure 5.1. We have drawn straight lines corresponding to $D_p = 4/3$ and $D_p = 5/3$ using the relation $\mathcal{A} \sim \mathcal{P}^{2/D_P}$ [56].[14]

II. It is a common assumption in turbulence research that turbulent dissipation is well represented by a single squared velocity derivative (dissipation surrogate). However, *except of their mean values* other properties of these two quantities are *different even in homogeneous and quasi-isotropic flows*. This problem has been formulated by [58] (see also [59], [1]). In case of a passive scalar it is much more difficult to realize the reasons for such a difference. However, there are clear indications that there exists a considerable difference between fractal and multifractal properties of the dissipation rate of passive scalar and its surrogate. For example it has been shown [60] that the fractal dimension (Kolmogorov capacity) of the carrier of the rate of dissipation of passive scalar is equal to 3, while its' value for an individual squared derivative is $5/3$ (D_∇). This result is in good agreement with the one obtained from a recent simulation of turbulent dispersion [61]. Some results adapted from [61] are shown in figure 5.2.

---

[14] Gifford [57] recognized that *two* values of D_p are better characterizing the data of Lovejoy [54] than the single value $D_p = 1.35$, reported by Lovejoy for all points.

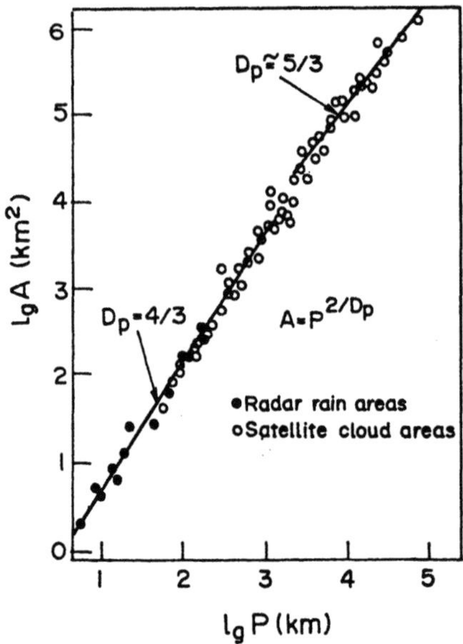

Fig. 5.1. Area A versus perimeter P for cloud and rain areas. Adapted from Lovejoy 1982 [54].

It is seen that the fractal dimension D_∇ reduces from 3 at the initial moment to about 5/3 at $t = 1$ sec (figure 5.2 a). The 'evolution' of D_∇ shown in fig. 5.2 a is linked to multifractality. At the same time the dimension of the perimeter of the projection of the cloud surface increases from 1 at $t = 0$ to about 4/3 at $t \geq 0.5$ sec (figure 5.2 b). A similar result for the dimension of the perimeter $(1.37 + 0.4)$ has been obtained in measurements of the Chernobyl spot of the radionuclides contamination [62].

6 Summary, discussion and some open problems

Summarizing we would like first to reiterate the main points of this communication.

$\mathcal{A}$. It is argued that regions with large fluctuations of turbulent energy are characterized by strong anisotropy and a local cascade of angular momentum, i.e. of a quantity of the type of Loytsianskii's invariant. These arguments - which have been supported by laboratory and numerical data on asymptotic properties of higher order intermittency exponents of turbulent energy - have been used for the analysis of diffusion of a puff of passive scalar. The result is a scaling law for the turbulent diffusivity $\mathcal{K} \sim \ell^{4/5}$, where ℓ is the characteristic

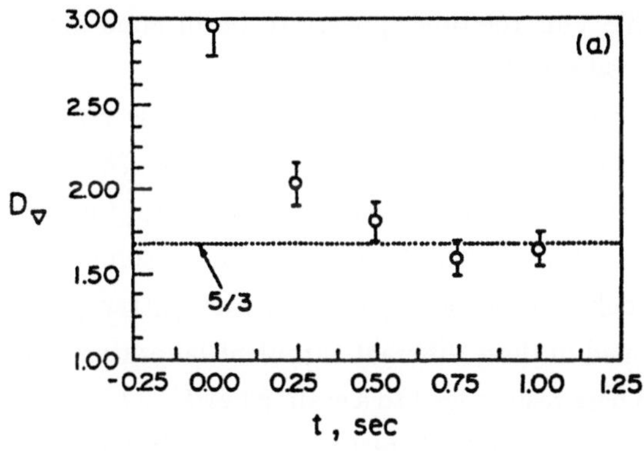

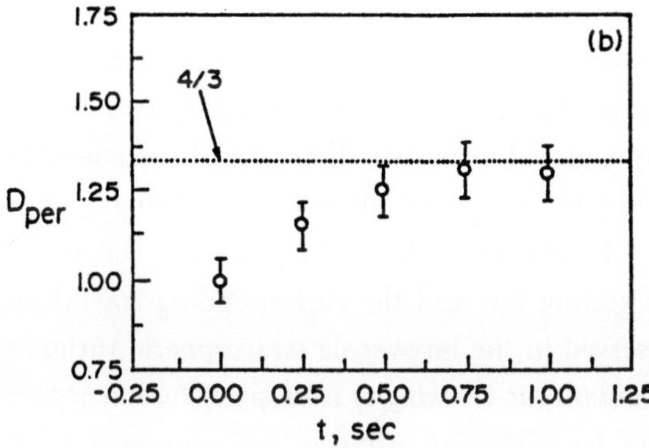

Fig. 5.2. a) Evolution of the fractal dimension $D_{\nabla'}$ of a cloud of tracer particles. b) Evolution of the fractal dimension of the perimeter of the cloud surface projection. Adapted from Stiasnie et al. 1993 [61].

scale of the puff. This relation appears to be in good agreement with a large number of observations [21].

$\mathcal{B}$. It is claimed that Kolmogorov turbulence is critical in respect to the localization effects of subregions with large helicity (*helical fractons*) and that the Kolmogorov cascade is renormalized in the *helical fractons*. The *quantitative* consequences of such a renormalization have been confirmed by the analysis of the asymptotic behavior of the higher order intermittency exponents of the field of helicity, obtained in three different turbulent laboratory flows (grid, boundary layer and jet). These results lead to a scaling law $\mathcal{K} \sim \ell^{8/7}$ in which the turbulent diffusion is controlled by helical fractons. This scaling law is in good agreement with a variety of observations in troposphere and in the ocean [22].

$\mathcal{C}$. It is shown that the asymptotic properties of the higher order intermittency exponents of turbulent dissipation in quasi-two-dimensional turbulence (arising as a result of helical instability of purely two-dimensional turbulence) are controlled by a global quasi-two-dimensional cascade of helicity [23]. Again this is confirmed by the results of laboratory modeling of quasi-two-dimensional turbulence [49]. In this case the scaling law for the turbulent diffusivity is $\mathcal{K} \sim \ell^{5/3}$. This scaling law and the corresponding fractal and spectral scaling relations are observed in the large scale stratospheric turbulence.

Thus local spontaneous breaking of isotropy of turbulent flow results in anomalous scaling laws for turbulent diffusion (as compared to the scaling law of Richardson) *and are observed, as a rule, in different atmospheric layers from the atmospheric boundary layer (ABL) to the stratosphere. The breaking of rotational symmetry is important in the ABL, whereas reflectional symmetry breaking is dominating in the troposphere locally and in the stratosphere globally.*

As has been already mentioned in the abstract the above results are of speculative nature (mainly due to use of dimensional arguments and scalings) [15] and leave several important questions open. Some of these questions are discussed below.

An important criterion of validity of results obtained via dimensional arguments is that different characteristics of the flow obtained in such a way should be consistent with the *same* governing parameter. For example, in the case when the governing parameter is $\mathcal{L}$ (see section 2) the turbulent energy

---

[15]Citing R. K. Kraichnan - *The wonderful thing about scaling is that you can get everything right without understanding anything* [63]. A similar statement is due to P. Bradshaw - *...it is clear that if a result can be derived by dimensional analysis alone... then it can be derived by almost any theory, right or wrong, which is dimensionally correct and uses the right variables* [65]. Also the citation from R. E. Normal should be seen as a warning: *It is increasingly clear that deterministic chaos and universal scaling theories can explain everything* [64].

spectrum should have the form

$$E_u(k) \sim \mathcal{L}^{2/3} k^{-3/5} \tag{6.1}$$

and the spectrum for $E_c(k)$

$$E_c(k) \sim \langle N \rangle \mathcal{L}^{-1/3} k^{-11/5}. \tag{6.2}$$

The available evidence does not allow to make a definite judgement about the existence of spectra (6.1) and (6.2). However, it seems that such spectra can be observed in appropriate conditions. Indeed, a spectrum $E_c(k))$ with the exponent '-11/5' was observed over almost two decades in the low wave number region in the experiments in the coastal region of the Baltic sea [66], figure 6.1. In [67] the reason for such a spectrum is seen in the possibility of energy supply over (almost) the whole range of scales (cf. section 2).

In case when the the governing parameter is $\tilde{\epsilon}$ (section 3) the situation is more serious, since the energy spectrum in this case takes the form

$$E_u(k) \sim \tilde{\epsilon}^{6/7} k^{-9/7} \tag{6.3}$$

and is *not compatible with the existing experimental evidence*. It has been claimed in [22] that in this particular case *the energy spectrum can be not compatible with (3.2) in the generally accepted sense due to complicated structure of helical fractons which are able to trap and detain the passive scalar within their interior.* This claim, of course, requires further elaboration, but even if it is true the spectrum of $E_c(k)$

$$E_c(k) \sim \langle N \rangle \tilde{\epsilon}^{-3/7} k^{-13/7} \tag{6.4}$$

should be compatible with the corresponding governing parameter. Again there exist no firm evidence on the existence of the spectrum (6.4).

Finally, in case when the governing parameter is ζ (see section 4, eq.(4.1)) there is an alternative parameter $\tilde{\zeta}$ [49] (eq (4.4)). As has been pointed in section 4 it is more likely that the relevant parameter is ζ as having clear physical meaning and compatible with the existing experimental evidence, though rather limited in the case of a passive scalar. In this respect the results of [49] should be seen as supporting the first choice, especially in view of the results for $E_c(k)$ obtained in [37] - [46]. Still, the possibility of (co)existence of both situations cannot be excluded totally and the issue remains open including the question about possible relation between ζ and $\tilde{\zeta}$. A trivial (but almost useless) answer to the last question follows again from the very dimensional argument, i.e.

$$\zeta = \langle | \, dh/dt \, | \rangle \sim \tilde{\zeta} = \langle | \frac{\partial \epsilon}{\partial z} | \rangle, \tag{6.5}$$

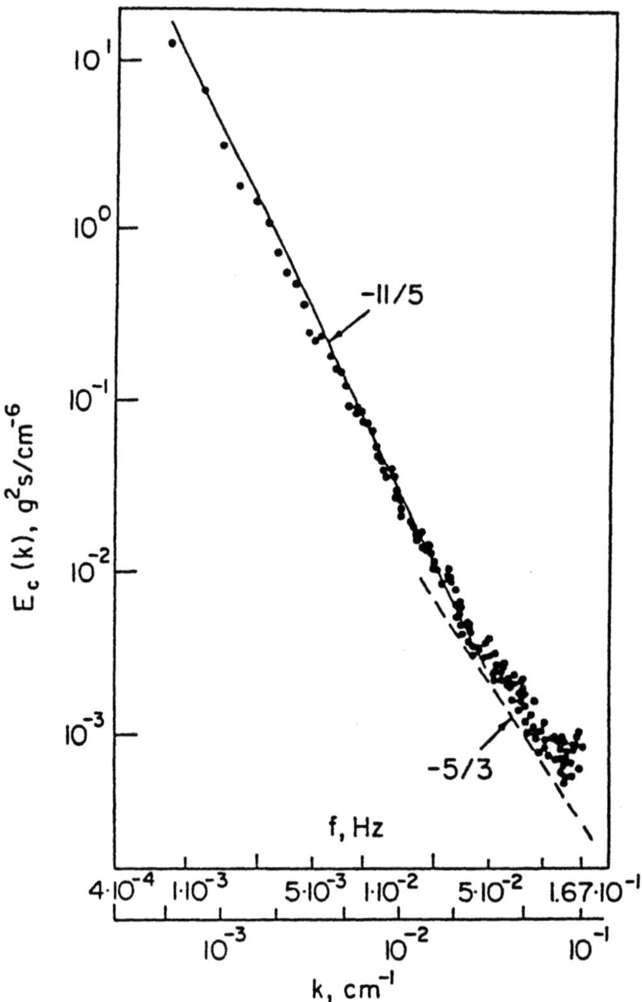

Fig. 6.1. Spectrum of dye concentration in experiments with continuous source (Ozmidov et al. 1971 [66]). Adapted from Monin & Ozmidov 1978 [67].

the physical meaning of which (if such can be found) is not clear. It is noteworthy that the '$-7/3$' turbulent energy spectrum can be obtained from totally different considerations as an exact solution of the kinetic equation for inertial-gravity waves [68], [69] and from the so called 2.5-dimensional averaged equations for rotating fluid [70]. The relation of these approaches to the discussed above properties of Q2D turbulence is not clear yet.

7 A note of warning

As mentioned, the above *interpretations* of the experimental observations are of speculative nature and further analysis is necessary to validate or disprove the claims made, since *the correspondence with the experimental results may occur for the wrong reasons* as happens from time to time in the field of turbulence.

The author is grateful to G. Falkovich, K. S. Gage, I. Hosokawa and A. Mahalov for useful information and to the authors of [61] for the permission to use the figure 5.2.

References

[1] Tsinober A., Kit E. and Dracos T. 1992 Experimental investigation of the field of velocity gradients in turbulent flows. *J. Fluid Mech.* **242**, 169 - 192.

[2] Kit E., Tsinober A. & Dracos T. 1993 Velocity gradients in a turbulent jet flow, in *Proceedings of the 4th European Conference on Turbulence*, ed. F.T.M. Nieuwstadt (Kluwer), *Appl. Sci. Res.*, **51**, 185 - 190.

[3] Richardson L. F. 1926 Atmospheric diffusion on a distance-neighbor graph, *Proc. Roy. Soc. London*, **A110**, 709 - 737.

[4] Taylor G. I. 1959 The present position in the theory of turbulent diffusion, In *Atmospheric Diffusion and Air Pollution, eds. F. N. Frenkiel and P. A. Sheppard*, pp. 101 - 112, Academic Press, New York London.

[5] Contrary to the velocity field experimental results for passive scalars are significantly different from 'theoretical' predictions. Among the reasons for

such 'misbehavior' of passive scalars in turbulent flows is the fact that the passive scalar carries the signature (at least in part) of the complex structure of the turbulent velocity field and that even in a purely laminar flow the passive scalar can behave in a turbulent manner (Lagrangian chaos).
For an overview and a partial list of references on 'misbehavior' of a passive scalar in turbulent flows see [45] and Holzer M. and Siggia E. 1994 Turbulent mixing of a passive scalar, *Phys. Fluids*, **6**, 1820 - 1837.

[6] Zaslavsky G. M. 1992 Anomalous transport and fractal kinetics, In *Topological Aspects of the Dynamics of Fluids and Plasmas, eds. H. K. Moffatt, G. M. Zaslavsky, P. Compte and M. Tabor*, pp. 481 - 491, Kluwer, Dordrecht/Boston/London.

[7] Monin A. S. & Yaglom A. M. 1971 & 1975 *Statistical fluid mechanics*, vol. **1** & vol **2**. MIT Press, Cambridge.

[8] Monin A. S. and Ozmidov R. V. 1985 *Turbulence in the ocean*, Reidel, Dordrecht.

[9] Hentschel H. G. E. and Procaccia I. 1983 Fractal nature of turbulence as manifested in turbulent diffusion, *Phys. Rev.*, **A27**, 1266 - 1269. These authors have reanalized the data on turbulent diffusion of smoke puffs compiled by Gifford (F. Gifford, Jr. 1957 *J. Meteor.*, **24**, 410 - 414). They found that the dependence of $\ell(t)$ is better approximated by the relation $\ell^2 \sim t^{3+\theta}$ with $0.45 > \theta > 0.15$ which corresponds to the exponent for $\mathcal{K}$ in the range $1.42 \div 1.37$.

[10] Bouchard J.- P. and Georges A. 1990 Anomalous diffusion in disordered media: statistical mechanisms, models and physical applications, *Phys. Rep.*, **195**, 127 - 293.

[11] Schlesinger M. F., Klafter J. and West B. J. 1986 Lévy walks with applications to turbulence and chaos, *Physica*, **140A**, 212 - 218.

[12] Schlesinger M. F., Klafter J. and West B. J. 1987 Lévy dynamics of enhanced diffusion: application to turbulence *Phys. Rev. Lett.*, **58**, 1100 - 1103.

[13] Kuzmin G. A. 1983 Ideal incompressible hydrodynamics in terms of vortex momentum density, *Phys. Lett.*, **A96**, 88 - 90.

[14] Oseledets V. I. 1988 On a new way of writing the Navier-Stokes equation. The Hamiltonian formalism, *Comm. Mosc. Math. Soc.*, **44**, 210 - 211.

[15] Moffatt H. K. 1983 Transport effects associated with turbulence with particular attention to the influence of helicity, *Rep. Prog. Phys.*, **46**, 621 -664.

[16] Drummond I. T., Duane S. and Horgan R. R. 1984 Scalar diffusion in simulated helical turbulence with molecular diffusivity, *J. Fluid Mech.*, **138**, 75 - 91.

[17] Cattaneo F., Hughes D. W. and Proctor M. R. E. 1988 Mean advection effects in turbulence, *Geophys. Astrophys. Fluid Dyn.*, **41**, 335 - 342.

[18] Chechkin A. V., Tur A. V. and Yanovsky V. V. 1993 Transport of passive admixture in helical turbulent medium, *In: Proc. Int. Conf. Physics in Ukraine, Kiev, 22-27 June 1993*, pp.50 - 53; also *Physica*, **A**, submitted.

[19] Borgas M. S. 1993 The multifractal lagrangian nature of turbulence, *Phil. Trans. R. Soc. Lond.*, **342**, 379 - 411.
Borgas M. and Sawford B. L. 1994 Stochastic equations with multifractal random increments for modeling turbulent dispersion, *Phys. Fluids*, **6**, 618 - 633.

[20] Tur A. and Levich E. 1992 The origin of organized motion, *Fluid Dyn. Res.*, **10**, 75 - 90.

[21] Bershadskii A., Kit E., Tsinober A. & Waisburd, H. 1994 Strongly localized events of energy, dissipation, enstrophy and enstrophy generation in turbulent flows, *IMA Conference on Multiscale Stochastic Processes Analysed Using Multifractals and Wavelets, Cambridge, 29 - 31 March 1993*; *Fluid Dyn. Res.*, **14**, (August 1994, in press).

[22] Bershadskii A., Kit E. and Tsinober, A. 1993 Self-organization and fractal dynamics in turbulence, *Physica* , **A 199**, 453 - 475.

[23] Bershadskii A., Kit E. and Tsinober A. 1993 Spontaneous breaking of reflectional symmetry in real quasi-two-dimensional turbulence on traveling waves and solitons, *Proc. Roy. Soc.* **A 441**, 147 - 155

[24] Meneveau C. 1991 Analysis of turbulence in the orthonormal wavelet representation, *J.Fluid.Mech.*, **232** (1991), 469 - 520.

[25] Hosokawa I. 1993 *Private communication* on unpublished results on generalized dimension D_q of the field $\mathbf{u}^2$ from direct numerical simulations of Navier-Stokes Equations, see Hosokawa I. and Yamamoto K. 1989 Fine structure of a directly simulated turbulence. *J. Phys. Soc. Japan* **58**, 20 - 23.

[26] Pasquill F. & Smith F. B. 1983 *Atmospheric Diffusion. Study of the dispersion of windborne material from industrial and other sources.* Ellis Horwood.

[27] Bershadskii A. and Tsinober A. 1993 On the influence of organized structures on turbulent diffusion in the ABL: final stage, *Bound. Layer Met., submitted.*

[28] Högström V.: 1964 An experimental study of atmospheric diffusion, *Tellus,* **16**, 205 - 251.

[29] Cramer H.E., Record F.A. and Vangan H.C.: 1958 The study of the diffusion of gases in the lower atmosphere, MIT Dep.Meteorology, Final Report No.AF 19(604) - 1058.

[30] Gifford F. A. 1983 Atmospheric diffusion in the mesoscale range: the evidence of recent plume width observations, *Sixth symposium on turbulence and diffusion, Boston, USA, 1983,* pp. 300 - 304.

[31] Okubo A. 1971 Oceanic diffusion diagrams, *Deep-Sea Research,* **18**, 789 - 806.

[32] Alexander S. 1986 Fractons, *Physica,* **140A**, 397 - 404.

[33] Pierrehumbert R.T. 1986 Universal short-wave instability of two-dimensional eddies in an inviscid fluid, *Phys. Rev. Lett.,* **57**, 2157-2159.

[34] Waleffe F. 1990 On the three-dimensional instability of strained vortices, *Phys. Fluids,* **A2**, 76-80.

[35] Brissaud A., Frisch U., Leorat J., Lesieur M. and Mazure A. 1973 Helicity cascades in fully developed isotropic turbulence. *Phys. Fluids,* **16**, 1366-1368.

[36] Moffatt H. K. and Tsinober A. 1992 Helicity in laminar and turbulent flows, *Annu. Rev. Fluid Mech.,* **24**, 281 - 312.

[37] Vinichenko N. K. and Dutton J. A. 1969 Empirical studies of atmospheric structure and spectra in the free atmosphere, *Radio Sci.,* **4**, 1115 - 1126.

[38] Gage K.S. and Nastrom G.D. 1986a Theoretical interpretation of atmospheric wavenumber spectra at wind and temperature observed by commercial aircraft during GASP., J. Atm. Sci., **43**, 729-740.

[39] Gage K.S. and Nastrom G. D. 1986b Spectrum of atmospheric vertical displacements and spectrum of conservative scalar passive additives due to quasi-horizontal atmospheric motions, *J. Geophys. Res.*, **D91**, 13211 - 13216.

[40] Pao Y.- H. and Goldburg A., eds. 1969 *Clear Air Turbulence and its detection*, Plenum, New York; see papers by G.K. Mather and by R. T. H. Collis, R. M. Endlich & R. L. Mancuso.

[41] Monin & Ozmidov 1985 [8], p. 218 make the following statement: *In most cases, however, large-scale turbulence in the ocean seem to be not purely three- or two-dimensional, but intermediate between the two. To verify this statement we calculated the slopes of 47 spectra of large-scale velocity fluctuations reported by the Woods Hole Oceanographic Institute (1965, 1966, 1967, 1970, 1971, 1974, 1975). At frequencies from about 5 to 0.005 cycleh^{-1} the mean slope of the spectra prove to be - 2.11. The slope ranged from - 1.87 to - 2.70. As usual, the spectra analyzed have pronounced peaks at the inertial and tidal periods. When estimated separately, the slopes of the spectra at frequencies above and below the inertial and tidal ones had slightly differing values: -2.34 in the former case and -1.92 in the latter.*

[42] Caughey S. J. 1977 Boundary layer turbulence spectra in stable conditions, *Boundary Layer Meteorology*, **11**, 3 - 14.

[43] Caughey S. J. and Palmer S. G. 1979 Some aspects of turbulence structure through the depth of the convective boundary layer, *Quart. J. Roy. Met. Soc.*, **105**, 811 - 827.

[44] Gage K.S. and Nastrom G. D. 1986c Horizontal spectra of atmospheric traces measured during the global atmospheric sampling program, *J. Geophys. Res.*, **D91**, 13201 - 13209.

[45] Sreenivasan K. R. 1991 On local isotropy of passive scalars in turbulent shear flows, *Proc. Roy. Soc.*, **A434**, 165 - 182.

[46] Prasad, R. R. and Sreenivasan, K. R. 1990 The measurement and interpretation of fractal dimensions of the scalar interface in turbulent flows, *Phys. Fluids*, **A2**, 792 - 807.

[47] Jaesh, Tong C. and Warhaft Z. 1994 On temperature spectra in grid turbulence, *Phys. Fluids*, **6**, 306 - 312.

[48] Warhaft Z. and Lumley J. L. 1978 An experimental of the decay of temperature fluctuations in grid-generated turbulence, *J. Fluid Mech.*, **88**, 659 - 688.

[49] Branover H., Bershadskii A., Eidelman A. and Nagorny M. 1993 Possibility of simulating geophysical flow phenomena by laboratory experiments, *Bound. Layer Met.*, **62**, 117 - 128.

[50] Morel P. and Larcheveque M. 1974 Relative dispersion of constant-level balloons in the 200-mb general circulation. J. Atm. Sci., **31**, 2189-2196.

[51] Mory M. and Hopfinger E.J. 1986 Structure functions in a rotationally dominated turbulent flow, *Phys. Fluids*, **29**, 2140-2146.

[52] Olson F. C. W. and Ichie T. 1959 Horizontal diffusion, *Science*, **130**, No. 3384. 1255.

[53] Corrsin S. 1959 Outline of some topics in homogeneous turbulent flow, *J. Geophys. Res.*, **64**, 2134 - 2150.

[54] Lovejoy S. 1982 Area-Perimeter Relation for Rain and Cloud Areas, Science, **216**, 185 - 187.

[55] Townsend A. A. 1966 The mechanism of entrainment in free turbulent flows, *J. Fluid Mechanics*, **26**, 689 - 715.

[56] Mandelbrot B.B. 1982 *The Fractal geometry of Nature*, p.110, Freeman.

[57] Gifford F. A. 1989 The shape of large tropospheric clouds, or "very like a whale", *Bull. Amer. Met. Soc.*, **70**, 468 - 475.

[58] Gibson C. H. and Masiello P. J. 1972 Observations of the variability of dissipation rates of turbulent velocity and temperature fields, in *Statistical Models and Turbulence*, 431 (ed. M. Rosenblatt and C. Van Atta, Springer, 1972).

[59] Sreenivasan K. R., Antonia R. A. and Danh H. Q. 1977 Temperature dissipation fluctuations in a turbulent boundary layer, *Phys. Fluids*, **20**, 1238 - 1247.

[60] Bershadskii A. and Tsinober A. 1993 On differences in fractal properties of rate of dissipation of energy and passive scalar and their surrogates, submitted.

[61] Stiassnie M., Hadad V. and Poreh M. 1993 Simulation of turbulent dispersion, submitted to *J. Fluid Mech.*.

[62] Bar'yakhtar V. G., Gonchar V. Yu. and Yanovsky V. V. 1993 Origin of the fractal structure of the Chernobyl spot of the radionuclides contamination, *Ann. Geophys.*, Suppl.II to vol. **11**, C306.

[63] Kadanoff L. P. 1990 Scaling and structures in the hard turbulence region of Rayleigh-Bénard convection. In *New Perspectives in Turbulence*, ed. L.Sirovich, p. 265.

[64] Normal R. E. 1993 Strange spatio-temporal patterns, self-organized universal crises and politico-dynamics via symmetric chaos, neutral networks and the fractal wavelet transform of the magic of names. *Nonlinear Science Today*, **3**, 13 - 14.

[65] Bradshaw P. 1994 Turbulence: the chief outstanding difficulty of our subject, *Experiments in fluids*, **16**, 203 - 216.

[66] Ozmidov R. V., Astok V. K., Gezentsvey A. N. and Yukhat M. K. 1971, Statistical characteristics of the concentration field of a passive impurity introduced into the sea, *Atmosph. Ocean. Phys.*, **7**, 636 - 641.

[67] Monin A. S. and Ozmidov R. V. 1978 Turbulence in the ocean. In *Ocean Physics*, **I**, 148 - 207, Nauka, Moscow (in Russian)

[68] Falkovich G. E. and Medvedev S. B. 1992 Kolmogorov-like spectrum for turbulence of inertial-gravity waves, *Europhys. Lett.*, **19**, 298 - 284.

[69] Falkovich G. 1992 Inverse cascade and wave condensate in mesoscale atmospheric turbulence, *Phys. Rev. Lett.*, **69**, 3173 - 3176.

[70] Mahalov, A. 1993 *Private communication.*

Conditionally-averaged dynamics of turbulence, new scaling and stochastic modelling

Evgeny A. Novikov

Institute for Nonlinear Science
University of California, San Diego
La Jolla, CA 92093, U.S.A.

Abstract. Conditional statistical characteristics of turbulence as functions of fixed vorticity (in particular characteristics of vortex stretching and twisting) are obtained from the Navier-Stokes equations (NSE) analytically and by direct-numerical simulations. A hierarchy of new scales for coherent structures in three-dimensional (3D) and 2D turbulence is obtained from a covariance analysis of NSE for local characteristics of motion (vorticity in 3D and vorticity gradient in 2D). Corresponding scales for gradient of a scalar field in 2D and 3D turbulence are obtained from the convection-diffusion equation. A complementary description of turbulence in terms of random processes, consistent with NSE, is also developed. The intermittency is described in terms of breakdown coefficients and associated infinitely-divisible probability distributions. Applications of these results to subgrid modelling for large-eddy simulations are considered.

1 Introduction and summary

The major request from the naval, aerospace, energy and environmental engineering communities to the theory of turbulence is to reduce the enormous number of degrees of freedom in turbulent flows to a level manageable by computer simulation. The general goal of this work is to advance the knowledge of the structure of turbulence, aiming at a subgrid-scale modelling for large-eddy simulations (LES). Among the tools being developed for achievement of this goal are:

1. Conditional averaging of the Navier-Stokes equations (NSE) and corresponding functional formalism.

2. Incorporation of experimentally and numerically observed coherent structures into the statistical description of turbulence.

3. Use of Markov processes with dependent increments, consistent with NSE.

4. Description of intermittency in terms of breakdown coefficients and associated infinitely- divisible probability distributions.

A number of new results were obtained, in particular:

1. A new scale for the experimentally and numerically observed three-dimensional (3D) "vortex strings" $l_s \sim L Re^{-3/10}$ (L-external scale $Re = \frac{LV}{\nu}$ — Reynolds number, V-characteristic velocity, ν— molecular viscosity). Corresponding scale for $2D$ cogerent vortices $l_c = L Re^{-1/4}$. Characteristic scales for coherent structures of a scalar field in $3D$ and $2D$ turbulence are also obtained.

2. Exponential behavior of conditionally-averaged rates of vortex stretching and dissipation as functions of fixed vorticity – a result of direct-numerical simulations (DNS) on a CM-5 parallel computer, using 6th-order finite differences on $(256)^3$ grid.

3. Probability distribution for $3D$ vectors of velocity increments, which is far from Gaussian and has an unusual form.

4. General probability distribution of the breakdown coefficients – characteristics of intermittency.

5. A hierarchy of subgrid-scale models, preliminary tested by DNS (for moderate Re) and by LES for isotropic and free-surface turbulence.

2 Conditionally averaged dynamics of turbulence

The dynamics and statistics of turbulent flows is better understood in terms of local characteristics (lc) which have an internal mechanism of amplification[1]. For three-dimensional (3D) turbulent flow the lc is the vorticity field and self-amplification is due to the vortex stretching[2,3]. For 2D turbulence the lc is the vorticity gradient[4,5,6]. We use the concept of self-amplification, because in both cases the deformation rate tensor, responsible for amplification, is expressed in terms of lc (vorticity in 3D and vorticity gradient in 2D). Conditional averaging of the Navier-Stokes equations (NSE), written in terms of lc with fixed lc at a point, transforms the major nonlinear amplification term into a linear term[4,5,6].

This allows, in particular, an analytical study of the conditionally-averaged 3D vorticity field:

$$\overline{\Omega}_i(\mathbf{r},\boldsymbol{\omega}) = [f_1(\boldsymbol{\omega})]^{-1} \int \omega_i' f_2(\mathbf{r},\boldsymbol{\omega},\boldsymbol{\omega}')d\boldsymbol{\omega}' \tag{1}$$

Here $\mathbf{r}$ is the distance from the point with fixed vorticity $\boldsymbol{\omega}$, f_1 and f_2 are the one-point and two-point probability density functions (*pdf*) for the vorticity field. We note that the conditionally-averaged NSE with fixed vorticity in n points corresponds to a hierarchy of "kinetic" equations for the n-point *pdf*[3,6]. For the Fourier-transform of (1) we have general expression[6]:

$$\tilde{\Omega}_i(\mathbf{k},\omega) = g(\omega,k,\mu)(\sigma_i - \mu n_i) + h(\omega,k,\mu)\epsilon_{ijk}\sigma_j n_k, \tag{2}$$

where $\sigma_i = \omega_i \omega^{-1}, n_i = k_i k^{-1}, \mu = \sigma_i n_i, g(-\mu) = g(\mu), h(-\mu) = -h(\mu)$. Here we used the local isotropy of turbulence and solenoidality of vorticity field, $\mathbf{k}$ is the wave-number vector with unit vector $\mathbf{n}$, $\boldsymbol{\sigma}$ is unit vector of fixed vorticity, μ is a scalar product of these unit vectors, ϵ_{ijk} is the unite antisymmetric tensor. Scalar g is a symmetric function of μ and scalar h is antisymmetric. Scalar h represents twisting of vortex lines, which is necessary for the statistical balance between vortex stretching and viscouse smoothing for high Reynolds number Re[3−6] (see also below discussion after formula (7)). This statistically important twist probably contributes to the helically-shaped explosion of "vortex strings" when they become unstable[7,8].

The NSE for 3D vorticity in incompressible fluid are:

$$\frac{\partial \omega_i}{\partial t} + v_k \frac{\partial \omega_i}{\partial x_k} = \frac{\partial v_i}{\partial x_k}\omega_k + \nu\Delta\omega_i + \phi_i, \quad \frac{\partial v_i}{\partial x_i} = 0, \tag{3}$$

$$\omega_i = \epsilon_{ijk}\frac{\partial v_k}{\partial x_j}, \quad \phi_i = \epsilon_{ijk}\frac{\partial f_k}{\partial x_j}, \quad \frac{\partial f_i}{\partial x_i} = 0 \tag{4}$$

Here v_i, ω_i and f_i are respectively the velocity, the vorticity and an external force, ν is the kinematic viscosity. The first term in the right-hand side of (3) represents the effect of vortex stretching, which is absent for 2D flow. We consider homogeneous and isotropic turbulence and average (3) conditionally with fixed vorticity. The nonlinear terms are reduced as follows[4−6]:

$$\overline{v_k \frac{\partial \omega_i}{\partial x_k}} = 0, \quad \overline{\frac{\partial v_i}{\partial x_k}\omega_k} = \overline{\frac{\partial v_i}{\partial x_k}}\omega_k = \alpha(\omega,t)\omega_i \tag{5}$$

$$\alpha(\omega,t) = \overline{\frac{\partial v_i}{\partial x_k}}\sigma_i\sigma_k, \quad \overline{D}_{ik} = \frac{1}{2}(\overline{\frac{\partial v_i}{\partial x_k}} + \overline{\frac{\partial v_k}{\partial x_i}}) = \frac{1}{2}\alpha(\omega,t)(3\sigma_i\sigma_k - \delta_{ik}) \tag{6}$$

The overline denotes conditional averaging with fixed ω. We note that conditional averaging, generally, does not commute with spatial and temporal derivatives[6].

As shown in (6), α is the eigenvalue of the conditionally-averaged tensor of deformation rates $(\overline{D}_{ik})$ in the direction of σ_i. δ_{ik} is the unit tensor. The effect of large- scale random forces on various characteristics of vorticity has been studied elsewhere[6,8,9]. In this section, we consider decaying turbulence without random forcing ($\phi_i = 0$).

Conditional averaging of equation (3) with the use of (5) and multiplication by σ_i gives the conditional balance for the magnitude of vorticity:

$$\frac{\overline{\partial \omega}}{\partial t} = (\alpha - \beta)\omega, \quad \beta(\omega, t)\omega \equiv -\nu \overline{\Delta \omega_i \sigma_i} \tag{7}$$

Here β is a coefficient of relaxation due to viscous smoothing. For turbulence with high Re the effects of vortex stretching and viscous smoothing are expected to be balanced for every ω[4−6]: $\alpha \approx \beta$. These coefficients are linear functionals of the conditionally-averaged vorticity field[6]:

$$\alpha = -\epsilon_{ijm}\sigma_i \int r_j \tilde{\Omega}_m \mu d\mathbf{k} = - \int \mu(1 - \mu^2)h d\mathbf{k}, \tag{8}$$

$$\beta\omega = \nu\sigma_i \int k^2 \tilde{\Omega}_i d\mathbf{k} = \nu \int k^2(1 - \mu^2)g d\mathbf{k}, \tag{9}$$

where (2) is used. Formula (8) shows that vortex stretching is linked with the twisting of vortex lines, represented by scalar h. The second term in field (2) corresponds to two coaxial distributed vortex rings with opposite signs of vorticity, which produce stretching of the central fluid element in the direction of σ_i (axis of rings).

Multiplication of equation (7) by ω gives the conditional balance of enstrophy:

$$\tau(\omega, t) \equiv \frac{1}{2}\frac{\overline{\partial \omega^2}}{\partial t} = q - d, \quad \text{where } q = \alpha\omega^2, \quad d = \beta\omega^2 \tag{10}$$

Here τ, q and d respectively represent the conditionally-averaged time derivative, production and dissipation of enstrophy. If we now multiply (10) by the *pdf* of the magnitude of vorticity $p(\omega, t)$ and integrate over ω, we recover unconditional (total) balance of enstrophy:

$$\frac{1}{2}\frac{\partial}{\partial t}\langle \omega^2 \rangle = \left\langle \frac{\partial v_i}{\partial x_k}\omega_i\omega_k \right\rangle - \nu\left\langle (\frac{\partial \omega_i}{\partial x_k})^2 \right\rangle \tag{11}$$

Here $\langle\ \rangle$ means unconditional statistical averaging. For the derivation of (11) from (10), we used the following general formulae[6]:

$$\left\langle \frac{\overline{\partial \psi(\omega)}}{\partial t} \right\rangle = \left\langle \frac{\partial \psi(\omega)}{\partial t} \right\rangle = \frac{\partial}{\partial t}\langle \psi(\omega) \rangle, \tag{12}$$

where in our case $\psi(\omega) = \omega^2$. In order to quantify the preceding conditional statistics, a series of direct numerical simulations (DNS) of isotropic turbulence has been performed on a CM-5 parallel computer, using 6th-order finite-differences on a $(256)^3$ grid[10]. Details of a similar numerical procedure are provided in Ref. [11]. The DNS were initialized with random noise in the lowest four wavenumber bins. The energy was held constant by uniformly rescaling the velocity field after each time step. After the enstrophy reached a constant value the velocity field was allowed to decay and we collected data. The energy spectra (that are plotted in Figure 1 in Ref. [10]) show good agreement with experimental data[12], especially for high wavenumbers, which is important for modeling vorticity in isotropic turbulence.

For evaluation of the conditionally-averaged characteristics, the data-sets were divided into 256 subsets, corresponding to equal intervals of ω. The results for α and β are presented in Figures 2a,b in Ref. [10] for different Taylor-microscale Reynolds numbers $R_\lambda \sim \sqrt{Re}$. The mutual balancing of stretching with dissipation is closer for the higher value of R_λ as was expected[4,6]. Most surprisingly, α and β grow exponentially. In fact, $\alpha \approx 0.13\,\omega_* \exp\,(0.16\,\omega/\omega_*)$, $(\omega_* = <\omega^2>^{1/2})$, with the same coefficients for $52 \leq R_\lambda \leq 80$.

We can speculate that this exponential growth of two opposing physical effects (stretching and diffusion) provides a statistical environment for local instabilities and strong fluctuations–formation of strong localized vortices ("vortex strings"), followed by a quick breakdown of such vortices when they become unstable. The conditionally-averaged vorticity production term $(\alpha\omega)$, when acting by itself, due to the exponential dependence of α will produce a singularity in the vorticity in a finite time. The diffusion term $(\beta\omega)$, when acted by itself, will cause quick decay of vorticity. For high Reynolds numbers these two effects will balance each other, but only statistically. In a particular realization, the α term can dominate locally in space and time, and a vortex string will grow until it becomes unstable under the influence of the large-scale motion[8]. Then the β term will take over and diffuse the vorticity. The vortex strings are expected to be twisted because of the twist in the conditionally-averaged vorticity field(2).

The *pdf* of vorticity (see Figure 3 in Ref. [10]) decreases exponentially for high ω, such that $p(\omega) \sim \exp\,(-1.8\omega/\tilde{\omega})$ and $\exp\,(-2.1\omega/\tilde{\omega})$ for $R_\lambda = 79.9$ and 51.8 respectively. The experimentally-measured attenuation $p(\omega) \sim \exp\,(-2.56\,\omega/\tilde{\omega})$ is more rapid than the DNS predictions probably because of the limited range of ω that had been measured[13]. The negative exponent in $p(\omega)$ is of order of magnitude larger than the positive exponent in $\alpha(\omega)$, so the finiteness of the unconditional production of enstrophy is insured.

We note that the simple analytical solution for the conditionally-averaged vorticity field assumes that α does not depend on ω[6]. With constant α, the production term $\alpha\omega$, when acting by itself, will not produce a singularity in the vorticity in a finite time. The preceding DNS results show for moderate Re that α depends exponentially on ω and the conditionally-averaged vorticity field has a more intricate structure (connected with the formation and destruction of vortex strings) than in the simple analytical solution. The coefficients $\alpha(\omega)$ and $\beta(\omega)$ provide the partial information (8), (9) about the conditionally-averaged vorticity field. The next step is to obtain the whole field, which is of great importance not only for understanding the structure of turbulence, but also for turbulence modeling in many applications (compare with simple subgrid-scale models in Refs. [11,14]), applied for isotropic and free-surface turbulence; see also next secton).

3 New scales

The natural starting point in turbulence scaling is the Kolmogorov internal scale[15–17]

$$l_\nu = \left(\frac{\nu^3}{\epsilon}\right)^{1/4} \sim L\, Re^{-3/4}, \quad \epsilon \sim \frac{V^3}{L} \tag{13}$$

where ϵ is the mean rate of energy dissipation. In the inertial range of scales

$$l_\nu << r << L, \tag{14}$$

we have classicial "2/3 law"[15–17]:

$$\langle [\mathbf{v}(\mathbf{x}+\mathbf{r}) - \mathbf{v}(\mathbf{x})]^2 \rangle \sim (\epsilon r)^{2/3} \tag{15}$$

The number of effective degrees of freedom is:

$$N \sim \left(\frac{L}{l_\nu}\right)^3 \sim Re^{9/4} \tag{16}$$

For example, a ship wake with $Re \sim 10^9$ requires $N \sim 10^{20}$. Thus, for this and many other applications DNS are impossible now and in the foreseeable future.

The majority of degrees of freedom is associated with small-scale motions in the inertial range (14). The traditional LES approach is to resolve as many degrees of freedom as possible with a given computational capacity and to model the rest of degrees of freedom. However, taking into account a delicate nature of turbulence, such an approach might be inconsistent with NSE. The alternative is to use a natural matching scale, dictated by NSE. Such a scale was found [8,9] from the balance of vorticity correlations in 3D turbulence and was associated with "vortex strings".

$$l_s \sim L\, Re^{-3/10} \tag{17}$$

This scale (with exact numerical coefficient[8]) was obtained from NSE by using a special functional formalism[18,19]. At this scale the effects of large-scale motion and viscosity are balanced and the nonlinear effect of vortex stretching gives zero contribution in the balance of vorticity correlations. The number of degrees of freedom, based on this scale, is:

$$N_s \sim \left(\frac{L}{l_s}\right)^3 \sim Re^{9/10} \tag{18}$$

Thus, for $Re \sim 10^9$ we have $N_s \sim 10^8$. In order to realize such a huge savings in numerical capacity, we need a modelling of the formation and destruction of vortex strings (see previous section).

One way of doing such modelling is to introduce in NSE (3) a vortex relaxation term:

$$-\tau_s^{-1}\left[\omega_i(\mathbf{x}) - S\int d\mathbf{r}\, m(\mathbf{r})\overline{\Omega}_i(-\mathbf{r},\hat{\omega}(\mathbf{x}+\mathbf{r}))\right],\tag{19}$$

$$\tau_s \sim L^{2/3}\epsilon^{-1/3}Re^{-1/5},\tag{20}$$

$$\hat{\omega}_i(\mathbf{x}+\mathbf{r}) = \frac{1}{1+\gamma(\mathbf{r})}\left[\omega_i(\mathbf{x}+\mathbf{r}) + \gamma(\mathbf{r})\overline{\Omega}_i(\mathbf{r},\omega(\mathbf{x}))\right]\tag{21}$$

Here τ_s is the relaxation time[8], S is the solenoidal projection operator, m and γ are weighing coefficients, depending on numerical scheme, and $\hat{\omega}$ is an intermediate field, designed for smooth relaxation. We plan to test this vortex relaxation in LES.

In 2D turbulence lc is the vorticity gradient and corresponding scale for coherent vortices is[9]:

$$l_c \sim L\, Re^{-1/4}\tag{22}$$

For passive scalar fields in 3D and 2D turbulence similar consideration is based on the correlation balance of the gradient of a scalar field[20]. Corresponding scales for coherent structures (apart from numerical coefficients) can be formally obtained from (17) and (22) by substitution diffusivity instead of viscosity[20]. A hierarchy of scales for more delicate properties of coherent structures in 3D and 2D turbulence is also obtained from corresponding correlation balances by using the same functional formalism[9,20].

4 Velocity increments and Markov modelling

A complimentary to the vorticity analysis is the description of turbulence in terms of velocity increments (vi):

$$u_i(\mathbf{x}',\mathbf{x}), = v_i(\mathbf{x}') - v_i(\mathbf{x}) = u_r n_i + \tilde{u}_i, \mathbf{x}' = \mathbf{x} + \mathbf{r}, n_i = r_i r^{-1}\tag{23}$$

Here $u_r = u_i n_i$ is the radial (longitudinal) component of vi, $\tilde{u}_i$ is the transversal vi (vector contained in the plane normal to the separation distance $\mathbf{r}$). The two components are physically different, even simply because of incompressibility. Loosely speaking u_r and $\tilde{u}_i$ signify correspondingly deformation along the vector $\mathbf{r}$ and rotation around a vector normal to $\mathbf{r}$.

In the inertial range we have the Kolmogorov result

$$\langle u_r^3\rangle = -\frac{4}{5}\,\epsilon r,\tag{24}$$

which can be written in tensor form[21]

$$\langle u_i u_j u_k\rangle = -\frac{4}{15}\,\epsilon\,(r_i\delta_{jk} + r_j\delta_{ki} + r_k\delta_{ij})\tag{25}$$

The Kolmogorov result was originally obtained for decaying turbulence. The same result was derived[18] (by using the above mentioned functional formalism[18,19]) for statistically stationary turbulence with large-scale random forcing. Statistical preference of negative u_r, emphasized by (24), corresponds to compression of fluid element in the direction of $\mathbf{r}$ and (because of incompressibility) expansion in normal directions. Since $\tilde{u}_i$ represents vortex, oriented normally to $\mathbf{r}$, we can interpret (24) as an inertial range manifestation of the vortex stretching (compare with local description of this effect in Section 2).

In LES we are dealing with filtered velocity field:

$$\{v_i(\mathbf{x})\} \equiv \int d\mathbf{x}' v_i(\mathbf{x}') f(\mathbf{x}, \mathbf{x}') \tag{26}$$

where f is a filter with a characteristic scale l, which is much larger than l_ν. The subgrid stress tensor can be expressed in terms of vi:

$$\tau_{ij} = \{v_i v_j\} - \{v_i\}\{v_j\} = \frac{1}{2} \int d\mathbf{x}' d\mathbf{x}'' u_i(\mathbf{x}'', \mathbf{x}') u_j(\mathbf{x}'', \mathbf{x}') f(\mathbf{x}, \mathbf{x}') f(\mathbf{x}, \mathbf{x}''), \tag{27}$$

According to the concept of conditional averaging[6], we need to average the product of vi in (27) conditionally with fixed filtered velocity. This will give us closed equations for filtered velocity in LES. Thus, subgrid-scale modelling is linked with the statistics of vi.

For statistically stationary turbulence, the continuity equation for the probability has the form (compare with Ref. [21]):

$$u_i \frac{\partial P}{\partial r_i} + \frac{\partial}{\partial u_i}(\overline{\alpha}_i P) = 0 \tag{28}$$

Here $P(\mathbf{u}, \mathbf{r})$ is the Eulerian probability density for the vector of vi $(PDVVI)$, which depends on probabilistic argument $\mathbf{u}$, distance $\mathbf{r}$ and parametrically may depend on absolute position in turbulent flow. $\overline{\alpha}_i$ is the operator or relative acceleration, conditionally averaged with fixed $\mathbf{u}$. Assuming that the relative velocity of fluid particles is Markovian with local relaxation and simplest forcing (diffusion in velocity space), we have[21,22]:

$$\overline{\alpha_i} = -\frac{c}{\tau(r)}(u_i + \gamma u_r n_i) - \frac{1}{2} m_{ij} \frac{\partial}{\partial u_j}, \tag{29}$$

$$\tau(r) = \langle u_k^2 \rangle \epsilon^{-1}, \quad m_{ij} = \epsilon \left[a n_i n_j + b \left(\delta_{ij} - n_i n_j \right) \right]. \tag{30}$$

Here c and γ are nondimensional constants, $\tau(r)$ is characteristic time of local relaxation, $\langle u_k^2 \rangle$ is Eulerian structural function of second order; a and b are nondimensional coefficients of diffusion in radial and transverse directions.

From incompressibility we have

$$\frac{\partial}{\partial r_i} \langle u_i u_j \rangle = 0 \tag{31}$$

By multiplying (28) by u_j, integrating over $\mathbf{u}$, and using (31), we get the constraint:

$$\langle \overline{\alpha}_i \rangle = 0, \tag{32}$$

which is clearly satisfied by the operator (29). Similarly, (25) gives tensor constraint:

$$-\frac{4}{3}\epsilon\delta_{jk} = \langle\overline{\alpha}_j u_k\rangle + \langle\overline{\alpha}_k u_j\rangle \tag{33}$$

which is also satisfied by (29) and imposes conditions on coefficients:

$$c(2+s) = (b+\frac{4}{3})(3+s), \; c(2\gamma - s) = (a-b)(3+s) \tag{34}$$

Here we used similarity for the second-order moment in the inertial range and incompressibility (31):

$$\langle u^2\rangle \sim r^s, s = \frac{2}{3} - \mu(\frac{2}{3}), \langle\tilde{u}^2\rangle = (2+s)\langle u_r^2\rangle, \tag{35}$$

$\frac{2}{3}$ corresponds to the classical Kolmogorov similarity (15), $\mu(k)$ is the intermittency exponent (see next section), and $\mu(\frac{2}{3})$ is known to be negative and small.
Consider probability for large vi:

$$u >> \langle u^2\rangle, \tag{36}$$

which corresponds to motions with large local shear of velocity, for example, to local jetlike motions ("streaks"), when one of the fluid particles is inside the jet and another is outside. The asymptotic solution of (28) with classical similarity reduces to the function of unusual argument[22]:

$$P(u_r, \tilde{u}, r) = (\epsilon r\tilde{u})^{-3/4} f\left[\frac{u_r^2 + \tilde{u}^2}{(\epsilon r\tilde{u})^{1/2}}\right] \tag{37}$$

$$f(z) = N_{\pm}^{-1} \exp\left\{-\theta_{\pm} z^{2/3}\right\} \tag{38}$$

Here constants N and θ are different for the cases $u_r > 0$ and $u_r < 0$, which is reflected by the subscript $\pm$. This asymptotics was obtained without Markovian assumption and corresponds to experimentally observed exponential behaviour of *pdf* for u_r. The intermittency corrections to this solution were also presented[22]. The global solution of (28) was obtained numerically[23].

The operator (29) also gives the Lagrangian description of turbulence[21,24,23], which corresponds, in particular, to Richardson's law:

$$\langle r^2(t)\rangle_L \sim \epsilon t^3 \tag{39}$$

Here $r(t)$ is the distance between two fluid particles and subscript L indicates the Lagrangian ensemble of averaging. The exact relations between Lagrangian and Eulerian descriptions[25,26] are used in this approach. The next step is to implement Markov modelling in LES.

5 Intermittency and infinitely-divisible distributions

The concept of scale similarity of random fields was developed more than a quarter century ago[27–29] (see also a more recent account of this theory[24] with some additional details, including the multifractal representation). The original purpose of this concept was to describe the phenonema of self-similar intermittency of turbulent flows which was observed experimentally[30]. However, the concept is quite general and its applications (and rediscoveries) have emerged in many areas of science, as diverse as biology and astrophysics. The scale similarity can be imbedded into the theory of infinitely-divisible probability distributions[31]. This gives us access to the well developed mathematical apparatus, which can be used not only for the description of experimental data but also for derivation of scale similarity from basic principles (from the Navier-Stokes equations in the case of turbulence).

We can distinguish between discrete and continuous self-similarity. In the discrete case there is a preferable scale factor leading to the logarithmically-periodic modulations[27,24] (this phenomena sometimes is called lacunarity[32,33]). Here we concentrate on the continuous self-similarity without preferable scale factor.

For simplicity consider a one-dimensional section of a non-negative scalar field (e.g., dissipation rate). This is in accord with experimental reading in time (with the aid of "frozen-flow" hypothesis). Thus ϵ_r is the dissipation (or similar quantity, see Ref. [29]) averaged over the segment r. For a turbulent flow which is locally isotropic in scales less than a certain external scale L, the two- and three-dimensional statistical characteristics have the same form (see obvious exceptions in Ref. [24]).

Consider inertial range of scales $l_* << r << L$, where l_* is an inertial scale, which can differ from the Kolmogorov internal scale because of the intermittency correction[27,34,35]. In the inertial range we shall single out three segments inserted in one another with length $r < \rho < l$ and introduce corresponding breakdown coefficients (bdc):

$$q_{r,l} \equiv \epsilon_r/\epsilon_l \leq l/r, \tag{40}$$

$$q_{r,l} = q_{r,\rho}q_{\rho,l} \tag{41}$$

In (40) we utilized the fact that the dissipation rate is nonnegative. The scale similarity is determined by the following conditions: (i) probability distribution for bdc depends only on the ratio of the corresponding scales and (ii) $q_{r,\rho}$ and $q_{\rho,l}$ are statistically independent [instead of (ii) we can use a less restrictive condition[24]]. More general conditions, which take into account exact relative positions of segments, are considered in Ref. [29] (see also discussion below). For the moments of bdc, from conditions of scale similarity and (41), we have

$$\left\langle q_{r,l}^p \right\rangle = a_p(l/r) = a_p(\rho/r)a_p(l/\rho), \tag{42}$$

$$a_p(l/r) = (l/r)^{\mu(p)}, \quad \mu(0) = 0 \tag{43}$$

Here $\langle\ \rangle$, as before, means statistical averaging and we used arbitrariness of ρ and normalization of probability. In Refs. [28] and [29] it was shown that

$$\mu(1) = 0, \quad 0 < \mu(2) \equiv \mu < 1, \tag{44}$$

45

$$\mu(p + \delta) - \mu(p) \leq \delta, \quad (\delta \geq 0) \tag{45}$$

$$\mu(p) \leq \mu + p - 2, \quad (p \geq 2) \tag{46}$$

Inequality (45) follows from (40), and (46) follows from (45) and (44). The probability density $W(q, l/r)$ for $q_{r,l}$ is uniquely defined by the set of $\mu(p)$ with integer p ($p = 0, 1, 2...$), because (46) ensures the fulfillment of the Carleman condition[36]:

$$\sum_{p=1}^{\infty} (a_{2p})^{-\frac{1}{2p}} = \infty \tag{47}$$

If function $\mu(p)$ has analytical continuation into the complex domain, than characteristic function of $ln(q_{r,l})$ has the form:

$$\psi(s, l/r) = \langle \exp{(isln q_{r,l})} \rangle = \int_o^{l/r} q^{is} W(q, l/r) dq = (l/r)^{\mu(is)} \tag{48}$$

The last equality in (48) can be inverted, giving:

$$W(q, l/r) = \frac{1}{2\pi q} \int_{-\infty}^{\infty} \exp{[-isln q + \mu(is)ln(l/r)]} ds \tag{49}$$

The standard way to construct a model is to make an assumption, say, about $W(q, 2)$. From (43) we obtain

$$\mu(p) = \log_2 \left[\int_o^2 q^p W(q, 2) dq \right] \tag{50}$$

Then, from (49) (or from a more long formula[29,24] when $\mu(p)$ is not analytical) we may calculate $W(q, l/r)$. In order for model to have physical and mathematical sense, we must ensure that *for arbitrary l/r* the probability density W is non-negative and properly normalized by integration over the finite interval $[0, l/r]$. Generally, it is not easy to verify these conditions analytically or even numerically. Thus, many models, proposed in the literature, remain questionable.

The solution to this problem comes from the observation that for arbitrary l/r and arbitrary integer n, equation (48) can be written in the form:

$$\psi(s, l/r) = \psi^n(s, (l/r)^{1/n}) \tag{51}$$

Thus, $ln q_{r,l}$ has the infinitely-divisible probability distribution[36]. Now we can use the Lévy-Baxter-Shapiro (LBS) theorem[36], which gives the general form of the characteristic function for the infinitely-divisible distribution, concentrated on interval $[0, \infty)$:

$$\chi(s) = \exp \left\{ ibs - \int_o^{\infty} \frac{1 - e^{isx}}{x} P(dx) \right\}, \quad b \geq 0 \tag{52}$$

Here P is a measure on the open interval $(0, \infty)$ such that $(1+x)^{-1}$ is integrable with respect to P. From (48) and (52), by using variable $z_{r,l} = -ln(\frac{r}{l}q_{r,l})$, $(0 \leq z_{r,l} < \infty)$ we get

$$\mu(p) = \kappa p - \int_o^\infty \frac{1 - e^{-px}}{x} F(dx), \tag{53}$$

$$\kappa = 1 - \frac{b}{ln(l/r)} \leq 1, \quad P(dx) = ln(l/r) \cdot F(dx) \tag{54}$$

Here measure F has the same (stated above) properties as P. From the first equality in (44) we have an additional condition on the measure:

$$\int_o^\infty \frac{1 - e^{-x}}{x} F(dx) = \kappa \leq 1 \tag{55}$$

All other properties of $\mu(p)$ and $W(q, l/r)$ are ensured by the theorem (52). Thus, we get a new approach for modeling, namely, by choosing measure F.

Consider, for example, measure with the density

$$F'(x) = Ax^{\alpha-1} \exp\left(-x/\sigma\right) \tag{56}$$

where A, α and σ are positive constants. From (53) and (55) we get

$$\mu(p) = \kappa \left[p - \frac{(p\sigma + 1)^{1-\alpha} - 1}{(\sigma + 1)^{1-\alpha} - 1} \right], \quad \alpha \neq 1 \tag{57}$$

$$\mu(p) = \kappa \left[p - \frac{ln(p\sigma + 1)}{ln(\sigma + 1)} \right], \quad \alpha = 1 \tag{58}$$

A particular case of (58) with $\kappa = \sigma = 1$ was considered in Ref.[29] and corresponds to the constant density $W(q, 2)$. The density $W(q, l/r)$ for arbitrary l/r was calculated analytically in Ref.[29] and it has all the necessary properties. Another case of (58) with $\kappa = 1$, $\sigma \approx 0.283$ was considered recently[37] and compared with experimental data on $\mu(p)$. Formula (58) with $\kappa = 1$ corresponds to gamma distribution[35,36] for $z_{r,l}$. A particular case of (57) with $\kappa = 1$, $\alpha = 1/2$ was also considered recently[38] and corresponds to the distribution of $z_{r,l}$ presented in Problem 14, Chapter 13, Ref. [36]. As far as I know, the general formula (57) and corresponding probability distribution has not been considered before, at least, in the literature on turbulence. Experimental data on $\mu(p)$ (see, for example, Ref. [39]) corresponds to a smooth curve, which can be easily fitted by choosing parameters in (57) or (58). A more sensitive criterion will be comparison with detailed experimental data on probability density $W(q, l/r)$ for a variety of scale factor l/r. Some measurements of $W(q, l/r)$ have been presented in the literature[40,41]. However, in order to find an optional model, we need a more systematic study. A particular interest is in the asymptotic behavior of $W(q, l/r)$ when q approaches its maximum value l/r. Physically it corresponds to local events when all dissipation in the interval l is concentrated in the subinterval $r < l$. Let us note that, having in mind the scale similarity conditions, the optimal position of subinterval r is when its center coincides with the center of interval l. To demonstrate this, consider the opposite situation when one of the boundaries of subinterval r coincides with a boundary of interval l. In this case, the complementary subinterval $l - r$ can be considered on the same footing as subinterval r, taking into account the local isotropy of the turbulent dissipation field. From definition (40) we get relation between corresponding bdc's:

$$q_{r,l} = \lambda - (\lambda - 1)q_{l-r,l}, \quad \lambda = l/r > 1 \tag{59}$$

Here λ is the scale factor for $q_{r,l}$, corresponding scale factor for $q_{l-r,l}$ is $\lambda/(\lambda-1)$. By squaring both sides of (59), statistical averaging and using (44), we have

$$\left\langle q_{r,l}^2 \right\rangle = \lambda(2 - \lambda) + (\lambda - 1)^2 \left\langle q_{l-r,l}^2 \right\rangle \tag{60}$$

Here the statistical averaging is over the whole dissipation field or over the joint probability distributions of two $bdc's$. Equation (43), with the definition of μ in (44), gives

$$\lambda^\mu = \lambda(2 - \lambda) + (\lambda - 1)^2 (\frac{\lambda}{\lambda - 1})^\mu \tag{61}$$

We see that, no matter the value of μ (experimentally $\mu \approx 0.2$), equation (61) cannot be satisfied for arbitrary λ. Exception is $\lambda = 2$. For the central position of subinterval we do not have this contradiction with scale similarity.

Returning to the asymptotic of $W(q, \lambda)$ where q approaches its maximum value λ, consider the case of a gap:

$$\int_{\lambda_1}^{\lambda} W(q, \lambda)dq = 0, \quad \lambda_1 < \lambda \tag{62}$$

From (43) and (62) we get:

$$\mu(p) = \log_\lambda \left\{ \int_o^\lambda q^p W(q, \lambda)dq \right\} \leq p \log_\lambda(\lambda_1) \tag{63}$$

Thus,

$$h = \lim_{p \to \infty} \frac{\mu(p)}{p} \leq \log_\lambda(\lambda_1) < 1 \tag{64}$$

The value of h is important for the description of turbulent velocity increments[22]. If there is no gap, then

$$1 \geq h \geq \lim_{p \to \infty} \frac{1}{p} \log_\lambda \left\{ \lambda_i^p \int_{\lambda_i}^\lambda W(q, \lambda)dq \right\} = \log_\lambda(\lambda_i) \tag{65}$$

for any $\lambda_i < \lambda$. Thus, $h = 1$. We hope that presented results, including connection between self-similarity and infinitely-divisible distributions, can serve as a basis for a more detailed experimental, analytical and numerical studies of turbulence and other phenomena with scale similarity.

After the paper[31] was submitted for publication, an anonimous referee indicated preprints[42,43] (the last one just published), which are relevant to the subject. The approach in Refs.[42,43] is more heuristic. Let us note, the Lévy-Khinchin representation in Ref. [42] does not take into account that the energy dissipation rate ϵ is non-negative. The non-negativity leads to important constraints (44)-(46) and to the Carleman condition (47) (see details in Refs. [29,24]). The LBS theorem (52) does the precise job, reflecting all these facts. The concrete model, studied in Refs. [42,43], corresponds to $h = 2/3$ and, thus, to a substantial gap (62) in the *pdf*. Such a gap contradicts the existing experimental data[40,41]. Let us stress that more detailed experimental studies of $W(q, \lambda)$ are needed in order to find an optimal model.

The effects of intermittency are important for LES, because the spottiness of small-scale turbulence drives numerics. An intermittency correction in terms of $\mu(p)$ for a simple LES model was obtained in Ref. [24].

6 Conclusion

Different aspects of the phenomena of turbulence are described above by different approaches: conditionally-averaged dynamics (vortex stretching and dissipation), correlation balances combined with functional formalism (new scaling), statistics of velocity increments, Markov modelling, breakdown coefficients (intermittency) and related infinitely-divisible distributions. Each of these aspects is an independent and fruitful area of research. By putting them together, we can see the connections and a more general picture is emerging.

Acknowledgements

This work is supported by the U.S. Department of Energy under Grant No. DE-FG03-91ER14188 with Dr. Oscar P. Manley as program manager and by the Office of Naval Research under Grants No. ONR-N00014-92-J-1610 and No. ONR-14-94-1-0040 with Dr. Edwin P. Rood as program manager.

References

[1] E.A. Novikov, *Fluid Dyn. Res.* **6**, 79 (1990).

[2] G.J. Taylor, *Proc. Roy. Soc. London* **A164**, 15 (1938).

[3] E.A. Novikov, *Dokl. Akad. Nauk SSSR* **12**, 299 (1967).*[Sov. Phys. Dokl.* **12**, 1006 (1968)].

[4] E.A. Novikov, in Proc. of the Monte-Verita Sympos. (1991), edited by T. Dracos and A. Tsinober, "New Approaches and Concepts in Turbulence" (Birkauser, Basel, p. 131, 1993).

[5] E.A. Novikov, *Phys. Lett.* **A162**, 385 (1992), *J. Phys.* **A25**, L657 (1992).

[6] E.A. Novikov, *Fluid Dyn. Res.* **12**, 107 (1993).

[7] S. Douady, Y. Couder and M.E. Brachet, *Phys. Rev. Lett.* **67**, 983 (1991), and video produced by these authors.

[8] E.A. Novikov, *Phys. Rev. Lett.* **71**(17), 2718 (1993).

[9] E.A. Novikov, *Phys. Rev. E* **49**(2), R975 (1994).

[10] E.A. Novikov and D.G. Dommermuth, *Modern Physics Letters B*, **8**, No. 23, 1395 (1994).

[11] D.G. Dommermuth and E.A. Novikov, in Proc. of the Sixth Inter. Conf. on Numerical Ship Hydro., pp. 239-270, Iowa City, 1993.

[12] G. Comte-Bellot and S. Corrsin, *J. Fluid Mech.* **48**, 273 (1971).

[13] A. Bershadskii, E. Kit and A. Tsinober, *Phys. Fluids* **A5** (7), 1523 (1993).

[14] T.S. Lund and E.A. Novikov, Annual Research Briefs, CTR, pp. 27-43, Stanford University (1992).

[15] L.D. Landau and E.M. Lifshitz, Fluid Mechanics (Pergamon, N.Y., 1959).

[16] A.S. Monin and A.M. Yaglom, Statistical Fluid Mechanics (MIT Press, Cambridge, MA, 1975), v. 2.

[17] W.D. McComb, The Physics of Fluid Turbulence (Claredon, Oxford, 1990).

[18] E.A. Novikov, Zh. Eksp. Teor. Fiz. **47**, 1919 (1964) [Sov. Phys. JETP **20**, 1290 (1965)].

[19] Appendix H in Ref. [17].

[20] E.A. Novikov, *J. Physics A*, **27**, L797 (1994).

[21] E.A. Novikov, *Phys. Fluids A* **1**, 326 (1989).

[22] E.A. Novikov, *Phys. Rev. E* **46**, R6147 (1992)

[23] G. Pedrizzetti and E.A. Novikov, *J. Fluid Mech.* **280**, 69 (1994).

[24] E.A. Novikov, *Phys. Fluids A* **2**, 814 (1990).

[25] E.A. Novikov, *Appl. Math. Mech.* **33** (5), 862 (1969).

[26] E.A. Novikov, *Phys. Fluids* **29** (12), 3907 (1986).

[27] E.A. Novikov, *Dokl. Akad. Nauk SSSR* **168**/6, 1279 (1966) [*Sov. Phys. Dokl.* **11**, 497 (1966)].

[28] E.A. Novikov, *Dok. Akad. Nauk SSSR* **184**, 1072 (1969) [*Sov. Phys. Dokl.* **14**, 104 (1969)].

[29] E.A. Novikov, *Prikl. Mat. Mekh.* **35**, 266 (1971) [*Appl. Math. Mech.* **35**, 231 (1971)].

[30] E.A. Novikov and R.W. Stewart, *Izv. Ser. Geofiz.* No. 3, 408 (1964) [*Izv. Akad. Nauk SSSR Geophys. Ser.* No. 3, 245 (1964)].

[31] E.A. Novikov, *Phys. Rev. E* 50(5), R3303 (1994).

[32] B.B. Mandebrot, "The Fractal Geometry of Nature", Freeman, N.Y. (1983)

[33] L.A. Smith, J.D. Fournier and E.A. Spiegel, *Phys. Lett.*, **A114**, 465 (1986).

[34] E.A. Novikov, *Appl. Math. Mech.* **27**, 1445 (1963).

[35] G. Paladin and A. Vulpiani, *Phys. Rev.* **A35**, 1971 (1987).

[36] W. Feller, "An Introduction to Probability Theory and its Applications", V. 2, Wiley (1991), reprint from (1966)-edition.

[37] Y., Saito, *J. Phys. Soc. Japan* **61**(2), 403 (1992).

[38] P.L. Vanyan, submitted to *Physica D* (1994).

[39] C. Meneveau and K.R. Sreenivasan, *Nucl. Phys. B, Proc. Suppl.* **2**, 49 (1987).

[40] C.W. Van Atta and T.T. Yeh, *J. Fluid Mech.* **71**, 417 (1975).

[41] A.B. Chhabra and K.R. Sreenivasan, *Phys. Rev. Lett* **68** (18), 2762 (1992).

[42] Z.S. She and E. Waymire, "Quantized Engergy Cascade and Log-Poisson Statistics in Fully Developed Turbulence," Preprint, 1994.

[43] B. Dubrulle, *Phys. Rev. Lett.* **73**(7), 959 (1994).

Observation of Anomalous Diffusion and Lévy Flights

Eric R. Weeks, T. H. Solomon [†], Jeffrey S. Urbach, and

Harry L. Swinney

Center for Nonlinear Dynamics, University of Texas at Austin
Austin, Texas 78712 USA

Abstract: Chaotic transport is studied experimentally in two-dimensional flow in a rapidly rotating annular tank. The flow consists of a chain of vortices sandwiched between two azimuthal jets. Automated image processing techniques are used to track the motions of neutrally buoyant tracer particles suspended in the flow. If the flow has periodic time dependence, the tracers typically follow chaotic trajectories, alternately sticking in vortices and flying for long distances in the jets. Probability distribution functions (PDFs) are measured for sticking and flight times. The flight PDFs are power laws, indicating in some cases that the particle motion can be characterized as Lévy flights (with a divergent second moment for flight times). The variance of an ensemble of particles is found to vary in time as $\sigma^2 \sim t^\gamma$, with $\gamma > 1$ (superdiffusion). The dependence of the variance exponent γ on the flight and sticking PDFs is studied and found to be consistent with calculations relating Lévy flights and anomalous diffusion ($\gamma \neq 1$). In a turbulent flow, Lévy flights no longer are present and the mixing appears to be normally diffusive ($\gamma = 1$). A review of previous experiments on anomalous diffusion is included.

1 Introduction

Mixing of passive scalars in fluid flows is governed by an interplay between advection and molecular diffusion. In the long time limit, the resulting transport can be described as enhanced or *normal* diffusion, characterized by a variance

$$\sigma^2(t) = \left\langle \left(x(t) - \langle x(t) \rangle \right)^2 \right\rangle \sim t^\gamma \qquad (1)$$

that scales linearly with time ($\gamma = 1$). The average can be over a scalar concentration field (e.g., for a dye) or an ensemble of tracer particles. Normal diffusion arises from disordered motion of the particles. For molecular diffusion, the disorder is due to Brownian motion, whereas for enhanced normal diffusion, advection of the particles by the flow (coupled with molecular diffusion) leads to a trajectory which, like Brownian motion, can be treated as a random walk, but with significantly enhanced transport rates. In most cases the Central Limit Theorem applies and leads to the variance growing linearly with time[1]. The Central Limit Theorem applies to random walks without long-term correlations in the motion of the particles, and when the motion is not dominated by rare events.

Long time transport in real fluid flows must be normally diffusive, while short-term transport is usually dominated by the velocity field and, in some cases, is *anomalous* ($\gamma \neq 1$) [1-3]. (Anomalous diffusion is possible for long-term transport only in theoretical models where molecular diffusion is neglected.) In flows with persistent vortices, for example, passive tracers can be trapped inside or near the vortices for long periods of time (sticking), resulting in *subdiffusion* [4,5], $\sigma^2(t) \sim t^\gamma$ with $\gamma < 1$. This occurs when the first moment of the distribution function for these sticking times diverges [2,6,7]. Subdiffusion has been found in simulations of chaotic maps [6] and Hamiltonian models [7].

In flows with coherent jets, passive particles can travel long distances ballistically (flights). In cases where the mean square length of flights in the jets diverges, these motions are termed *Lévy flights*. Lévy flights are characterized by probability distribution functions for flight lengths which are power law rather than exponential, with $P(l) \sim l^{-\mu}$ $(1 < \mu \leq 3)$. (In some cases, the term Lévy *flight* is reserved for instantaneous jumps, with the term Lévy *walk* describing motions with finite velocity. We use the term flight even in the latter case.) Lévy flights can lead to transport which is *superdiffusive* [8-11], with $\gamma > 1$. In model flows with both persistent vortices *and* coherent jets, competition between sticking and flights can lead to either subdiffusion or superdiffusion [12]. Lévy flights and superdiffusion have been found in several model systems [13-18].

Although anomalous diffusion has been studied theoretically for several decades, observations of anomalous transport in laboratory systems have only recently been made. In fluid flows, anomalous diffusion has been observed previously for mixing in a linear chain of time-independent vortices [4,5] and in turbulent capillary waves (surface waves) [19,20]. Direct evidence of Lévy flights has not been obtained in previous experiments, although studies of superdiffusion in polymer-like micelles [9,10] and in laser cooling [21] have used Lévy flight models to explain the experimental data.

In this article, we present experimental measurements of transport in several simple flows composed of both vortices and jets [8,11]. The flow is generated in an annular tank that rotates rapidly to insure two-dimensionality. Automated image processing techniques are used to track the motions of numerous neutrally-buoyant tracer particles in the flow. These digitized trajectories are then ana-

---

[1] The Central Limit Theorem states that the distribution function for a sum of a large number of random variables has the form of a gaussian, with the variance linearly proportional to the number of terms in the sum. This fails when the second moment of the distribution of the random variables is infinite. The variance of the distribution of the sum can be interpreted as the variance of an ensemble of random walkers. For further discussion see [1].

lyzed to determine the variance of a spreading distribution and the probability distribution functions (PDFs) for both sticking and flight times. These experiments provide the first *direct* observations of Lévy flight motion in any laboratory system. The exponents for the variance and the PDFs are compared with theoretical predictions relating Lévy flights and superdiffusion.

Section 2 is a review of previous experiments on anomalous diffusion. Several models of anomalous diffusion in Hamiltonian systems are discussed in Sect. 3. Our experiments are described in Sect. 4, and the results of these experiments are discussed in Sect. 5.

2 Previous Experiments

Mixing in fluids is caused by both molecular diffusion and advection due to the velocity field. Often the transport is anomalously diffusive for short time scales, but crosses over to normal diffusion in the asymptotic limit [22]. As the crossover time can be rather large in some hydrodynamic systems, these intermediate time scales with anomalous scaling are of great interest. (For example, in oceanic flows with a length scale $L \approx 10$ km and a typical diffusion constant $D \approx 10^{-5} \mathrm{cm}^2/\mathrm{s}$, the time scale set by diffusion is $\tau_\mathrm{d} = L^2/D \approx 10^9$ yr, while time scales of interest are typically ~ 1 yr.) Several experiments have examined the intermediate time scales where advection dominates the transport.

2.1 Subdiffusion in a Linear System of Vortices

Pomeau, Pumir, and Young proposed that the transport of a passive impurity in a linear lattice of two-dimensional vortices, bounded by rigid walls, would be subdiffusive [23]. Cardoso and Tabeling tested this result experimentally [4,5]. They studied a linear array of vortices of alternating signs which were constrained to be two-dimensional and stationary. This system provides a simple example of transport dominated by sticking near vortices.

The experimental arrangement was a Plexiglas cell 350 mm by 35 mm, with a fluid layer of electrolyte less than 3 mm thick in the cell. A longitudinal current through the electrolyte and a series of permanent magnets with alternating poles resulted in a stationary one-dimensional array of vortices with alternating signs, bounded by the walls of the cell. A fluorescent dye with electrical conductivity and density matched to the electrolyte was used to examine the transport properties. The vortex turnover times were on the order of a few seconds, whereas the diffusive time scales for large scale mixing of the dye were between 100 and 200 hours. Since the velocity field was time independent, mixing in the system was dependent on molecular diffusion across streamlines (see Sect. 3).

Dye injected close to the edge of a vortex initially mixed between the vortices and along the walls of the cell, but not into the centers of the vortices. Dye invaded the vortex array around the vortices with the variance growing subdiffusively: $\sigma^2(t) \sim t^\gamma$ with $\gamma = 2/3$, in agreement with the theoretical predictions

[23]. For times greater than about an hour, the normal diffusion limit was observed as slight three-dimensional mixing effects began to dominate. (Mixing into the centers of the vortices took place on this time scale.)

2.2 Superdiffusive Transport by Capillary Waves

The Faraday instability, which occurs when a liquid with a free surface is periodically oscillated vertically, leads to capillary waves at the fluid-air interface. The waves oscillate at half the driving frequency. For thin layers of fluid in wide containers, the pattern has square symmetry, independent of the shape of the container walls [24,25]. If the waves have weakly turbulent time dependence, particles floating at the surface can follow erratic trajectories, resulting in anomalous transport.

In experiments of Ramshankar and Gollub [19,20], a 8 cm square cell of Plexiglas 2 cm deep was filled to a depth of 1 cm with distilled water. The apparatus was shaken at 320 Hz, yielding 160 Hz oscillating waves with an average wavelength $\lambda = 2.7$ mm. Slightly above onset of the Faraday instability, the capillary waves become disordered in space and time, enhancing the transport. This occurs at a forcing amplitude $\epsilon \sim 0.0004$, with $\epsilon = (A - A_C)/A_C$ being the nondimensional forcing amplitude where A_C is the amplitude at the wave onset. The transport was studied by considering both mixing of fluorescent dye and small particles floating on the surface. Studies were done at both low forcing amplitude and high forcing amplitude, but in no case was it possible to reach asymptotic time scales dominated by molecular diffusion.

At low forcing values, the motion was found to be highly correlated, with a correlation length of about 6 λ. Through tracking of individual particles, the motion was found to be described by fractional Brownian motion (fBm), a generalized description of Brownian motion [26]. Fractional Brownian motion characterizes the scaling of the distribution $P(\tau)$ of displacements $\delta X(\tau) = X(t+\tau)-X(t)$, the distance a particle has traveled in time τ. For fBm, this distribution scales as $b^{-H}P(b\tau) \sim P(\tau)$. With $H = 1/2$, normal Brownian motion is recovered. Experimentally with $\epsilon = 0.063$, it was found $H \approx 0.6$.

From the statistical properties of fBm the variance (1) can be related to the scaling exponent H:

$$\sigma^2(t) \sim t^{2H} \quad , \tag{2}$$

with normal diffusion occuring for Brownian motion with $H = 1/2$. Superdiffusive scaling was observed in the experiments of [19,20], in agreement with the prediction from fBm, providing an elegant link between the random walk description of transport and the anomalous diffusion properties. However, too few trajectories were observed to be able to determine if the trajectories were Lévy flights.

For higher values of ϵ ($\epsilon \sim 0.35$), corresponding to much stronger forcing, the correlated motion was destroyed. The particles moved according to normal Brownian motion, and normal diffusion was observed with both particle and dye measurements. This is similar to the normal diffusion limit that we observe in experiments with weak turbulence (see Sect. 5.3).

2.3 Superdiffusion in Polymer-Like Micelles

Anomalous diffusion has been observed in recent experiments on certain complex fluids with no flow [9,10]. These studies considered motion of fluorescent tracers attached to long flexible cylindrical micelles, which behave as polymers. However, unlike polymers, the micelles often break and recombine, which results in enhanced transport of tracers attached to the micelles. Due to the high micellar concentration, micelles are constrained to move along their length (reptation). This results in normal diffusion, with small diffusion constants. However, shorter micelles have more freedom of movement and thus have larger diffusion constants. Initially, transport is dominated by the more prevalent longer micelles. For intermediate times, the breaking and recombination of micelles allows tracers to spend time moving on shorter micelles. Tracers move at correspondingly faster rates and farther distances while on a short micelle. This motion is describable as a Lévy flight and leads to superdiffusion. For long enough times, the tracers have visited micelles of all length scales and the transport is again normally diffusive.

The micelles of interest are amphiphilic molecules of CTAB (cetyl trimethyl ammonium bromide), which are dissolved in a saline solution. In this solution they form long flexible micelles of varying length. The action of micelle breaking and recombination results in a equilibrium distribution of micelle lengths given by

$$P(L) = L_0{}^{2\sigma-2} L^{1-2\sigma} \exp\left(-L/L_0\right) \tag{3}$$

with σ being a positive parameter close to zero which allows for a larger probability for longer micelles to break (see [9] for details), and L_0 being on the order of 500 nm. (The micellar radius is ~ 2.5 nm.) Transport properties were studied with the technique of fluorescent recovery after fringe-pattern photobleaching [9,27]. Fluorescent probes are incorporated into the micelles, and illuminated by a laser flash which destroys their fluorescent properties. The diffusion of new fluorescent tracers into the bleached region allows measurement of the transport, and thus characterizes the motion of the micelles.

Simple relationships can be obtained between the statistical properties of the micelles and the expected transport. The diffusion constant is related to the length of a micelle by

$$D(L) \sim L^{-2\beta} \tag{4}$$

with $\beta \geq 1$, allowing for deviations from ideal reptation behavior. The probability to move a distance l before a micelle breaks or recombines then becomes

$$P(l) \sim l^{-\mu} \tag{5}$$

with[2] $\mu = 1 + 2(1 - \sigma)/\beta$. For $\mu \leq 3$, the motion of a tracer attached to a micelle is a Lévy flight. The variance is expected to scale as t^γ with $\gamma = 2/(\mu - 1)$. These arguments apply for intermediate time scales during which

---

[2] Note that we define μ differently than [9,10], to match with the definitions used in Sect. 3

the tracers are exploring smaller and smaller micelles. (Initially, the majority of tracers are on the more common long micelles, although this depends on tracer concentration. For a sufficiently high concentration of tracers, micelles of all lengths are saturated with tracers and normal diffusion is observed.)

The experiments of [9,10] yielded diffusion exponents ranging from $\gamma \approx 1$ (normal diffusion) to $\gamma \approx 1.33$ (superdiffusion), depending on the concentration of micelles, concentration of the saline solution, and the temperature. The micellar concentration and salinity strongly affect the parameters σ and β, as well as the length scales where (4) and (5) are expected to hold. Increasing the temperature reduced the superdiffusive effect, as the breaking time decreases for higher temperatures [28] and thus the intermediate time scale in which anomalous diffusion occurs is shorter than the observations. These are the first experiments to use a Lévy flight description to interpret measured superdiffusive mixing, although direct measurements of the probability distribution function given by (5) were not possible.

2.4 Anomalous Diffusion in Other Systems

Hints of anomalous diffusion have been seen in the motion of ocean tracers [29], although there were not enough tracers present to quantify adequately the transport. The other systems discussed in this subsection are not fluid flows. Studies of the photoconductivity of amorphous materials [30,31] yielded subdiffusion, arising from a broad distribution of local trapping times. Lévy flights have recently been proposed to explain laser cooling of atoms [21].

A recent experiment has used long-term particle tracking techniques similar to those discussed in Sect. 4 to examine motion of low density lipoprotein receptors on the surface of human skin fibroblasts [32]. Subdiffusion was observed with a crossover to normal diffusion at longer times; this has been interpreted as a random walk in a two-dimensional system with random obstacles [33].

Study of gelatin polymers [34] has shown behavior that was interpreted as anomalous diffusion, similar to the work discussed in Sect. 2.3. In this case, the long time limit behavior was related to subdiffusive motion, similar to a random walk occuring on a fractal lattice with the same fractal dimension as the polymers themselves [34].

3 Anomalous Diffusion in Hamiltonian Model Systems

We will not attempt to survey the work on transport in Hamiltonian systems; for a review see [3,12]. Hamiltonian systems are of interest to fluid studies because of their connection to two-dimensional fluid flow (such as the work of Cardoso and Tabeling discussed in Sect. 2, and our measurements presented in Sect. 5). Two-dimensional velocity fields for incompressible flows are describable by the stream function $\psi(x, y, t)$. The equations describing the motion of a passive scalar particle are

$$dx/dt = -\partial\psi/\partial y \quad , \quad dy/dt = \partial\psi/\partial x \quad , \tag{6}$$

which are Hamilton's equations of motion with the streamfunction as the Hamiltonian and x and y as the conjugate coordinates [35]. The motion of a tracer particle corresponds to a trajectory in phase space for the equivalent Hamiltonian system.

For a two-dimensional, time-independent Hamiltonian system, phase space trajectories fall on closed curves. In two-dimensional, time-independent flows, therefore, tracer particles follow streamlines. Ideally, no mixing occurs, although in real fluid systems, molecular diffusion can result in enhanced transport which is either diffusive [36] or subdiffusive [4,5] (as seen in Sect. 2.1). If the streamfunction (Hamiltonian) is time-dependent, the particle (phase space) trajectories can be chaotic: neighboring tracer particle paths can diverge exponentially with time [35,37]. This is termed *chaotic advection*, and can occur *even if the flow is laminar*. The phase space of periodic Hamiltonian systems typically has both ordered and chaotic regions. Enhanced diffusive mixing of tracers in the chaotic regions can be either normal or anomalous, depending on the structure of the phase space [15,16].

As discussed in Sect. 1, competition between sticking (near vortices) and flights (in jet regions) can give rise to anomalous diffusion [12]. The sticking and flight mechanisms have both been studied in Hamiltonian systems. Sticking is often associated with island chains near closed, ordered regions of phase space [7,38,39]. In some cases sticking within a hierarchy of island chains in chaotic maps leads to Lévy flights in phase space [40,41].

Several theoretical studies of Hamiltonian systems have predicted sticking times with probability distribution functions (PDFs) having the form of a power law: $P_S \sim t^{-\nu}$ [17,18,38-40]. Studies have also predicted power law flight PDFs: $P_F \sim t^{-\mu}$ [3,14,18,40-43]. As discussed above, flights with $\mu \leq 3$ are Lévy flights and lead to superdiffusion. Several calculations have been done relating the exponents μ and ν to the anomalous diffusion exponent γ. For flights with a constant velocity, separated by sticking events with no motion, Klafter and Zumofen have shown that [44]

$$\gamma = 2 + \nu - \mu \qquad \nu < 2, \quad 2 < \mu < 3 \quad , \tag{7a}$$

$$\gamma = 4 - \mu \qquad \nu > 2, \quad 2 < \mu < 3 \quad . \tag{7b}$$

An additional study by Zaslavsky for random walks in Hamiltonian systems found [40]

$$\gamma = 2\nu/(\mu - 1) \quad . \tag{8}$$

This formula arose from considerations of fractal space-time behavior arising from the island chain hierarchies in some Hamiltonian systems.

Equations (7) and (8) do not, in general, apply to systems with exponential sticking PDFs, which have a finite first moment for sticking times. Such systems can be considered random walks with jumps at discrete time intervals (the first moment of the sticking PDF). For each jump, the random walker undergoes a flight motion of constant velocity with a length given by the flight PDF. Calculations have shown that for those conditions [14,42,45]

$$\gamma = 2 \qquad \mu \leq 2 \qquad\qquad (9a)$$
$$\gamma = 4 - \mu \qquad 2 < \mu < 3 \qquad\qquad (9b)$$
$$\gamma = 1 \qquad \mu \geq 3 \qquad\qquad (9c)$$

with (9b) being equivalent to (7b).

The theoretical predictions above are compared to experimental data in Sect. 5.2.

4 Experimental Setup

The experiments are conducted using an annular tank that rotates at 1.5 Hz (Fig. 1). This rotation rate is fast enough to ensure the two-dimensionality of the flow [46] by the Taylor-Proudman theorem [47]. The fluid is contained between two cylinders with inner radius and outer radius of 10.8 cm and 43.2 cm, respectively. The Plexiglas lid is flat and rotates rigidly with the tank. The bottom is slightly conical, resulting in an inner height that varies from 17.1 cm at the inner radius to 20.3 cm at the outer radius. The conical bottom models to first order the *beta effect* due to the curvature of the earth, an effect which is important for oceanic and atmospheric flows, but is unimportant for the current experiments. The annulus is completely filled with a mixture of water and glycerol (38% glycerol by weight) with a kinematic viscosity $\nu = 0.03$ cm^2/s.

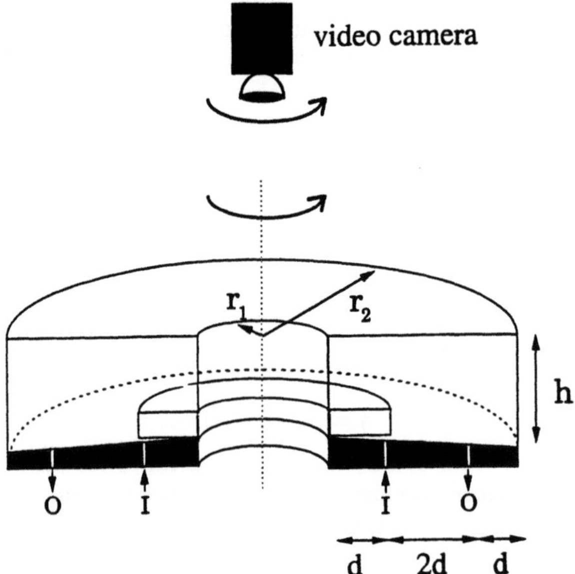

Fig. 1. The rotating annulus. Two concentric rings of holes are on the bottom. Flow is produced by pumping fluid in through the inner ring of holes (marked I) and out through the outer ring (marked O). A Plexiglas ring sits above the inner ring of holes. The flow is observed through a video camera that rotates overhead.

Fluid is pumped through two concentric forcing rings of 120 holes, each 0.26 cm in diameter. Fluid is pumped at a rate of 45 cm^3/s into the annulus through the inner ring (at 18.9 cm) and out through the outer ring (at 35.1 cm). The radial flow couples with the Coriolis force to produce a strong azimuthal jet between the two rings of holes. Jet velocities are typically 4 cm/s, two orders of magnitude greater than the forcing radial velocity. The jet is in the opposite direction of rotation of the tank.

Above the forcing rings there exist sharp velocity gradients. At the forcing rates used in this experiment, the shear layer above the outer ring of holes is unstable to a chain of five, six, or seven vortices [46]. The instability at the inner shear layer is inhibited by a 6.0 cm tall annular Plexiglas barrier with inner (outer) radius of 10.8 (19.4) cm that is inserted above the inner ring of holes. (Without the barrier, the flow would be composed of two vortex chains, one above each forcing ring [46].) The vortex chain rotates relative to the tank at approximately half the speed of the azimuthal jet as seen in the annulus frame of reference. In a reference frame moving with the vortices, the vortex chain is sandwiched by azimuthal jets going in opposite directions (Fig. 2).

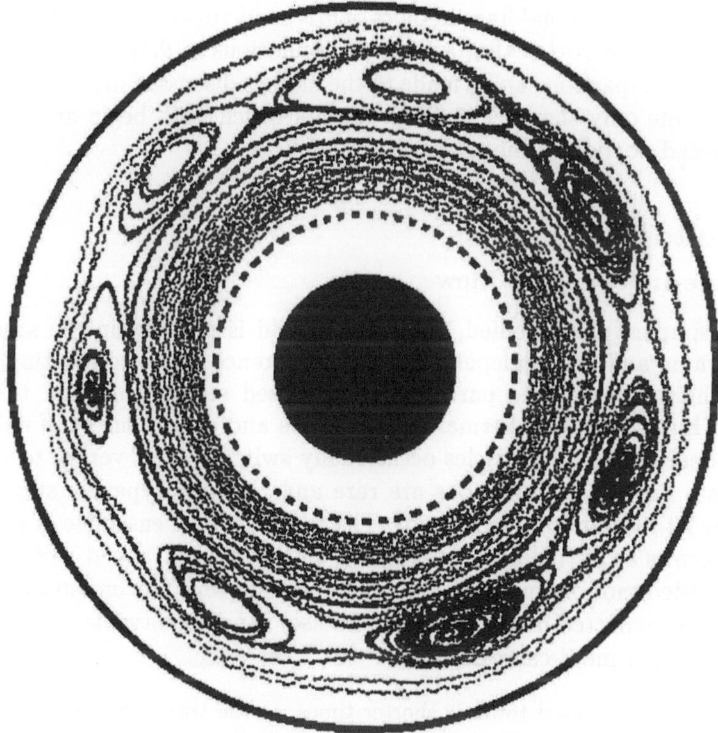

Fig. 2. A flow with seven vortices sandwiched between two azimuthal jets is revealed by the trajectories of 20 particles tracked for 300 s in a reference frame co-rotating with the vortices. The vortex chain rotates with a period of 73.2 seconds in the reference frame fixed to the annulus. The pictured flow is the modulated wave flow discussed in Sect. 5.2. The dashed circle shows the location of the Plexiglas barrier (see Fig. 1).

Two measurement techniques are used. Hot film probes mounted in the lid above the outer ring of holes are used to characterize the velocity field. In addition, transport was measured using novel long-term particle tracking techniques [48]. Neutrally buoyant, 1 mm diameter wax particles are suspended in the fluid, and illuminated from the side of the annulus. (The ability of these particles to follow the flow passively is discussed in [11].) A CCD camera mounted above the annulus (Fig. 1) and co-rotating with the vortex chain is used to view the particles. The signal from the camera is sent to a computer that processes and stores the images in a compressed format onto a hard drive. Software is then used to post-process the images and identify the particles. In this manner up to 100 particles can be simultaneously tracked for long periods of time. A typical experimental run of 6 hours contains up to 400 particle trajectories of lengths ranging from 100 to 2000 s. Duration of individual trajectories is limited by slight deviations from an ideal two-dimensional flow, carrying particles out of the illuminated region [11].

Trajectories are analyzed to determine the variance and probability distribution functions. The motion of particles is analyzed as a random walk in the angular coordinate. Flight events are defined to be motions between vortices that show up as diagonal lines in plots of $\theta(t)$, and sticking events are motions of particles within a vortex that appear as oscillations in $\theta(t)$ (see Fig. 5). In this way, direct comparison can be made to the models discussed in Sect. 3. Measured PDFs include only the sticking/flight events which both begin and end during the recorded portion of the trajectory[3].

5 Results

5.1 Time-Independent Flow

For the simplest flow studied, the velocity field is periodic in the annulus reference frame and time-independent in the reference frame co-rotating with the vortex chain. Ideally, the particles follow closed streamlines, and there is no mixing. However, slight thermal imperfections and other sources of noise result in non-ideal behavior. Particles occasionally switch from a vortex to the jet or vice-versa, but these transitions are rare and particles typically stay within a vortex or jet for hundreds of seconds. The variance of an ensemble of particles in the jets grows as t^2 (ballistic separation), and particles trapped within a vortex have bounded motion and only mix within the vortex. The crossover to normal diffusion (as expected in the presence of noise) is not observable within the time scales the experiment can examine.

---

[3] These PDFs are biased towards shorter times, as the trajectory durations seen are limited to ~ 25 minutes. This results in a change in the exponent of the power laws of observed PDFs. We have corrected for this by generating artificial trajectories with power law PDFs, and then chopping them at finite times to match the experimental observations, as discussed in [11]. The value of the "true" PDF exponent is taken to be that of the chopped artificial trajectories that matches the experimental PDF. This results in a typical correction of 0.3 to the exponent of the power law.

5.2 Flows with Chaotic Advection

The three flows exhibiting chaotic advection that we have studied will be classified in terms of the dynamics of the flow in the co-rotating frame of the vortices, not by the dynamics in the annulus reference frame in which the velocity spectra are obtained. The three flows, each obtained for the same total pumping rate, 45 cm^3/s, are:

(a) *Time-periodic flow.* This flow is produced by reducing by 50% the pumping through both inner and outer forcing rings of one 60° sector. As the chain of six vortices moves around the annulus, each vortex moves through this sector, which perturbs the vortices with a period equal to the rotation period of the vortex chain, 70.0 s. The velocity power spectrum is periodic both in the frame of the annulus (which is also the reference frame of the perturbation), as Fig. 3(a) illustrates, and in the frame co-rotating with the vortices. In any other reference frame the velocity field is quasiperiodic with two frequencies.

(b) *Chaotic flow.* This flow is produced by shutting off the pumping in one 60° sector. The flow consists of a chain of vortices (varying between five and six vortices over long periods of time), but the velocity spectrum now contains broadband noise (Fig. 3(b)). Thus, in contrast to the laminar, regular, velocity fields of (a) and (c), the velocity field for this flow is nonperiodic.

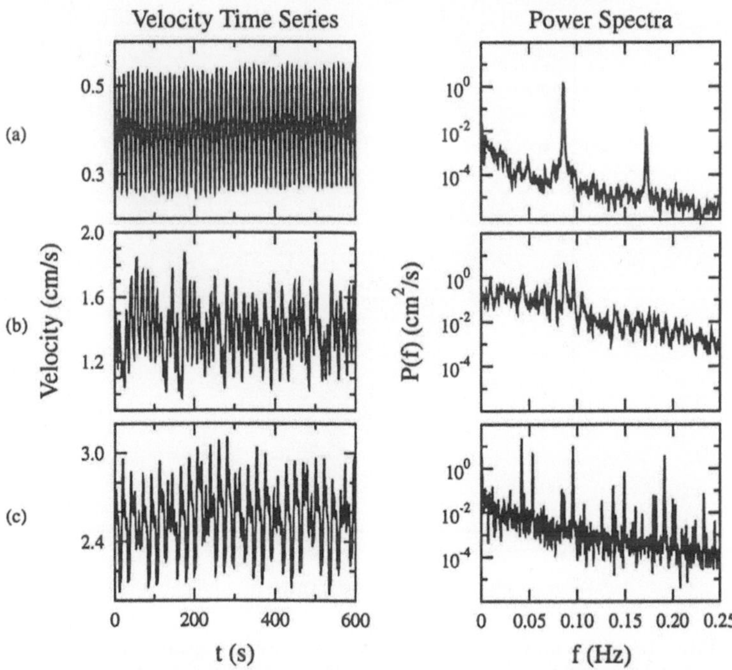

Fig. 3. Velocity time series and power spectral density obtained from measurements of the azimuthal velocity component at r = 35.1 cm. (a) periodic flow; (b) chaotic flow; (c) modulated wave flow.

(c) *Modulated wave (time-periodic) flow.* Spectra for this flow with seven vortices (rather than six, as in (a)) contain two incommensurate frequencies, as Fig. 3(c) shows. The spectrum perhaps contains broadband fluid noise in addition to the inevitable measurement noise present (cf. Fig. 3(a)). In the reference frame moving with the vortices, the motion is time-periodic, similar to (a); however, the periodicity arises from an instability in (c) rather than a perturbation as in (a). In the vortex reference frame the instability of the modulated wave flow has a frequency of 0.00033 Hz and a mode number of 3 (measured from particle tracking). In contrast, the periodic flow in (a) has a frequency of 0.014 Hz and a mode number of 1, from the perturbation.

The advection time scale for these flows is set by the vortex turnover time, $\sim 23\,$s. This time can be measured from Fig. 4, a graph showing the time a particle takes to move halfway around a vortex plotted against the angle moved through (with respect to the center of the annulus, in the reference frame moving with the vortex chain). This graph was obtained by taking the difference in time and angle between successive pairs of maxima and minima in $\theta(t)$ for a sticking event; see Fig. 5(b). Small angles correspond to particles near the center of the vortex. The average time for all of the particles shown in Fig. 4 is 11.7 s, giving a vortex turnover time of 23.4 s.

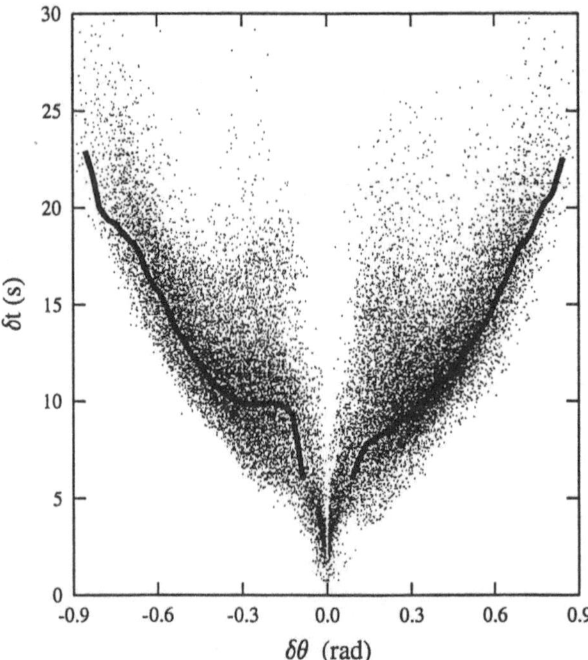

Fig. 4. Time to move halfway around a vortex plotted against angular displacement (measured relative to the center of the annulus, not the center of the vortex). The scattered points are the measured points for the seven vortex state (modulated wave flow), and the solid line is drawn through the regions of highest density. The plot extends the full width of a vortex, $2\pi/7 = 0.898$ rad.

Chaotic advection is observed in these three flows. Particles frequently make transitions to and from vortices. This is seen in Fig. 5, showing the motion of one particle in the modulated wave flow. Flight events of various lengths appear as diagonal lines in the graph of $\theta(t)$, and sticking events are oscillations about a fixed θ (Fig. 5(b)). These three flows all have flight and sticking events of durations ranging from ~ 20 s (comparable to the vortex turnover time) up to ~ 900 s.

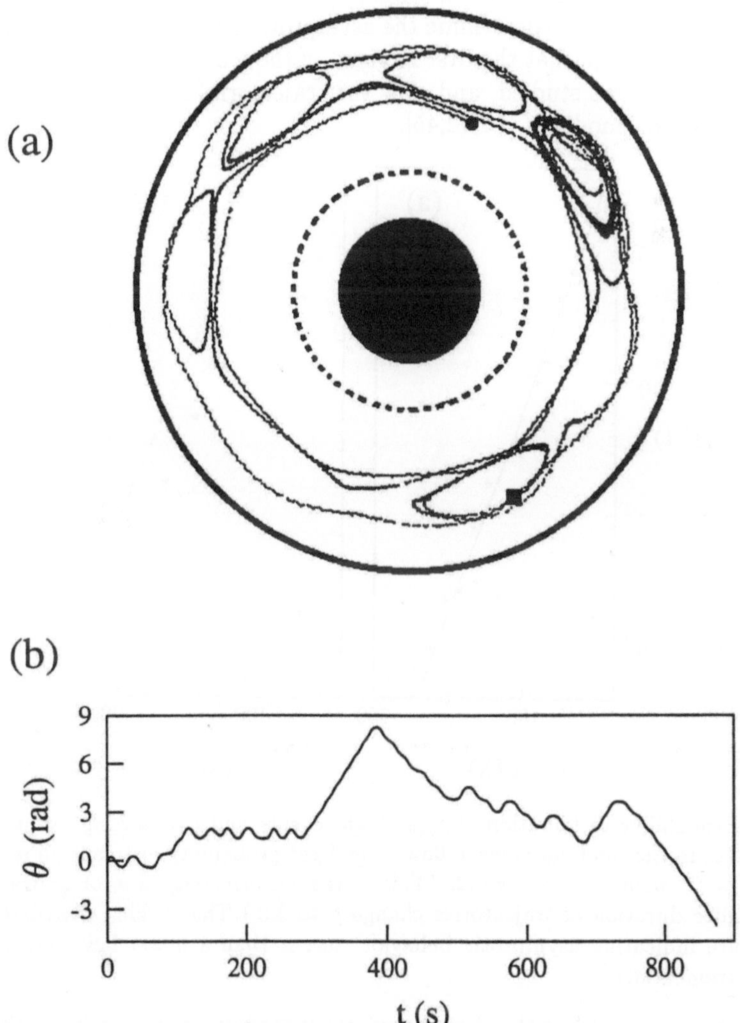

Fig. 5. (a) The trajectory of a single particle in the modulated wave flow. The particle starts at the square and ends at the circle. Hyperbolic fixed points are visible between some of the vortices, where the motion of the particle is sensitive to transitions into and out of vorticies. (b) The azimuthal coordinate $\theta(t)$ is shown for the trajectory in (a). Diagonal lines are flight events, and the oscillations are sticking events. From 500 s to 700 s, the particle hops between four vortices, resulting in staircase-like motion.

64

The flight PDFs for the three cases are all power laws (e.g. Fig. 6(a)), with exponents of 2.3 for the periodic flow, 2.6 for the chaotic flow, and 3.2 for the modulated wave flow[4]. The first two flows thus are examples of Lévy flights; the modulated wave flow is not a Lévy flight despite the power law behavior.

The sticking PDFs are harder to characterize; Fig. 6(b) shows a typical PDF. The sticking PDFs are perhaps power laws at intermediate times with a crossover to exponential behavior at long times, but the data are inconclusive. Statistics are scarce at long times due to the slight deviations from ideal two-dimensional flow, and it is difficult to determine the asymptotic behavior. It appears that the tails are steep enough that the first moment of the true sticking PDF is finite for all three of the flows studied, and thus the trajectories can be considered using the random walk model of [14,42,45].

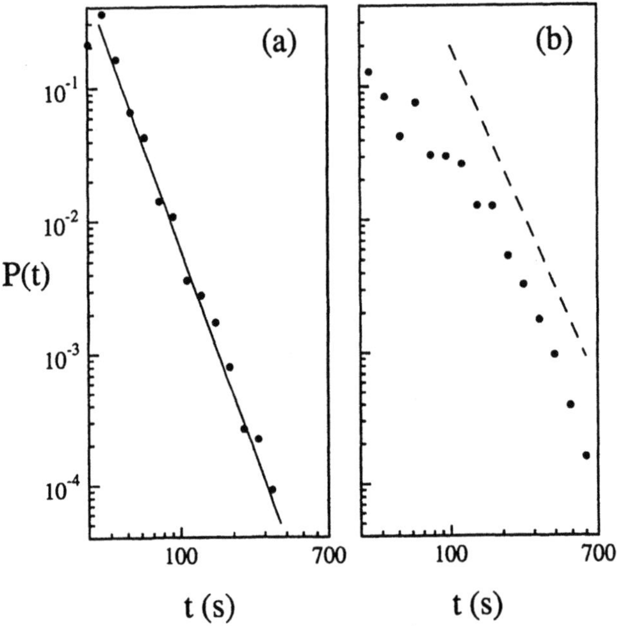

Fig. 6. Probability distributions for (a) flight events and (b) sticking events, shown for particles in the modulated wave flow. The flight probability distribution is a power law, $P_F \sim t^\mu$ with $\mu = 3.6 \pm 0.2$. (This is the uncorrected value of μ; corrections for the finite duration of trajectories change μ to 3.3.) The sticking distribution has a curvature indicating asymptotic behavior steeper than a power law (but does not appear exponential).

The variance $\sigma^2(t)$ for the three flows, calculated using the methods discussed in [11], is shown in Fig. 7(a). Figure 7(b) shows the slope of the variance, which for long times yields the exponent γ. The periodic and chaotic flows have plateaus in γ at 1.65±0.15 and 1.55±0.15 respectively. The modulated wave flow does not

---

[4] These are the corrected power laws for the PDFs (see footnote 3); the uncorrected slopes of the PDFs are 2.6, 2.9, and 3.7 for these flows.

appear to have a plateau region wide enough to determine γ, although it appears superdiffusive for the time scales resolvable by the experiment. For times less than the vortex turnover time (~ 24 s), the flows all appear ballistic ($\gamma \approx 2$) as the particles have not explored the finite extent of a vortex. The crossover to normal diffusion is not observable in the time the particles are visible.

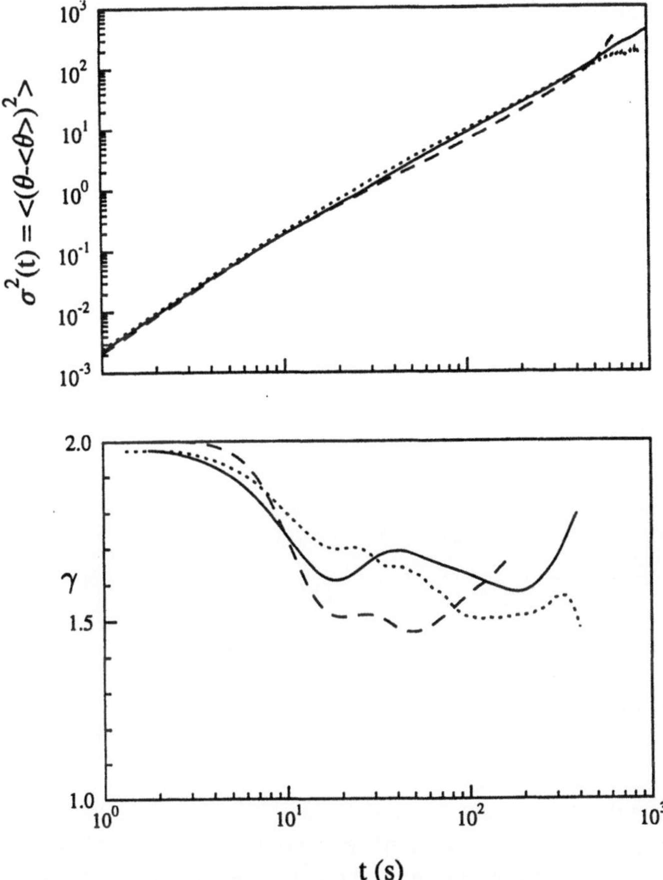

Fig. 7. (a) Variance of the azimuthal displacement of an ensemble of particles for periodic flow (solid line), chaotic flow (dotted line), and modulated wave flow (dashed line). (b) Slope of the variance for the different flows. The periodic flow and the chaotic flow have plateaus that yield the exponents for power law growth, $\gamma = 1.65 \pm 0.15$ (periodic flow) and $\gamma = 1.55 \pm 0.15$ (chaotic flow). The modulated wave flow does not appear to have a plateau.

Figure 8 shows the flight length compared to flight time. The constant slope for flights in the two directions indicates that the motion occurs with approximately constant velocity; hence the motion can be compared with calculations for random walks with constant velocity (Sect. 3). As mentioned above, the calculations done for sticking PDFs with finite first moments appear to be the relevant ones for comparison. Equation (9) yields $\gamma = 1.7$, 1.4, and 1.0 for the periodic, chaotic, and modulated wave flows respectively. The exponents fit within the experimental error for the periodic and chaotic flows (cf. Fig. 7). However, the modulated wave flow does not appear to be normally diffusive in the time scale visible in the experiment, in conflict with the predicted result. It is possible that the discrepancy arises because the convergence to the asymptotic behavior is slower for a power law flight PDF with an exponent close to 3. The second moment of the flight PDF is finite for this case, but it is still large, which affects the time scales the asymptotic behavior is reached.

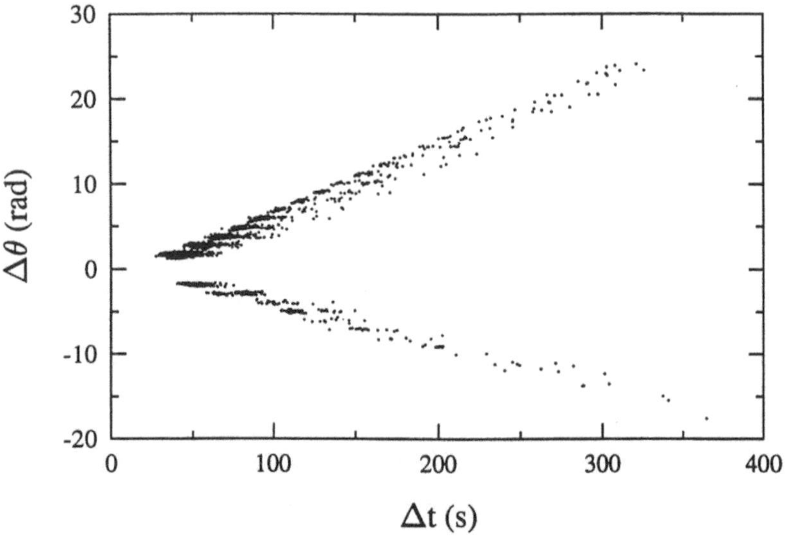

Fig. 8. Flight length $\Delta\theta$ plotted against flight duration Δt. The approximately linear relationship shows that flights have roughly constant velocity. The horizontal bands differ in spacing by $\pi/3$, which is the angular spacing between vortices (for the periodic flow shown). The spread of times within a horizontal band is due to the slowing down of particles which pass close by the hyperbolic points.

It is difficult to compare our results with (8), as the sticking PDFs seen in the experiment do not appear to have reliable power law behavior over enough decades to accurately determine ν, the exponent for the power law. If a power law is assumed, for the periodic flow we find a reasonable fit is $\nu = 1.6 \pm 0.3$ [11], yielding $\gamma = 2.4 \pm 0.5$, larger than the measured value, 1.65 ± 0.15. It is probable that our particles cannot probe the small scale island chains necessary to see the prediction of (8); without definite power law behavior it is difficult to make a comparison.

5.3 Weakly Turbulent Flow

The final flow studied is a weakly turbulent flow. To generate this flow, only the outer ring of forcing holes is used. Water is used ($\nu = 0.009\,\mathrm{cm}^2/\mathrm{s}$) to increase the Reynolds number, rather than the water-glycerol mixture of the other flows. In addition, the azimuthal symmetry is broken by reversing the flow through alternating 60° sectors. This results in a flow with no stable vortices or jet regions; particles move in an erratic fashion (Fig. 9(a)) and the power spectrum for the velocity field has no strong peaks. The graph of $\theta(t)$ shows no long-term flight events (Fig. 9(b)); compare with Fig. 5.

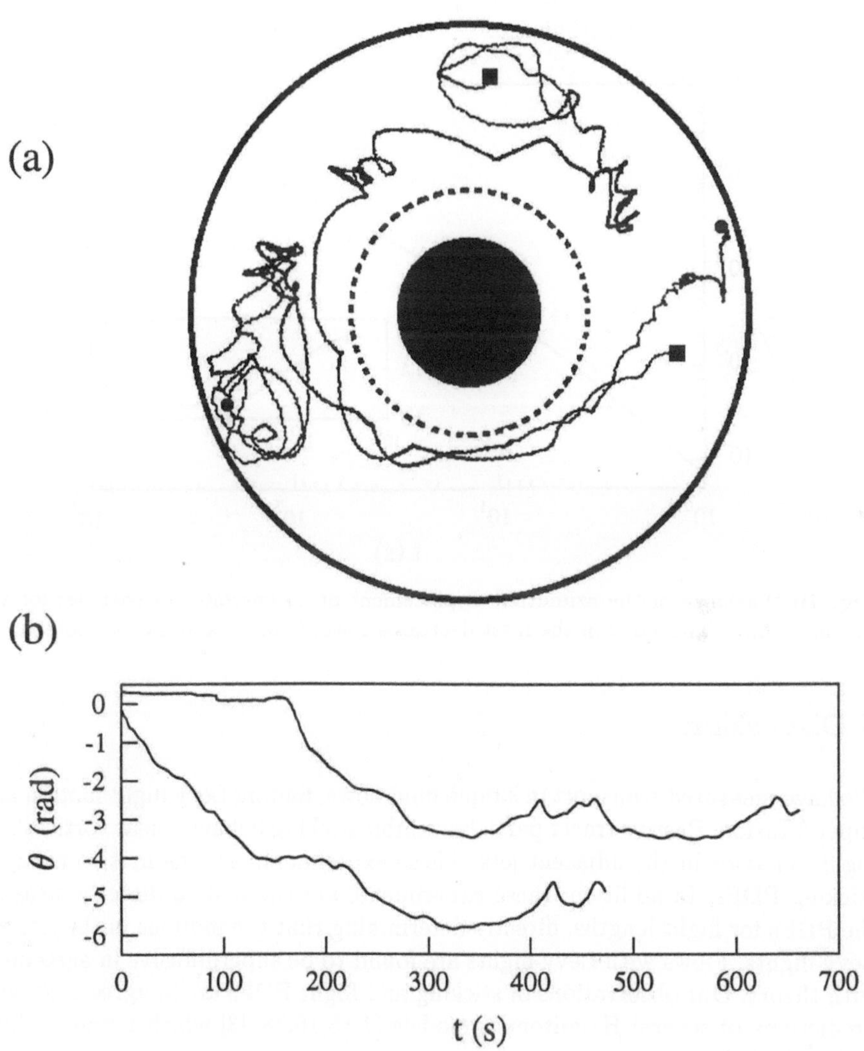

Fig. 9. (a) Two tracer particle trajectories in the turbulent flow, viewed in the annulus reference frame. The ends of one erraticle are marked with squares, the other by cicircles; both particles start at the right side of the annulus. (b) The azimuthal displacement for the particles in (a); the upper trace is for the particle marked with circles.

By defining a flight as the angular distance traveled between successive extrema in $\theta(t)$, the results can be loosely compared with the theoretical predictions of Sect. 4. The resulting flight PDF is an *exponential* [11] with a decay time of 15 s. Interpreting the results as a random walk with a finite second moment, we should have normally diffusive transport by the Central Limit Theorem. In addition, turbulent flows are predicted to show normal diffusion (at time scales large compared to the eddy turnover times) [49]. The experiment shows the exponent γ decreasing to 1.0 (Fig. 10), although we were unable to see clearly the asymptotic limit; this appears in agreement with the random walk prediction and the prediction for turbulence.

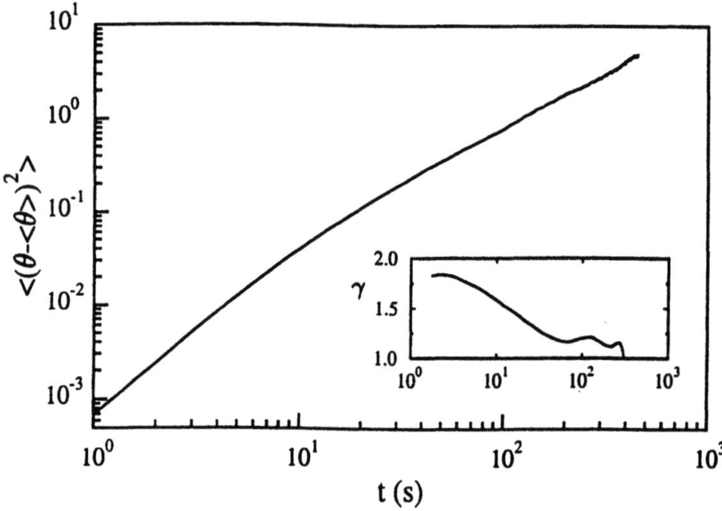

Fig. 10. Variance of the azimuthal displacement of an ensemble of particles for the turbulent flow. The slope in the inset decreases towards unity and has no plateau.

6 Discussion

We have measured transport in simple fluid flows, finding Lévy flight motion and superdiffusion. Passive tracer particles exhibit sticking behavior near vortices and flight behavior in the adjacent jets. These experiments are the first to measure sticking PDFs. In addition, these experiments are the first to directly measure the PDFs for flight lengths, directly determining that the motions in the jets are Lévy flights. Flows with Lévy flights are found to be superdiffusive in agreement with theory. Our observations of sticking and flight PDFs are in agreement with predictions of several Hamiltonian models [7,15,16,38-43] which found sticking and flight behavior.

Theoretical and numerical studies of anomalous diffusion are abundant, but experimental studies are rare (cf. Sect. 2). The present observations demonstrate that concepts such as Lévy flights and anomalous diffusion are relevant to studies of mixing. We hope that these experiments will motivate further work to understand anomalous diffusion in laboratory settings.

Acknowledgments

We thank M. Pervez for help with particle tracking, D. del-Castillo-Negrete for daily discussions, and J. Klafter, P. J. Morrison, M. F. Shlesinger, G. M. Zaslavsky, and G. Zumofen for helpful discussions. This research is supported by the Office of Naval Research Grant No. N00014-89-J-1495.

References

†   Permanent address: Department of Physics, Bucknell University, Lewisberg, Pennsylvania 17837 USA

1. J. P. Bouchaud, A. Georges, Anomalous diffusion in disordered media: statistical mechanisms, models and physical applications, Phys. Rep. **195** 127-293 (1990)

2. B. D. Hughes, M. F. Shlesinger, E. W. Montroll, Random walks with self-similar clusters, Proc. Natl. Acad. Sci. USA **78** 3287-3291 (1981)

3. M. F. Shlesinger, G. M. Zaslavsky, J. Klafter, Strange kinetics, Nature **363** 31-37 (1993)

4. O. Cardoso, P. Tabeling, Anomalous diffusion in a linear system of vortices, Eur. J. Mech. B/Fluids **8** 459-470 (1989)

5. O. Cardoso, P. Tabeling, Anomalous diffusion in a linear array of vortices, Euro. Lett. **7** 225-230 (1988)

6. T. Geisel, S. Thomae, Anomalous diffusion in intermittent chaotic systems, Phys. Rev. Lett. **52** 1936-1939 (1984)

7. G. Petschel, T. Geisel, Unusual manifold structure and anomalous diffusion in a Hamiltonian model for chaotic guiding-center motion, Phys. Rev. A **44** 7959-7967 (1991)

8. T. H. Solomon, E. R. Weeks, H. L. Swinney, Observation of anomalous diffusion and Lévy flights in a two-dimensional rotating flow, Phys. Rev. Lett **71** 3975-3978 (1993)

9. J. P. Bouchaud, A. Ott, D. Langevin, W. Urbach, Anomalous diffusion in elongated micelles and its Lévy flight interpretation, J. Phys. II **1** 1465-1482 (1991)

10. A. Ott, J. P. Bouchaud, D. Langevin, W. Urbach, Anomalous diffusion in "living polymers": a genuine Lévy flight?, Phys. Rev. Lett. **65** 2201-2204 (1990)

11. T. H. Solomon, E. R. Weeks, H. L. Swinney, Chaotic advection in a two-dimensional flow: Lévy flights and anomalous diffusion, Physica D **76** 70-84 (1994)

12. V. V. Afanasiev, R. Z. Sagdeev, G. M. Zaslavsky, Chaotic jets with multifractal space-time random walk, Chaos **1** 143-159 (1991)

13. T. Geisel, J. Nierwetberg, A. Zacherl, Accelerated diffusion in Josephson Junctions and related chaotic systems, Phys. Rev. Lett. **54** 616-619 (1985)

14. T. Geisel, A. Zacherl, G. Radons, Chaotic diffusion and 1/f-noise of particles in two-dimensional solids, Z. Phys. B - Condensed Matter **71** 117-127 (1988)

15. A. A. Chernikov, B. A. Petrovichev, A. V. Rogal'sky, R. Z. Sagdeev, G. M. Zaslavsky, Anomalous transport of streamlines due to their chaos and their spatial topology, Phys. Lett. A **144** 127-133 (1990)

16. B. A. Petrovichev, A. V. Rogal'sky, R. Z. Sagdeev, and G. M. Zaslavsky, Stochastic jets and nonhomogeneous transport in Lagrangian turbulence, Phys. Lett. A **150** 391-396 (1990)

17. J. B. Weiss, E. Knobloch, Mass transport and mixing by modulated traveling waves, Phys. Rev. A **40** 2579-2589 (1989)

18. G. M. Zaslavsky, D. Stevens, H. Weitzner, Self-similar transport in incomplete chaos, Phys. Rev. E **48** 1683-1694 (1993)

19. R. Ramshankar, D. Berlin, J. P. Gollub, Transport by capillary waves. Part I. Particle trajectories, Phys. Fluids A **2** 1955-1965 (1990)

20. R. Ramshankar, J. P. Gollub, Transport by capillary waves. Part II: Scalar dispersion and structure of the concentration field, Phys. Fluids A **3** 1344-1350 (1991)

21. F. Bardou, J. P. Bouchaud, O. Omile, A. Aspect, C. Cohen-Tannoudji, Subrecoil laser cooling and Lévy flights, Phys. Rev. Lett. **72** 203-206 (1994)

22. T. H. Solomon, J. P. Gollub, Chaotic particle transport in time-dependent Rayleigh-Bénard convection, Phys. Rev. A **38** 6280-6286 (1988)

23. W. Young, A. Pumir, Y. Pomeau, Anomalous diffusion of tracer in convection rolls, Phys. Fluids A **1** 462-469 (1989)

24. A. B. Ezerskii, P. I. Korotin, M. I. Rabinovich, Random self-modulation of two-dimensional structures on a liquid surface during parametric excitation, Pis'ma Zh. Eksp. Teor. Fiz. **41** 129-131 (1985); JETP Lett. **41** 157-160 (1985)

25. N. B. Tufillaro, R. Ramshankar, J. P. Gollub, Order-disorder transition in capillary ripples, Phys. Rev. Lett. **62** 422-425 (1989)

26. B. B. Mandelbrot, J. W. Van Ness, Fractional Brownian motions, fractional noises and applications, SIAM Review **10** 422-437 (1968)

27. R. Messager, A. Ott, D. Chatenay, W. Urbach, D. Langevin, Are giant micelles living polymers?, Phys. Rev. Lett. **60** 1410-1413 (1988)

28. S. J. Candau, F. Merikhi, G. Waton, P. Lemaréchal, Temperature-jump study of elongated micelles of cetyltrimethylammonium bromide, J. Phys. France **51** 977-989 (1990)

29. A. R. Osborne, A. D. Kirwan, Jr., A. Provenzale, L. Bergamasco, Fractal drifter trajectories in the Kuroshio extension, Tellus **41A** 416-435 (1989)

30. G. Pfister, H. Scher, Dispersive (non-Gaussian) transient transport in disordered solids, Adv. Phys. **27** 747-798 (1978)

31. H. Scher, E. W. Montroll, Anomalous transit-time dispersion in amorphous solids, Phys. Rev. B **12** 2455-2477 (1975)

32. R. N. Ghosh, W. W. Webb, Automated detection and tracking of individual and clustered cell surface low density lipoprotein receptor molecules, Biophys. J. **66** 1301-1318 (1994)

33. M. J. Saxton, Anomalous diffusion due to obstacles: a Monte Carlo study, Biophys. J **66** 394-401 (1994)

34. S. Z. Ren, W. F. Shi, W. B. Zhang, C. M. Sorensen, Anomalous diffusion in aqueous solutions of gelatin, Phys. Rev. A **45** 2416-2422 (1992)

35. H. Aref, Stirring by chaotic advection, J. Fluid Mech. **143** 1-21 (1984)

36. T. H. Solomon, J. P. Gollub, Passive transport in steady Rayleigh-Bénard convection, Phys. Fluids **31** 1372-1379 (1988)

37. J. M. Ottino, Mixing, chaotic advection and turbulence, Ann. Rev. Fluid Mech. **22** 207-253 (1990)

38. J. D. Meiss, E. Ott, Markov-tree model of intrinsic transport in Hamiltonian systems, Phys. Rev. Lett. **55** 2741-2744 (1985)

39. J. D. Meiss, E. Ott, Markov tree model of transport in area-preserving maps, Physica D **20** 387-402 (1986)

40. G. M. Zaslavsky, Fractional kinetic equation for Hamiltonian chaos, Physica D **76** 110-122 (1994)

41. T. Geisel, A. Zacherl, G. Radons, Generic 1/f noise in chaotic Hamiltonian dynamics, Phys. Rev. Lett. **59** 2503-2506 (1987)

42. J. Klafter, A. Blumen, M. F. Shlesinger, Stochastic pathway to anomalous diffusion, Phys. Rev. A **35** 3081-3085 (1987)

43. X.-J. Wang, Dynamical sporadicity and anomalous diffusion in the Lévy motion, Phys. Rev. A **45** 8407-8417 (1992)

44. J. Klafter, G. Zumofen, Lévy Statistics in a Hamiltonian System, Phys. Rev. E **49** 4873-4877 (1994)

45. M. F. Shlesinger, J. Klafter, Lévy walk representations of dynamics processes, in Perspectives in Nonlinear Dynamics, ed. M. F. Shlesinger, R. Cawley, A. W. Saenz, W. Zachary, (World, Singapore, 1986) pp. 336-349

46. T. H. Solomon, W. J. Holloway, H. L. Swinney, Shear flow instabilities and Rossby waves in barotropic flow in a rotating annulus, Phys. Fluids A **5** 1971-1982 (1993)

47. J. Pedlosky, *Geophysical Fluid Dynamics* (Second Edition) (Springer-Verlag, New York, 1987)

48. M. S. Pervez, T. H. Solomon, Long-term tracking of neutrally bouyant tracer particles in two-dimensional fluid flows, Exp. Fluids, in press.

49. I. Mezic, S. Wiggins, Birkhoff's ergodic theorem and statistical properties of chaotic dynamical systems with applications to fluid mechanical dispersion and mixing, submitted to Physica D.

Chaotic Lagrangian Motion on a Rotating Sphere

Vered Rom-Kedar[1], Yona Dvorkin[2] and Nathan Paldor[2].

[1] The Department of Applied Mathematics and Computer Science, The Weizmann Institute of Science, P.O.B. 26, Rehovot 76100, Israel.

[2] The Department of Atmospheric Sciences, The Hebrew University of Jerusalem, Jerusalem 91904, Israel.

Abstract: We study the motion of Lagrangian particles launched on a geopotential surface of a rotating sphere (e.g. floats in the deep ocean) where the latter is zonally perturbed by some travelling pressure wave (e.g. tidal waves). The motion of these particles is described by a near integrable, two-degrees-of-freedom Hamiltonian system. For some regions in parameter and phase space, the system may be reduced to a one-and-a-half-degrees-of-freedom Hamiltonian system and standard tools may be applied to prove the existence of chaotic homoclinic behavior, hence of the phenomena of anomalous transport associated with the homoclinic chaos. In other regions zonally localized structures and homoclinic tangles with back-flows, associated with the three dimensionality of the energy surfaces appear, as well as resonant behavior of unstable periodic orbits.

1. Introduction

The simplest possible two-dimensional motion on a rotating sphere, due only to Coriolis force (i.e. in the absence of any body force), is called the inertial motion. It may be relevant for describing the motion of Lagrangian ("passive") particles in the atmosphere (such as weather balloons, pollutants or satellites in the upper atmosphere) or for describing the motion of floats in the deep ocean. Each of these applications involves different scales of velocities, and various simplifying assumptions on the dynamics which need to be carefully examined (in particular, for the atmospheric applications the role of the interaction of changes in pressure and particle velocity). In the context of atmospheric and oceanographic flows, the possible solutions of the inertial (i.e. force free) motion have been described in various textbooks[1] for some specific cases (e.g. small latitudinal extent). The introduction of the general equations of motion (namely when the scale of the motion is allowed to encompass the entire globe), their integrals, and the full analysis of their possible solutions has been introduced only recently by Paldor and Killworth[2]. The influence of weak zonally travelling pressure waves, perturbing the geopotential surface from its mean spherical shape, was considered by Paldor and Boss[3]. They have demonstrated that chaotic trajectories exist and cause dispersion of lines of Lagrangian particles. In fact, many different waves exist in the real atmosphere - tidal (thermal or planetary) waves are examples of eastward propagating waves while Rossby waves propagate westward. The effect of these two waves on particle dispersion will be shown to be qualitatively different. Here we set the ground for a complete analysis of these equations, and include results regarding mixing and transport for some particularly simple cases. We concentrate on the mathematical aspects of the analysis, and at this point, use the physical circumstance as a motivation for posing the mathematical problems. The application of our results to observations on the physical system is deferred to a follow-up study.

Mathematically, the equations we consider are near integrable two-degrees-of-freedom Hamiltonian system. Most of our conclusions are based upon known results from various works in dynamical system theory. The use of several different tools in understanding one system seems to be unique, and brings up the issue of matching between the different techniques in the intermediate regimes. The new concept/method we introduce here is the representation of colored energy surfaces: we plot a two dimensional cross-section of the three dimensional energy surfaces and tag them according to the behavior of the angular variable. This geometrical representation enables an immediate interpretation of the generic qualitative behavior and aids in identifying the occurrence of all singular behavior.

It has been observed that the chaotic zone in two- and four-dimensional area preserving maps which is associated with the existence of homoclinic tangles exhibits anomalous transport in which Levy flights alter the decay rate of the autocorrelators and lead to non-diffusive behavior[4]. Moreover, such a behavior has been observed experimentally in a rotating system in which the inertial motion may be realized (see Weeks et al., this proceeding). The analysis of our system reveals the existence of homoclinic chaos of various types, with two types which are "generic" - the usual homoclinic tangle associated with one-and-a-half d.o.f. Hamiltonian systems and a homoclinic tangle with a back-flow in the angular direction. It follows that the phenomena of anomalous transport is present in our system in the former case. The appearance of the Lèvy flights in the second case is yet to be explored.

This paper is organized as follows: in section 2 we describe the structure of the integrable system and its symmetries. In section 3 we construct the energy surfaces and present the division to the parameter ranges according to the expected behavior under perturbation. In section 4 we present the Melnikov analysis results for the simplest, infinite wave length perturbation case. We discuss the implication of this result on maximal mixing and maximal chaotic zone.

2. The phase space geometry

The non-dimensional Lagrangian momentum equations for the eastward and northward velocity components (u, v) and the rate of change of the longitude and latitude coordinates (λ, ϕ) are given by:

$$\frac{d\lambda}{dt} = \frac{u}{\cos \phi} \tag{2.1a}$$

$$\frac{du}{dt} = v \sin \phi (1 + \frac{u}{\cos \phi}) - k\epsilon \frac{A(\phi)}{\cos \phi} \cos(k\lambda - \sigma t). \qquad (2.1b)$$

$$\frac{d\phi}{dt} = v \qquad (2.1c)$$

$$\frac{dv}{dt} = -u \sin \phi (1 + \frac{u}{\cos \phi}) - \epsilon A'(\phi) \sin(k\lambda - \sigma t) \qquad (2.1d)$$

Equations (2.1) have been nondimensionalized using the radius of the earth for length scale and twice the frequency of the earth's rotation for the frequency scale, so that a nondimensional time unit corresponds to $24h/2\pi \approx 1.901$ hours. Order one velocities in the nondimensional variables turn out to be nearly 1 km/second dimensional velocities, so the relevant nondimensional velocities for atmospheric flows are of order 0.01 , For satellites, the relevant scales are of order 1 and for oceanographic flows of order 0.001. $A(\phi)$ represents the latitude dependent amplitude of a pressure wave (divided by the density), which is assumed to be even:

$$A(\phi) = A(-\phi) \quad A(0) = 1, \quad A'(0) = 0. \qquad (2.2)$$

$A(\phi)$ may represent, for example, the amplitude of a daily tidal forcing ($k = 1, \sigma = 1$), for which the amplitude $A(\phi)$ may be either taken as constant ($A(\phi) = 1$) or to vary with latitude like $A(\phi) = \cos^3 \phi$, see Paldor and Boss[3] for discussion. Higher or negative wavenumbers may be associated with other waves present in the atmosphere (e.g. Rossby waves). For the atmospheric flows the relevant scale for the pressure wave is that of the order of the kinetic energy, namely ϵ is $O(10^{-4})$. Finally, we note that the polar coordinates introduce apparent singularity at the poles; $\cos \phi = 0$ there, but u vanishes there as well, so no real singularity is encountered when the solutions pass near the poles.

When no waves are present the motion is integrable; Two integrals of motion were derived in Paldor and Killworth[2] for the inertial trajectories, corresponding to the angular momentum D and the kinetic energy E:

$$D = \cos \phi (\cos \phi + 2u) \qquad (2.3a)$$
$$E = u^2 + v^2. \qquad (2.3b)$$

Using (2.3a) u may be expressed in terms of D and ϕ:

$$u = \tfrac{1}{2}(\frac{D}{\cos \phi} - \cos \phi) \qquad (2.4)$$

When $\epsilon = 0$ the equation for λ decouples, D is conserved so it is considered a parameter, and the phase flow of the two dimensional system obtained for (ϕ, v) depends on the value of D as follows:

When $|D| > 1$ the origin is stable (elliptic) and is surrounded by periodic orbits which visit both hemispheres. Denote the maximal latitude reached by a periodic orbit by $\phi_{\max} = \phi_{\max}(E, D)$.

When $|D| < 1$ the origin is unstable (hyperbolic) and two stable, elliptic fixed points are created at

$$(\phi_{ell}, v_{ell}) = \pm(\arccos \sqrt{|D|}, 0). \tag{2.5}$$

The origin has two homoclinic orbits†, one in each hemisphere: $q_h^{\pm}(t; \phi_{hmax}) = \pm(\phi_h(t; \phi_{hmax}), v_h(t; \phi_{hmax}))$ where ϕ_{hmax} is the maximal latitude reached by the homoclinic orbit and is given by

$$\cos \phi_{hmax} = |D|. \tag{2.6}$$

While not included here for lack of space, we note that Eqs. (2.1) provide example in which all unperturbed solutions may be found analytically - the homoclinic orbits to the origin $q_h(t; \phi_{hmax})$ for $|D| < 1$, the associated homoclinic solution for the longitude position $\lambda_h^{\pm}(t; \phi_{hmax})$, the period of the periodic orbits in ϕ, $P_\phi(\phi_{max}; \phi_{hmax}) = P_\phi(E, D)$ and the longitude position after completion of one period in the ϕ variable. In particular, exact predictions for the location of resonances may be found easily.

The homoclinic orbits separate the phase space to three distinct regions - regions R_1 and R_2 contain periodic orbits which are restricted to one hemisphere (north and south respectively) whereas region R_3 contains periodic orbits with $\phi_{max} > \phi_{hmax}$. The area of the bounded regions is:

$$\mu(R_1(\phi_{hmax})) = \mu(R_2(\phi_{hmax})) = \sin \phi_{hmax} - \phi_{hmax} \cos \phi_{hmax}. \tag{2.7}$$

It follows from (2.1) that when $\epsilon \neq 0$

$$\tfrac{dD}{dt} = -2\epsilon A(\phi)k \cos(k\lambda - \sigma t), \tag{2.8}$$

hence there is a major difference between the $k = 0$ and $k \neq 0$ cases:

Forcing with infinite wavelength ($k = 0$):

This forcing corresponds to a perturbation which does not vary with longitude. If the amplitude of the pressure wave is independent of latitude ($A(\phi) = const.$) then the perturbation merely changes periodically the radii of the spherical geopotential and hence has no effect on the motion along it. Thereby, only latitude dependent amplitude ($A'(\phi) \neq 0$) is of interest. Since the angular momentum is conserved in this case, using (2.4) it follows that equations (2.1c ,2.1d) are independent of λ and may be analyzed using the standard tools of a one-and-a-half degrees of freedom Hamiltonian system, depending on the

---

† These are solutions which are asymptotic to the origin as $t \to \pm\infty$.

parameters D, σ and on $A'(\phi)$:

$$\frac{d\lambda}{dt} = \frac{1}{2}\left(\frac{D}{\cos^2\phi} - 1\right) \tag{2.9a}$$

$$\frac{dD}{dt} = 0 \tag{2.9b}$$

$$\frac{d\phi}{dt} = v \tag{2.9c}$$

$$\frac{dv}{dt} = \frac{1}{8}\sin(2\phi)\left(1 - \frac{D^2}{\cos^4\phi}\right) + \epsilon A'(\phi)\sin(\sigma t) \tag{2.9d}$$

and the Hamiltonian:

$$H(\phi, v, t; D) = \frac{v^2}{2} + \frac{1}{8}\left(\frac{D}{\cos\phi} - \cos\phi\right)^2 - \epsilon A(\phi)\sin(\sigma t). \tag{2.10}$$

The equations and Hamiltonian written for $k = 0$ clearly demonstrate the relevance of the restriction to latitude dependent amplitude of the perturbation. More details regarding this case are presented in section 4.

<u>*Forcing with finite wavelength ($k \neq 0$):*</u>

Introducing the wave velocity c:

$$c = \frac{\sigma}{k} \tag{2.11}$$

and defining θ, the conjugate variable of D, by:

$$\theta = \tfrac{1}{2}(\lambda - ct), \tag{2.12}$$

(2.1) is replaced by:

$$\frac{d\theta}{dt} = \frac{1}{4}\frac{D}{\cos^2\phi} - \tfrac{1}{2}(c + \tfrac{1}{2}) = \frac{\partial H}{\partial D} \tag{2.13a}$$

$$\frac{dD}{dt} = -2k\epsilon A(\phi)\cos(2k\theta) = -\frac{\partial H}{\partial\theta} \tag{2.13b}$$

$$\frac{d\phi}{dt} = v = \frac{\partial H}{\partial v} \tag{2.13c}$$

$$\frac{dv}{dt} = \frac{1}{8}\sin(2\phi)\left(1 - \frac{D^2}{\cos^4\phi}\right) - \epsilon A'(\phi)\sin(2k\theta) = -\frac{\partial H}{\partial\phi}, \tag{2.13d}$$

where H is the Hamiltonian given by:

$$H(\phi, v, \theta, D) = \frac{v^2}{2} + \frac{1}{8}\left(\frac{D}{\cos\phi} - \cos\phi\right)^2 - \frac{c}{2}D + \epsilon A(\phi)\sin(2k\theta). \tag{2.14}$$

Using the symmetries of (2.13), it is easy to show that the relevant parameters ranges for (2.13) are:

$$k > 0, \qquad \epsilon \geq 0, \qquad -\tfrac{1}{2} \leq c < \infty, \tag{2.15}$$

as all other values may be reduced to the above. Mathematically, the natural period for θ is π/k, however, physically, it is π. When considering dispersion of particles, this distinction is important, hence we keep $\theta \in [0, \pi)$. By the evenness of $A(\phi)$, the Poincaré map in θ at θ_0 is symmetric with respect to:

i. $\phi \to -\phi, \quad v \to -v$ for all θ_0.

ii. $\phi \to -\phi, \theta \to -\theta, t \to -t$ when $\theta_0 = j\pi/2k + \pi/4k$ for some integer j.

However, (2.13) is *not symmetric* with respect to reflections in D: when $D \to -D$ the equation for θ changes in a non invariant manner (though for the unperturbed case, the system in (ϕ, v) is symmetric with respect to reflections in D). This asymmetry is due to the unidirectional rotation of the earth.

3. The Energy Surfaces

Equations (2.13) constitute a weakly coupled two-degrees-of-freedom Hamiltonian system, so their solutions lie on three dimensional energy surfaces in (v, ϕ, D, θ) phase space, given by level sets of H of (2.14). The structure of the energy surface serves as a backbone for understanding the solution structure in the different regions of parameter and phase space.

Below, the construction of the energy surfaces for $c > 0$ (Figure 1) and $c < 0$ (Figure 2) is described. Then we discuss the implications of the surface structure on the dynamical behavior in the different regimes in phase space and the analysis of these regimes.

3.1 Construction of the Energy Surfaces

We construct the surfaces at the cross-section $\theta = 0$, so that the $O(\epsilon)$ term vanishes. The actual energy surfaces are three dimensional, and may be represented symbolically as the cross-sections presented in Figure 1 or Figure 2, multiplied by a circle on which the variable θ varies. Other cross-sections on the circle, corresponding to different values of θ may be viewed as changing the value of H by an $O(\epsilon)$ amount and as slight deformation of the surface if $A(\phi)$ is not constant. This effect is especially significant near values of H for which the singularities of the energy surface change.

To present the motion in the θ direction in a compact fashion we color the regions on the energy surface for which $\frac{d\theta}{dt}$ is positive along the unperturbed orbits by light shading and the regions on which it is negative by dark shading. This coloring scheme enables us to read of the structure of the four-dimensional flow from the two dimensional energy surface plot.

The energy surfaces of Figure 1 and Figure 2 are plotted for increasing values of H - picking the typical structure in each regime of H values as described below.

To find the minimal relevant H value, we use (2.14) to conclude that at the origin

$$H_o(D) = H(0,0,0,D) = \tfrac{1}{8}(D-1)^2 - \tfrac{c}{2}D, \qquad (3.1)$$

Hence, the minimal value of H for which an energy surface includes the origin is given by:

$$H_{\min} = \min_D H_o(D) = -\tfrac{1}{2}c(1+c). \qquad (3.2)$$

The value of D at the origin for a given H value is:

$$D_\pm(H) = (1+2c) \pm 2\sqrt{2H + c(1+c)}. \qquad (3.3)$$

It follows that the minimizing D value (i.e. the value of D at H_{min}) is:

$$D_r = 1 + 2c. \qquad (3.4)$$

Consider the case $c > 0$. Then, $D_+(H) > 1$ for all $H > H_{\min}$, and $D_-(H) \geq 1$ for $H \in [H_{\min}, H_{-c/2}]$ where

$$H_{-c/2} = \min_H \{D_{-\text{sign}(c)} = 1\} = -c/2. \qquad (3.5)$$

From (2.14), for $\theta = 0$,

$$\tfrac{v^2}{2} = H - \tfrac{1}{8}\left(\tfrac{D}{\cos\phi} - \cos\phi\right)^2 + \tfrac{c}{2}D, \qquad (3.6)$$

so, for $D \in (D_-(H), D_+(H))$ the r.h.s. of (3.6) is positive at $\phi = 0$, hence the energy surfaces for these values of H are as depicted in Figure 1 a. Moreover, for $c > 0$, (3.6) has a real solution for some D value iff $H \geq H_{\min}$ so we need not consider smaller values of H.

To color Figure 1 a, we investigate the behavior of $\frac{d\theta}{dt}$. It follows from (2.13a) that

$$\left.\frac{d\theta}{dt}\right|_{\phi=0} = \tfrac{1}{4}(D - 1 - 2c)$$

hence $\frac{d\theta}{dt}$ is positive at $\phi = 0$ for $D > 1 + 2c$ and negative for $D < 1 + 2c$. Moreover, $\frac{d\theta}{dt}$ changes sign when $\frac{d\theta}{dt} = 0$, and this may happen iff $\phi = \phi_r$ where

$$\cos^2 \phi_r = \tfrac{D}{2c+1}. \qquad (3.7)$$

Hence, $\frac{d\theta}{dt}$ may change sign along orbits only if $0 < D < D_r$ where $D_r = 1 + 2c$ as in (3.4). Substituting (3.7) (and $\epsilon = 0$) in (2.14) we restrict D so that v^2 is non-negative at $\phi = \phi_r$; since

$$v^2 = 2H + \tfrac{c(1+c)D}{1+2c},$$

and the r.h.s. vanishes for $D = D_\Delta(H)$ where

$$D_\Delta(H) = -\tfrac{2H}{c}\left(1 + \tfrac{c}{1+c}\right), \qquad (3.8)$$

we find that $\frac{d\theta}{dt}$ vanishes on a given energy surface, H, for $\phi = \phi_r$ only when $D \in [\max\{0, D_\Delta(H)\}, D_r]$ for $c > 0$ and when $D \in [0, D_r]$ for $c < 0$.

It follows that for $c = 0.5$, $H_{\min} = -0.375$, $H_{-c/2} = -0.25$, hence Figure 1 a falls into this regime. Now we describe briefly the changes in the energy surfaces for $c > 0$ as H is further increased from its minimal value H_{min}. At $H = H_{-c/2}$, $D_-(H) = 1$, namely the bifurcation point lies on this energy surface. Increasing H further brings the separatrix and the elliptic points onto the energy surface. Equations (2.5) and (2.14) imply, in this case, that the elliptic points on the energy surface H have angular momentum:

$$D_{ell}(H) = \begin{cases} \frac{-2H}{c} & H > 0 \\ \frac{-2H}{c+1} & H < 0. \end{cases} \tag{3.9}$$

Also, equation (3.3) implies that $|D_-(H)| < 1$ for $H_{-c/2} \leq H \leq H_{\max}$ where

$$H_{\max} = \max_H \{H | D_-(H) \geq -1\} = H_o(-1) = \tfrac{1}{2}(c+1), \tag{3.10}$$

hence the separatrices emanating from $(0, 0, D_-(H), \theta)$ and the elliptic points are contained in the energy surface with energy H for such H values, as depicted in Figure 1 b-e.

The difference between the first three figures Figure 1 b,c,d is the different behavior of $\frac{d\theta}{dt}$ in the vicinity of the separatrices; Indeed, as long as $D_-(H) < D_\Delta(H)$, $\frac{d\theta}{dt}$ may not vanish along the separatrix, as shown in Figure 1 b. Since $D_-(H) = D_\Delta(H)$ when $\phi_r = \phi_{hmax}$, we use (2.6) and (3.7) to conclude that at that point $D_-(H) = D_\Delta(H) = D_c$, where

$$D_c = \tfrac{1}{2c+1}, \tag{3.11}$$

and this determines the critical value of H:

$$H_c = H_o(D_c) = \tfrac{1}{8}\left(\tfrac{1}{(1+2c)^2} - 1\right). \tag{3.12}$$

Hence, for $H \in (H_{-c/2}, H_c)$, $\frac{d\theta}{dt}$ is negative for all (ϕ_h, v_h) as depicted in Figure 1 b, for $H = H_c$, $\frac{d\theta}{dt}$ vanishes at one point along the separatrix, namely, at $\phi = \phi_{hmax}$, and otherwise it is negative, and for $H > H_c$, there is a region of ϕ values, $\phi_r < \phi < \phi_{hmax}$, for which $\frac{d\theta}{dt}$ is positive whereas for $\phi < \phi_r$ $\frac{d\theta}{dt}$ is negative, as shown in Figure 1 c ($H_c(c = 0.5) = -3/32 \approx 0.094$). When $H = 0$, $D_{ell} = D_\Delta(H) = 0$, corresponding to the degenerate behavior at the poles. When $0 < H < \frac{1}{8}$, $D_-(H) > 0$ and $D_{ell} < 0$, all orbits with $0 < D < 1$ have regions of back-flow where $\frac{d\theta}{dt} > 0$, while the periodic orbits with $D < 0$ have none. The intersection of the energy surface with $D = 0$ surface consists of interior orbits, corresponding to periodic motions restricted to one hemisphere, passing through the poles with $v < \frac{1}{2}$. At $H = \frac{1}{8}$ the separatrix reaches $D = 0$, connecting the equator and the poles (Figure 1 d). For $\frac{1}{8} < H$ The intersection

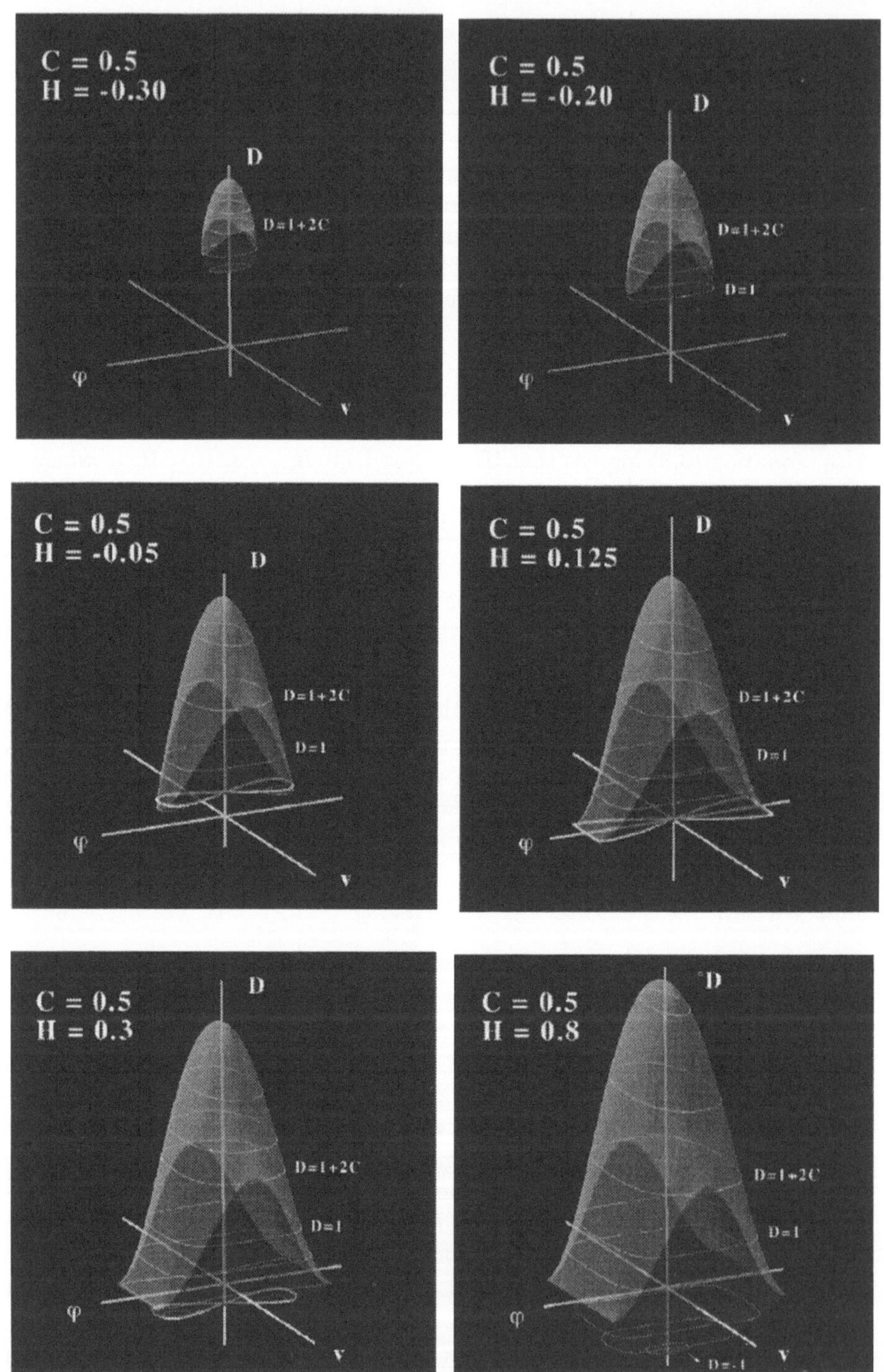

Figure 1. The energy surfaces for $c > 0, \theta = 0$.

of the energy surface with $D = 0$ surface contains exterior orbits, corresponding to periodic motions going through both hemispheres, passing through both the north and south poles with $v > \frac{1}{2}$. These motions correspond to high kinetic energies and are not expected to be encountered in observed oceanic or atmospheric flows. For $\frac{1}{8} < H < H_{\max}$ the separatrix and the elliptic points have negative angular momentum $(0 > D_-(H), D_{ell}(H) > -1)$ where $\frac{d\theta}{dt}$ does not change sign (Figure 1 e). When $H > H_{\max}$ the energy surface includes only exterior periodic orbits, and the periodic orbits with $0 < D < D_r$ have regions of back-flow (Figure 1 f).

Similar analysis for the $c < 0$ case results in Figure 2. Clearly new structures of the energy surface appear. Most notably, for $0 < H < H_{min}$, there exist an energy surface consisting of two connected components (Figure 2 a). At $H = H_{min}$ the two components coalesce at the origin (Figure 2 b), and for $H_{min} < H < \min\{-c/2, 0.125\}$ there exist two homoclinic orbits emanating from $D_\pm(H)$ with $0 < D_\pm < 1$ (Figure 2 c). For $0.125 < H < H_{max}$ the lower homoclinic orbit has $D_-(H) \in (0, -1)$ (Figure 2 d,e) whereas for $H_{min} < H < -c/2$ the upper homoclinic orbit has $D_+(H) \in (D_r, 1)$ (Figure 2 c,d). The behavior near $D = D_r$ for $H \approx H_{min}$ is of special interest.

3.2 Regimes of analysis:

The series of energy surfaces of Figure 1, Figure 2 and the structure of the unperturbed motion on them reveals that there are three types of typical structures which emerge for small non vanishing pressure waves. Moreover, the critical values of H and D for which a certain degenerate behavior appears arise naturally as the limiting values for which the boundary separating different subregions is approached.

Typical Structures

A. KAM surfaces: The KAM surfaces correspond, in the unperturbed case, to the motion on tori composed of the periodic motion in ϕ on the presented cross-section and the periodic motion in θ, with irrationally related period. By KAM theory, even for nonzero perturbation (yet sufficiently small) most of these tori persist. These tori are two dimensional surfaces which divide the energy surfaces, hence their persistence guarantees stability in the D direction (as is the standard case for two-degrees-of freedom systems).

B. Resonances: The coupling between the rationally related periodic motion in the angles ϕ and θ creates resonances. When $\frac{d\theta}{dt}$ is nonvanishing these resonances are necessarily of rotational nature in θ, namely θ covers the interval $[0, \pi]$. When $\frac{d\theta}{dt}$ changes sign along an orbit, (i.e. the orbits containing two-colors in Figure 1 or Figure 2) there exists the possibility of oscillatory resonance, which, under the perturbation, results in localized structures with respect to the travelling wave. Moreover, one expects that the size of these oscillatory resonances will be the most significant (in future work we plan to indicate the values of D on which strong resonances are to occur on the energy surface).

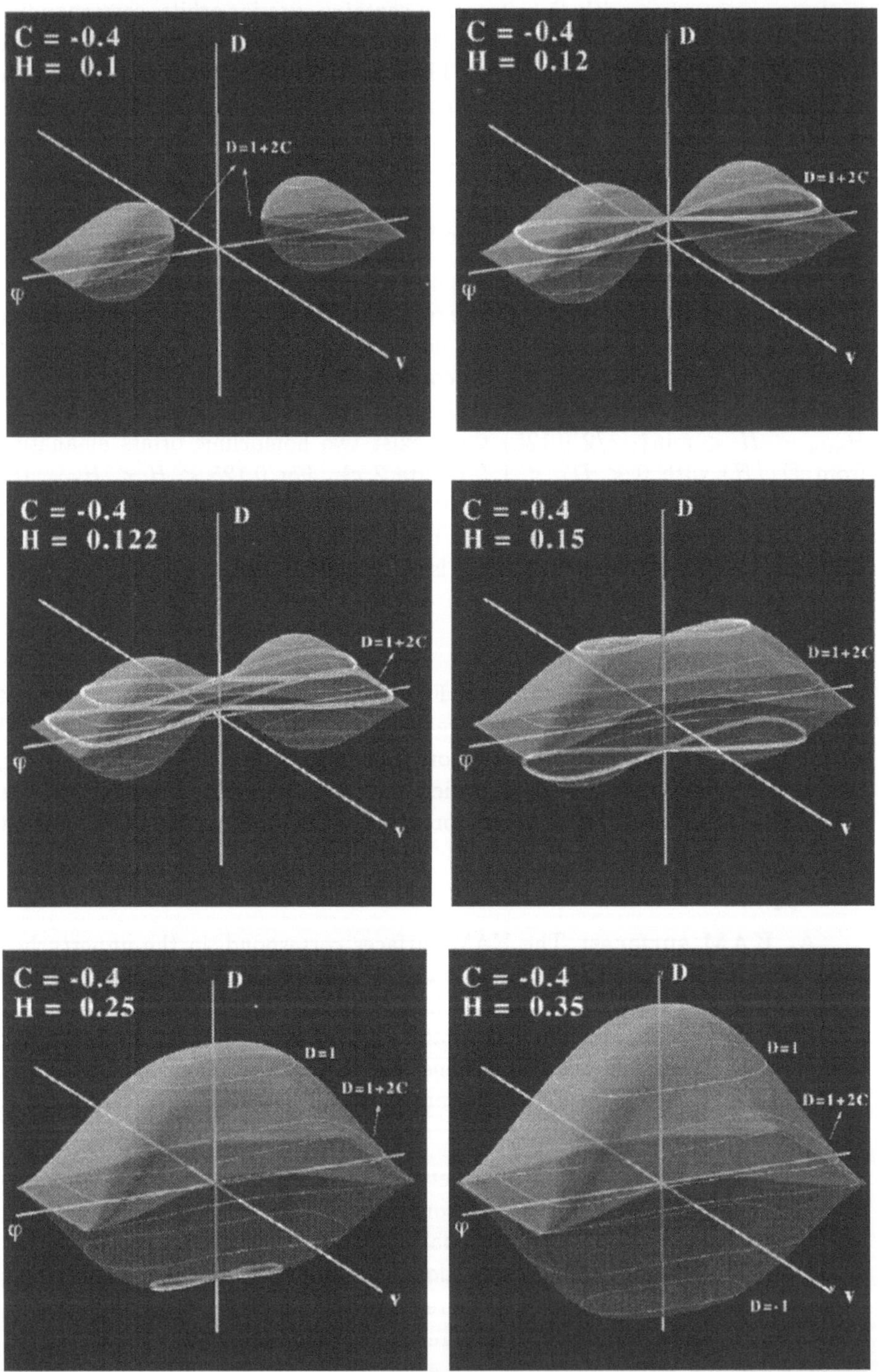

Figure 2. The energy surfaces for $c < 0, \theta = 0$.

C. Near separatrix behavior:

i. *Monotonic θ dependence:* when $\frac{d\theta}{dt}$ is bounded away from zero over an invariant region which includes the separatrix, the system (2.13) is reduced, as in Marsden and Holmes[5], to a one-and-a-half degrees-of-freedom Hamiltonian system, where θ replaces time. The reduction and the analysis of the reduced system will be presented elsewhere. From Figure 1 and Figure 2 we identify the intervals in H for which the unperturbed solutions in θ near the separatrix are monotonic in t (for these intervals the neighborhood of the separatrix has only one color shading in the figure):

I. $c > 0$:

Ia. For $H \in (-c/2, H_c)$ the component $\theta(t)$ is monotonic near the separatrix emanating from $D_-(H)$, as depicted in Figure 1 b. For these values of H, $D_-(H) \in (D_c, 1)$.

Ib. For $H \in (1/8, \frac{1}{2}(c + 1))$, the component $\theta(t)$ is monotonic near the separatrix emanating from $D_-(H)$, as depicted in Figure 1 f. For these values of H, $D_-(H) \in (-1, 0)$.

II. $c < 0$

IIa. For $H \in (H_{\min}, -c/2)$, the component $\theta(t)$ is monotonic near the separatrix emanating from $D_+(H)$, as depicted in Figure 2 b,c. For these values of H, $D_+(H) \in (D_r, 1)$.

IIb. For $H \in (1/8, \frac{1}{2}(c + 1))$, the component $\theta(t)$ is monotonic near the separatrix emanating from $D_-(H)$, as depicted in Figure 2 c,e. For these values of H, $D_-(H) \in (-1, 0)$.

ii. *Non monotonic θ dependence:* when $\frac{d\theta}{dt}$ changes sign along orbits in the vicinity of the separatrix (but not at the origin) the system is fully three dimensional on each energy surface. The methods developed by Wiggins[6] may be used to analyze the behavior of the separatrices. The geometrical interpretation of the results in terms of the transport in the four dimensional system is challenging. Such a behavior occurs in the following regimes:

I. $c > 0$:

Ic. For $H \in (H_c, 1/8)$, the component $\theta(t)$ is non-monotonic near the separatrix emanating from $D_-(H)$, as depicted in Figure 1 d,e. For these values of H, $D_-(H) \in (0, D_c)$.

II. $c < 0$

IIc. For $H \in (H_{\min}, 1/8)$, the component $\theta(t)$ is non-monotonic near the separatrix emanating from $D_-(H)$, as depicted in Figure 2 b,d. For these values of H, $D_-(H) \in (0, D_r)$.

Degenerate behavior

The interesting degenerate cases are listed below:

A. Hyperbolic resonance: When $\frac{d\theta}{dt}$ vanishes on the unstable fixed point $(\phi, v) = (0, 0)$, a strong resonance (1:1) occurs between the inertial trajectories and the forcing. Recent results by Kovacic and Wiggins and their extensions by Haller and Wiggins and Kaper and Kovacic[7] may be used to analyze this case. In these works the authors concentrated on finding criteria and proving theorems regarding the existence of transverse homoclinic orbits of various geometrical nature (analogously to finding the Melnikov integral in the one-and-a-half d.o.f. Hamiltonian systems). Such a behavior occurs in the following regime:

II. $-0.5 < c < 0$

IId. For $\epsilon = 0$ and $H = H_{\min}$, $\frac{d\theta}{dt} = 0$ at $D_-(H) = D_+(H) = D_r$, as depicted in Figure 2 b.

B. Behavior near Parabolic point: For $\epsilon > 0$ and all values of the other parameters, the behavior near the origin at $|D| = 1$ is mathematically non-trivial as it involves a perturbation of a parabolic point. It is expected to find exponentially small splitting of the separatrices near this point[8]. In terms of the energy surfaces, this region corresponds to the behavior near $H = H_{\min}$ and $H = H_{\max}$ where the energy surfaces change their topology.

C. Behavior on the $D = 0$ surface. For $\epsilon > 0$ and all values of the other parameters, the behavior near the separatrix for $D = 0$ is non-trivial, as, in the spherical coordinate system, in the limit $D \to 0$ the separatrices are discontinuous in v (reflecting the fact that v changes sign at the poles).

D. Zero wave speed. When $c = 0$, $\frac{d\theta}{dt} = 0$ at the elliptic fixed points $((\phi, v) = (\pm \arccos \sqrt{D}, 0))$, hence strong resonances are expected to occur along this surface. Moreover, since $D_r = D_c = 1$, these resonances end near $D = 1$ where a **parabolic resonance** appears.

E. Critical wave speed. When $c = -\frac{1}{2}$, the hyperbolic resonance (case IIb) occurs for $D_r = 0$, combining the two degenerate phenomena (A + C) which were discussed above.

4. Forcing with infinite wavelength ($k = 0$)

It follows from (2.9) that the (ϕ, v) system may be treated as a one-and-a-half d.o.f. Hamiltonian systems depending on the parameters D, σ and ϵ, and D may be taken to be positive as the (ϕ, v) system is invariant under $D \to -D$.

Interpretation in terms of the motion in the (λ, ϕ) has no such invariance; when $D < 0$ or $D > 1$ λ is monotonic in t, and when $0 < D < 1$, λ changes its direction when $\phi = \arccos \sqrt{D}$. In particular, the system (2.9) possesses a two dimensional surface of fixed points given by: $\{D, \lambda | (\phi, v, D, \lambda) = (\arccos \sqrt{D}, 0, D, \lambda), 0 \le D \le 1\}$. This situation is highly degenerate, especially at $D = 1$, hence additional perturbation, coupling the λ coordinate to the (ϕ, v) system (e.g. small k) is expected to cause strong resonances in addition to the $k = 0$ resonances discussed next.

Mathematically, one views the perturbed system as three dimensional in (ϕ, v, t) and considers the dynamics of the Poincaré map, a map found by sampling the solutions at $t = 2j\pi/\sigma + t_0$ for fixed t_0. For $D > 1$ the solutions to the unperturbed problem are periodic in t. In the Poincaré map, the orbits with periods which are rationally related to $2\pi/\sigma$ (these are called resonant orbits) appear as periodic orbits of the mapping, and those which are irrationally related trace a continuous curve. For small enough perturbation "most" of these curves survive (the ones which have periods sufficiently far from being resonant). When resonance of order n/m occurs, i.e., when the period of the periodic motion in the (ϕ, v) plane, $P_\phi(\phi_{\max}, D)$, satisfies:

$$P_\phi = \frac{n}{m}\frac{2\pi}{\sigma} \tag{4.1}$$

a chain of islands in the (ϕ, v) plane appears. Namely, a stable periodic orbit emerges, crossing the equator $2m$ times as t covers the interval $[0, 2n\pi/\sigma]$ and orbits which are sufficiently close to this solution - namely belong to the resonance band - oscillate about it (the width of the resonance band is proportional to $\sqrt{\epsilon}$ and is exponentially decreasing with n/m). It follows that strong resonances near the elliptic points occur when $\sigma \to 0$ near $|D| = 1$ and as $\sigma \to 1$ near $D = 0$. The latter limit is especially interesting as the elliptic points approach the separatrices as $D \to 0$.

For $|D| < 1$ the separatrix breaks down enabling orbits to spend several rounds in the north hemisphere and then hop to the south hemisphere and vice versa. The breakup of the separatrix is quantified by calculating the distance between the stable and unstable manifolds of the fixed point $(\phi, v, D) = (0, 0, \cos\phi_{hmax})$ along the ϕ-axis. In the Poincaré map $t = t_0$ this distance is given by:

$$d(t_0; \sigma, \phi_{hmax}) = \epsilon\frac{8M(t_0;\sigma,\phi_{hmax})}{|\sin(2\phi_{hmax})(1-\frac{1}{\cos^2\phi_{hmax}})|} + O(\epsilon^2) \tag{4.2}$$

Where $M(t_0; \sigma, \phi_{hmax})$ is the Melnikov function[9,5], a periodic function in t of period $T = 2\pi/\sigma$:

$$M(t_0; \sigma, \phi_{hmax}) = 2\cos(\sigma t_0)\int_0^\infty vA'(\phi)\sin(\sigma t)\Big|_{q_s(t)}dt = M_0(\sigma, \phi_{hmax})\cos(\sigma t_0). \tag{4.3}$$

As long as $M_0(\sigma, \phi_{hmax})$ does not vanish the Melnikov function has simple zeroes, and the stable and unstable manifolds intersect transversely. It is customary to present the amplitude of the Melnikov function, $M_0(\sigma, \phi_{hmax})$. However, the normalization factor in (4.2) reflects the changes of the vector field on the homoclinic loop with ϕ_{hmax}, hence **a true measure to the size of the chaotic zone must include this factor**. Moreover, this factor changes the nature of the dependence on ϕ_{hmax} as $M_0(\sigma, \phi_{hmax})$ is monotonically increasing with ϕ_{hmax} whereas the maximal distance, $d(\sigma, \phi_{hmax})$ is not. In fact d attains a global maxima at $\phi_{hmax} \approx 0.4, \sigma \approx 0.25$. As σ is increased the "maximal chaos latitude" slightly increases as well. The width of the stochastic layer is monotonically increasing with $d(\sigma, \phi_{hmax})$ (though not linearly), hence its magnitude measures, roughly, the extent of the chaotic region.

The most rapid mixing between states bounded to one hemisphere and states visiting both hemispheres occurs when the flux per unit time, given by the lobe area[10] divided by the period, normalized by the homoclinic loop area is maximal:

$$F(\sigma, \phi_{hmax}) = \frac{\text{Lobe area}}{\text{period} \cdot \text{homoclinic loop area}}$$
$$= \frac{2|M_0(\sigma,\phi_{hmax})|/\sigma}{\frac{2\pi}{\sigma}\mu(R_1)} = \frac{|M_0(\sigma,\phi_{hmax})|}{\pi(\sin\phi_{hmax}-\phi_{hmax}\cos\phi_{hmax})} . \tag{4.4}$$

Computing this function, we find that $F(\sigma, \phi_{hmax})$ attains its maxima at $\sigma \approx 0.25$ and $\phi_{hmax} \approx 0.5$. Hence the most rapid exchange of bounded to unbounded motion occurs for these values. To obtain more accurate estimates of the transport rates between the North and South hemisphere, the TAM may be employed[11].

5. Summary

Using the angular momentum and the Hamiltonian we have constructed energy surfaces, colored according to the zonal direction of propagation. This construction enabled us to classify readily the different regions in phase space in which the behavior is qualitatively different. In this initial study we have discussed the behavior in one particular case of infinite wavelength perturbation. For a fixed small strength of the pressure wave, we have identified the parameters and latitudes for which the chaotic zone is maximal and the values for which the mixing is most intense. The application of these results to observations on the dispersal of passive particles in the atmosphere and ocean is left for a sequel study.

Acknowledgement

The authors are indebted to Drs. M. Engel and A. Sigalov of the Center for Visualization of Dynamic Systems of the Hebrew University of Jeruslaem for their assistance in producing the figures contained in this paper. V. Rom-Kedar thanks for the hospitality of the Mathematics department of the University Of Chicago where part of this work has been carried out.

References

[1] W. Von-Arx [1974]. *An Introduction to Physical Oceanography*. Addison-Wesley Publishing Company Inc., Massachusetts.

[2] N. Paldor and P.D. Killworth [1988] Inertial Trajectories on a Rotating Earth, J. Atm. Sc., Vol. , No. 24, pp 4013-4019.

[3] N. Paldor and E. Boss [1992] Chaotic Trajectories of Tidally Perturbed Inertial Oscillations, J. Atm. Sc., Vol 49, 23, pp 2306-2318.

[4] E. Knobloch and J.B. Weiss, [1987], Mass Transport and Mixing by Modulated Traveling Waves. Phys. Rev. A, **36**, 1522, observed non-diffusive decay of autocorrelator. See Zaslavsky et al., this procceding, for the role of Lèvy flights in this process and for other references.

[5] See review and reference in J. Guckenheimer and P. Holmes [1983]. *Non-Linear Oscillations, Dynamical Systems and Bifurcations of Vector Fields*. Springer-Verlag, New York.

[6] S. Wiggins [1990]. *Introduction to Applied Nonlinear Dynamical Systems and Chaos*. Springer-Verlag: New York, Heidelberg, Berlin.

[7] G. Kovacic and S. Wiggins [1992] Orbits Homoclinic to resonances, with an application to chaos in a model of the forced and damped sine-Gordon equation, Physica D, 57 pp 185-225, G. Haller and S. Wiggins [1993] Orbits Homoclinic to resonances: the Hamiltonian case, Physica D, 66 pp 298-346, G. Haller and S. Wiggins [1994] N-pulse homoclinic orbits in perturbations of resonant Hamiltonian systems, to appear in Arch. of rational Mech., T. Kaper and G. Kovacic [1994] Multi-bump orbits homoclinic to resonance bands, preprint.

[8] P. Holmes, J. Marsden and J. Scheurle [1983] Exponentially small Splittings of Separatices with applications to KAM theory and Degenerate Bifurcations, Cont. Math. **81**, 213-244.

[9] V.K. Melnikov [1963] *On the Stability of the Center for Time Periodic Perturbations*, Trans. Moscow Math. Soc., **12**, 1.

[10] V. Rom-Kedar, A. Leonard and S. Wiggins [1990] An Analytical Study of Transport, Mixing and Chaos in an Unsteady Vortical Flow, JFM, vol. 214, pp. 347-394.

[11] V. Rom-Kedar [1994] Homoclinic Tangles - Classification and Applications, Nonlinearity, 7, 441-473.

Lévy walks and lattice gas hydrodynamics

F. Hayot and L. Wagner

Department of Physics, The Ohio State University, Columbus, Ohio 43210

Abstract. We describe an algorithmic implementation of Lévy walks in lattice gas hydrodynamics, which we use to discuss the effects of turbulence in flow around an infinite cylinder. We describe our results on pressure around the cylinder, drag, and coherence of the Von Karman vortex street.
Keywords. Lévy walk, lattice gas hydrodynamics, turbulence, flow around a cylinder.

1. Introduction

We wish to discuss an implementation of the main features of Lévy walks in lattice gas hydrodynamics[1]. The main features are those highlighted in the paper by Shlesinger,West and Klafter[2] in a description of enhanced turbulent diffusion (Richardson's law[3]). The situation we will study is turbulent channel flow, in particular the case where turbulent flow is incident on an infinite cylinder. We thus limit ourselves to two dimensional situations. Before describing the implementation, we need to briefly recall the main features of lattice gas hydrodynamics.

Lattice gas hydrodynamics is a fluid mechanical description at the microscopic or kinetic level. There are two different versions, a Boolean one and one which takes the form of a Boltzmann equation. The first deals with integer numbers, and averaging over lattice regions is required in order to describe real number fields such as fluid velocity. The implementation is purely algorithmic. How and why, after coarse graining, macroscopic fluid behavior emerges, has been amply shown[4](in the low Mach number limit). The second one operates at a less deep microscopic level, deals with real numbers, and amounts to solving a Boltzmann equation numerically. Since in this case no coarse graining takes place, there is no intrinsic statistical noise, which is a great advantage when one aims at obtaining quantitative results.

2. Lattice gas hydrodynamics

Let us now describe the Boolean version of lattice gas hydrodynamics, though the results described later will make use of a Boltzmann equation. The reason for doing so is that the introduction of Lévy walks is somewhat easier to describe in the Boolean case. The lattice gas is a collection of point particles, which move about from site to site on a two dimensional hexagonal regular lattice. At each site there are six possible directions for a particle to move into. In one unit of time the particle moves a distance of one unit between sites. Thus, though the direction of velocity can change, its magnitude is fixed at unity. Besides moving in the direction indicated by their velocities, at each time step also particles, at the same site, can undergo two-, three- or four- particle collisions, which are momentum conserving, and conserve the number of particles[4]. (Energy conservation is trivial in the simple models with moving particles we consider, because it is equivalent to particle number conservation.) The algorithm is such at at any given time step, at any given site, there can be only one or no particle moving in a given direction. The configuration of particles can therefore be represented by a set of binary numbers. This is the Boolean character, which assures numerical stability, but requires coarse- graining for defining velocity and density fields.

The simplest Boltzmann model is a discrete version of the so- called BGK (Bhatnagar, Gross, Krook) model[5], for which the evolution equation is the following:

$$f_i(\mathbf{x} + \mathbf{c_i}, t + 1) - f_i(\mathbf{x}, t) = (-1/\tau)(f_i(\mathbf{x}, t) - f_i^0(\mathbf{x}, t)), \qquad (1)$$

Here $i = 1, ..., 6$ numbers the six directions on the hexagonal lattice, of unit vectors denoted $\mathbf{c_i}$, and f_i is the number density of particles in the ith direction. The coefficient τ, which will be of order unity is the relaxation time to the local distribution f_i^0, which is chosen appropriately for our purposes, as a function of the local macroscopic density and velocity. The model we are using contains also rest particles which satisfy a separate equation with a separate local distribution function. The details of the model can be found in reference 6, where it is made clear which choices are made for the local functions and why these are advantageous. The kinematic viscosity in the model is linearly related to τ. We have used the model extensively to discuss drag and pressure distribution for flow around a cylinder at Reynolds numbers of the order of 100, and found quantitative agreement with experiment[7].

3. Lévy walk implementation

3.1. *Description*

Let us now describe how the Lévy walk part is transplanted into the Boolean algorithm of lattice gas hydrodynamics. The two principal features of the Lévy walk we retain is that there is exchange of momentum over large distances (the probability density distribution falls off algebraically at large distances), and that there is a characteristic time associated with each exchange with is distance dependent. First recall that an update in lattice gas hydrodynamics is a combination of both a translation of particles into the directions of their velocity vectors followed by a momentum and energy conserving collision, if allowed. In between two successive updates the Lévy walk part of the algorithm proceeds as follows.

Imagine pressure driven channel flow. A distance l is drawn from an algebraic probability density distribution of the form

$$p(l) \propto l^{-z} \qquad (2)$$

The value of z is 1.35 for the results reported here, and is basically the only parameter. If z gets too large, large distances contribute less, and therefore the exchanges do not change the laminar flow.

Once l is drawn, a point S is uniformly chosen through the channel, and the particle configurations at two sites, at equal distances l from S in, say, a direction transverse to the flow are exchanged. This operation corresponds to momentum exchange over a distance of $2l$, such that overall momentum is conserved. For a chosen l, the number of exchanges, or correspondingly the number of points S chosen, is made to depend on the value of l. The algorithm is such that the larger l is, the fewer exchanges over distance $2l$ take place. For a given l, as many different points S are drawn until the required number of exchanges has taken place. (If S is too close to the boundary for the exchange of chosen l to take place, a new S is drawn.) We typically use for the number of exchanges a linear dependence on l, of the form $N(l) = N_0 - l$. Since the Lévy walk implementation takes place within one update time, the characteristic time associated with each distance l is inversely proportional to the number of corresponding exchanges. In other words, as required, the characteristic time associated with long jumps is longer than for small jumps.

For the BGK model (equation (1)) the Lévy walk implementation is similar. The system is updated at time intervals one unit apart according to equation (1), and within each such time interval the Lévy walk algorithm is implemented, namely choices of l (and thus $N(l)$), and of positions S are made.

3.2. *Discussion*

The Lévy walk exchanges act to maintain the system in a stationary out of equilibrium state. They achieve this because they take place on time scales smaller than the relaxation time τ. Though the long range part of the exchanges might be represented in, say equation (1), through the introduction of some nonlocal long range force, there is also the characteristic , length dependent, time associated with each exchange. The time over which quantities in equation (1) evolve is of order unity, and the Navier- Stokes equations are obtained, at long times and large scales, through a Chapman- Enskog expansion, for which there is local momentum and particle number conservation. However, in the algorithm, the exchanges take place on time scales smaller than the unit update time. On these time scales there is only global momentum conservation, no longer a local one. (We have checked that the particle number i. e. mass transport which our procedure entails has a negligible effect.) We do not know how to formalize this.

4. Results and comparison with experiment

The experimental situation concerns flow impinging on an infinite cylinder, when the flow is turbulent. Turbulent intensity is of the order of several percent, as measured by the ratio of fluctuating to average velocities. The other quantity characterizing the turbulence is the integral scale, and its ratio to cylinder diameter. To model this situation, we implement the Lévy walk algorithm, described above in the Boolean case, for the Boltzmann model given by equation (1). What

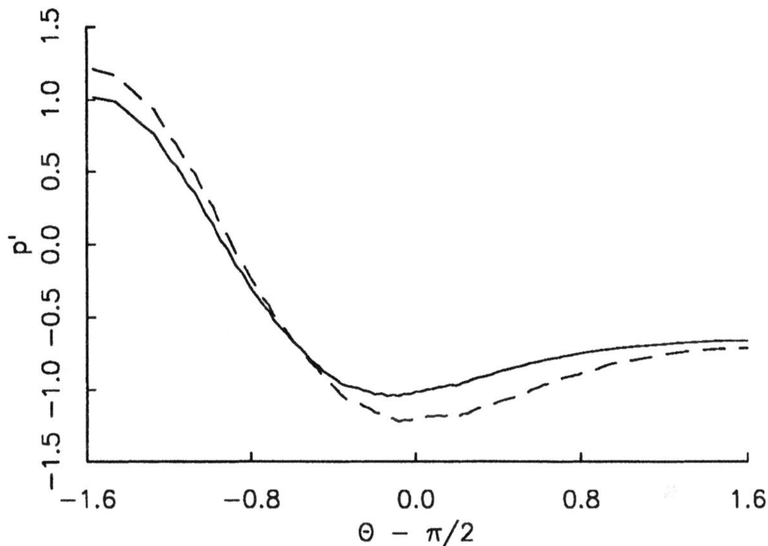

Figure 1: Dimensionless pressure $p' = (p - p_\infty)/\frac{1}{2}\rho U^2$ where p_∞ and U are pressure and velocity far from the obstacle, as a function of $\theta - \pi/2$, where θ is the angle around the cylinder measured from the stagnation point. Laminar flow corresponds to the solid line, turbulent flow, with $l_{max}/D = 7$ to the dash-dotted line . Reynolds number $Re = 76.8$.

replaces the integral scale is the maximum length l_{max} over which exchanges can take place. (These are truncated Lévy walks.) The comparison with experiment is qualitative, because the experiments we are aware of are done at very high Reynolds number, often approaching the region of the so- called drag crisis. Our simulations are done at a Reynolds number of about 100. The quantities we will be discussing are pressure around the cylinder and drag. The question is what changes relative to the laminar case occur in these quantities, when turbulence is present, as a function of the ratio of integral scale to cylinder diameter. We will also consider the issue of coherence of the Von Karman street, when exchanges take place over larger and larger distances, transporting momentum into and out of the wake, and thus interfering with the coherent structure of the vortices. Let us discuss these items in succession:

4.1. *Pressure around the cylinder*

Since on the hexagonal lattice, the cross- section of the cylinder is a polygon, and not a circle, the pressure around the cylinder is measured a few units away from it, in order to minimize this effect. A comparison of laminar pressure and turbulent pressure is shown in figure 1. for our model . The ordinate is the reduced pressure p', and the abscissa is proportional to the angle θ around the cylinder measured from the stagnation point. Over part of the forward region

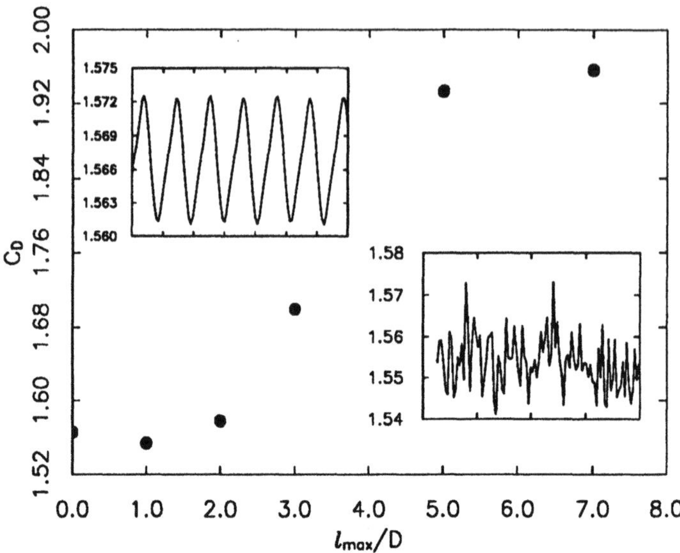

Figure 2: Drag coefficient C_D as a function of l_{max}/D. The inserts show the time series of drag in laminar flow at Re=77, and for flow with incident turbulence at $l_{max} = D$. Whereas the laminar time series is very regular, the second insert shows that even weak turbulence is strong enough to mask the small amplitude oscillations the drag undergoes as vortices are shed.

the turbulent pressure is higher than the laminar one, but around the back region the opposite is true, the greatest difference occuring close to , but above $\theta = \pi/2$. We also find (as expected) that the difference between laminar and turbulent pressure diminishes as the "integral scale" parameter l_{max}/D diminishes. These results are in qualitative agreement with those of Bearman[8] for flat plates, who observed that turbulence diminishes the base pressure (behind the plate), and also that, as the size of the plate increases (for a given integral scale) the difference between turbulent and laminar base pressure diminishes.

4.2. Drag

Drag on the cylinder is affected by transport of momentum into and out of the wake, which is a result of the Lévy exhanges. These exchanges lead to a clear decrease in the size of the region of negative streamwise velocity component right behind the cylinder. From a particle point of view, this means that there are fewer particles which impinge onto the cylinder hitting it from the back against the principal thrust of the flow. As a result, the drag is expected to increase as l_{max}/D increases sufficiently. This is what is observed in figure 2 which shows the drag coefficient C_D as a function of l_{max}/D. As sooon as l_{max}/D exceeds a value of 2, the drag starts increasing. It saturates when the size of the exchanges becomes large enough to be insensitive to the dimension of the vortex

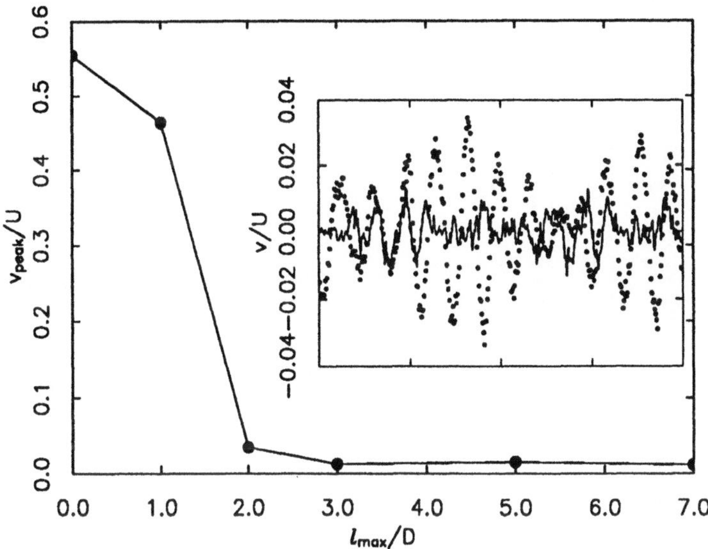

Figure 3: Transverse velocity ratio v_{peak}/U on the cylinder axis as a function of l_{max}/D. The time series for v/U is shown in the insert for $l_{max}/D = 2$(dots) and $l_{max}/D = 3$(full line). U is the average incoming velocity far from the cylinder.

street itself. According to a wake entrainment argument[8], turbulent drag should increase as $(l_{max}/D)^2$ relative to the laminar case, until it saturates. Our result on turbulent drag is in qualitative agreement with this.

4.3. *Coherence of Von Karman street*

It is natural to ask to what extent entrainment into and out of the wake, caused by the presence of turbulence in the incident flow, affects the coherence of the Von Karman street. As a measure of that coherence, we choose the component v of velocity perpendicular to the streamwise direction, on the axis of the cylinder. In the laminar case, this is a strong component, which at the Reynolds number we are considering varies periodically in time with a well defined frequency, which determines the value of the Strouhal number. Our results illustrated in figures 3. and 4. show two things: one that v_{peak},the maximum value of v, decreases rapidly as the parameter l_{max}/D increases, and becomes very small at $l_{max}/D = 3$, the value where also the drag starts deviating significantly from the laminar value (cf. figure 2.). Secondly, that though the intensity diminishes considerably, the frequency spectrum of v continues to be dominated by the Strouhal frequency component (because of lack of statistics our power spectrum data are not of good quality.) There thus may be no significant broadening of the Strouhal peak in the spectrum as the range of exchanges increases. We have noticed however that a contour plot of pressure, which in the laminar case shows clearly the presence of vortices, changes completely as l_{max}/D becomes

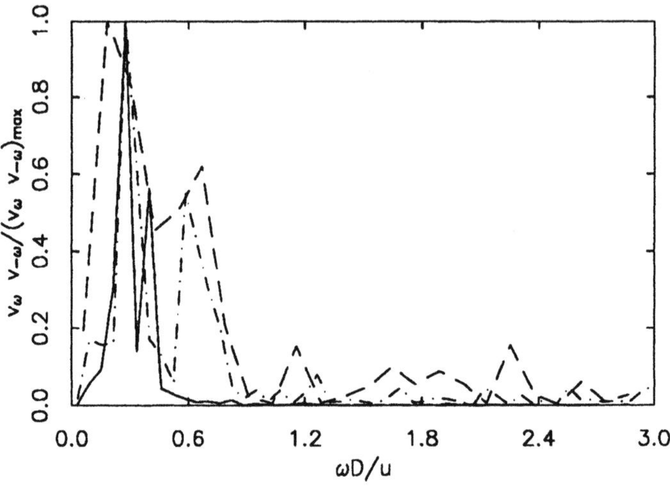

Figure 4: Power spectrum of v at a distance of $3.5D$ behind the cylinder on the cylinder axis for $l_{max} = 2D$ (solid line), $= 5D$ (dashed line), and $= 7D$ (dotted line), as a function of the dimensionless quantity (the Strouhal number) $\omega D/U$. All curves have been rescaled by their maximum value.

large enough, without any trace left of vortex structure. Correlations between transverse velocity components on and off the axis show the same phenomenon[9]. Speaking loosely, one might say that as the range of exchanges increases the Von Karman street gradually loses its spatial coherence, but in such a way that the rate of production of vortices (the Strouhal number) remains well defined, at least up to $l_{max}/D = 2$. For clearly higher values the vortices cease to exist.

There is a caveat here: these results depend sensitively on the precise relationship between $N(l)$ and l, whereas those on pressure for instance do not. It is therefore worth repeating here that as far as the coherence of the Von Karman street is concerned, our model studies only indicate what could happen and what are the parameters to measure. As to experiment, we do not know of a systematic investigation of the robustness of the Von Karman street with the integral scale of turbulence. There are indications that its robustness is not affected by turbulence[10], though an experiment on turbulent shear flow past a cylinder seems to show that vortex shedding can be disrupted at large integral scales of turbulence[11].

Acknowledgements: We are very grateful to Stephane Zaleski for his interest in this work and for helpful discussions.

This work was supported by the Department of the Navy, Office of Naval Research, under Grant No. N00014-92-J-1271, and benefitted from computer time provided by The Ohio Supercomputer Center.

References

[1] U. Frisch, B. Hasslacher, and Y. Pomeau, Phys. Rev. Lett. **56**, 1505 (1986)

[2] M. F. Shlesinger, B. J. West, and J. Klafter,Phys. Rev. Lett. **58**, 1100 (1987)

[3] L. F. Richardson, Proc. Roy. Soc. London ,Series A **110**, 709 (1926)

[4] U. Frisch, D. d'Humieres, B. Hasslacher, P. Lallemand, Y. Pomeau, and J.P. Rivet, Complex Syst. **1**, 648 (1987)

[5] P. Résibois and M. De Leener, *Classical Kinetic Theory of Fluids*, John Wiley & Sons (1977)

[6] H. Chen, S. Chen, and W. H. Matthaeus, Phys. Rev. A **45**, R5339 (1992)

[7] Lukas Wagner, Phys. Rev. E **49**, 2115 (1994); Lukas Wagner, Phys. of Fluids A, to be published

[8] P. W. Bearman, J. Fluid Mech. **46**, 177 (1971)

[9] Lukas Wagner and F. Hayot, J. of Stat. Phys., to be published.

[10] D. Surry, J. Fluid Mech. **52**, 543 (1972)

[11] M. Kiya and H. Tamura, J. of Fluids Eng. **111**, 126 (1989)

PART 2:

MATHEMATICAL APPROACHES

Definition of stable laws, infinitely divisible laws, and Lévy processes

Jean-Pierre Kahane[1]

[1] Département de Mathématiques, Université de Paris-Sud,
91405 Orsay Cedex, France

The main references are the books of Paul Lévy 1925, 1937, 1948. We shall discuss stable laws first and how to construct them through Poisson "bricks", then processes with stationary and independent increments ("processus additifs") and how to construct them through Poisson processes. We shall find infinitely divisible laws on the way.

The *law*, or distribution, of a real random variable X, is just a measure μ on the line defined as

$$\mu(I) = \mathrm{P}\,(X \in I) \tag{1}$$

(probability that X belongs to I). It is well defined by the characteristic function

$$\varphi(u) = \mathrm{E}\,(e^{iuX}) \tag{2}$$

(E means expectation, or mean value). Actually,

$$\varphi(u) = \int_{\mathbb{R}} e^{iux}\mu(dx), \tag{3}$$

is the Fourier transform of the measure μ. Characteristic functions are often easier to work with than distributions.

By definition the laws of X and cX ($c > 0$ fixed) belong to the same *type*. The laws of X and $cX+d$ ($c > 0$ and d real) belong to the same generalized type. The type of law of X is *stable*, meaning stable under addition of independent copies, if, given X_1 and X_2, independent copies of X, and positive constants c_1 and c_2, there exists a copy X_3 of X and a constant c such that

$$c_1 X_1 + c_2 X_2 = cX_3. \tag{4}$$

In short, we say that the law of X is stable, or that X is stable. Quasi-stable is a notion related to the generalized type ; we say that the law of X, or X itself, is quasi-stable, if

$$c_1 X_1 + c_2 X_2 = cX_3 + d, \tag{5}$$

whatever $c_1, c_2 > 0$. If c_1 and $c_2 > 0$ are given and (4) is assumed to hold, the law of X, or X itself, is said to be semi-stable. It is interesting to explore quasi-stable and semi-stable laws and the corresponding processes. However, we shall restrict ourselves to stable laws.

The characteristic function of cX is $\varphi(cu)$, and the characteristic function of $c_1 X_1 + c_2 X_2$ is the product $\varphi(c_1 u)\varphi(c_2 u)$. The law of X is stable if and only if, whatever c_1, $c_2 > 0$, there exists $c > 0$ such that

$$\varphi(c_1 u)\varphi(c_2 u) = \varphi(cu). \tag{6}$$

Clearly, $\varphi(u)$ cannot vanish at any point. We write

$$\varphi(u) = e^{-\psi(u)}, \tag{7}$$

so that the real part of $\psi(u)$ is positive, and the solution of the functional equation

$$\psi(c_1 u) + \psi(c_2 u) = \psi(cu), \tag{8}$$

is

$$\psi(u) = c\left(\frac{u}{|u|}\right)|u|^\alpha; \tag{9}$$

that is,

$$\begin{cases} \psi(u) = c_+\, u^\alpha, & \text{for } u > 0 \\ \psi(u) = c_-\, (-u)^\alpha, & \text{for } u < 0, \end{cases} \tag{10}$$

with $\Re c_+ > 0$, $\Re c_- > 0$ ($c_+ = c(+1)$, $c_- = c(-1)$).

Classical examples are the normal (Gaussian) law ($\alpha = 2$), the Cauchy law ($\alpha = 1$), the law of the time when Brownian motion is stopped at a given level ($\alpha = 1/2$). Actually, these are the only cases when the law is defined by an explicit density; in other cases we have to satisfy ourselves with the characteristic function.

In fact, (10) is only a necessary form for $\psi(u)$. We want moreover $e^{-\psi(u)}$ to be a characteristic function (Fourier transform of a non-negative function). This requires

$$0 < \alpha \le 2. \tag{11}$$

We shall see that (10) provides a solution whenever c_+ and c_- are well chosen, but not all c_+'s and c_-'s work; $c_+ = c_- > 0$ provides always a solution.

The case $\alpha = 2$ is apart; then, necessarily $c_+ = c_- > 0$, that is, $\psi(u) = cu^2$ with $c > 0$. For $0 < \alpha < 2$ a new idea is needed.

We are going to consider functions $\psi(u)$ such that $e^{-c\psi(u)}$ is a characteristic function whatever $c > 0$. Since the product of two characteristic function is a characteristic function again, the sum of two ψ-functions, and, more generally, any combination of ψ-functions with positive coefficients, is again a ψ-function; in other words, the ψ-functions form a convex cone. Moreover, this cone is closed under pointwise convergence, because a pointwise limit of characteristic functions, when continuous, is a characteristic function. In order to build this closed cone we only need the extreme points; these are the "bricks" that we shall introduce now.

If X is a Poisson variable with parameter (mean number) a, then

$$\mathrm{E}\left(e^{iuX}\right) = \exp\left(a(e^{iu} - 1)\right); \tag{12}$$

therefore the ψ-function corresponding to yX is

$$\psi(u) = a(1 - e^{iuy}). \tag{13}$$

The ψ-function corresponding to $y(X - E(X))$ is

$$\psi(u) = a(1 - e^{iuy} + iuy). \tag{14}$$

Eqs. (13) and (14) are the bricks that we are looking for.

At first, let us look at the closed convex hull of functions (13) ($a > 0$, $y > 0$). It consists of ψ-functions of the form

$$\psi(u) = \int_{\mathbb{R}^+} (1 - e^{iuy}) \, \nu(dy), \tag{15}$$

where ν is any locally bounded measure on $(0, \infty)$, such that both integrals

$$\int_0^1 y\nu(dy) \quad \text{and} \quad \int_1^\infty \nu(dy),$$

are finite. Choosing

$$\nu(dy) = \frac{dy}{y^{1+\alpha}} \quad (0 < \alpha < 1), \tag{16}$$

gives

$$\psi(iu) = \int_0^\infty (1 - e^{-uy}) \frac{dy}{y^{1+\alpha}} = cu^\alpha, \tag{17}$$

for some $c > 0$ and all $u > 0$. Therefore, using analytic continuation,

$$\begin{cases} \psi(u) = c\, e^{-i\frac{\pi}{2}\alpha}\, u^\alpha, & \text{for } u > 0, \\ \psi(-u) = c\, e^{i\frac{\pi}{2}\alpha}\, u^\alpha . \end{cases} \tag{18}$$

These are the ψ-functions of positive stable X. Let us emphasize the formula

$$E(e^{-ux}) = e^{-cu^\alpha} \quad (u > 0), \tag{19}$$

and the condition

$$0 < \alpha < 1. \tag{20}$$

We turn to bricks (14) with $a > 0$ and $y > 0$. They generate ψ-functions of the form

$$\psi(u) = \int_{\mathbb{R}^+} (1 - e^{iuy} + iuy) \, \nu(dy), \tag{21}$$

whenever integration makes sense, that is,

$$\int_0^1 y^2 \, \nu(dy) < \infty \quad \text{and} \quad \int_1^\infty y\, \nu(dy) < \infty. \tag{22}$$

Choosing

$$\nu(dy) = \frac{dy}{y^{1+\alpha}} \quad (1 < \alpha < 2), \tag{23}$$

gives

$$\psi(iu) = \int_0^\infty (1 - e^{-uy} - uy) \, \nu(dy) = -cu^\alpha, \tag{24}$$

for some $c > 0$ and all $u > 0$; therefore,

$$\begin{cases} \psi(u) = c\, e^{i\frac{\pi}{2}(2-\alpha)}\, u^\alpha, & (u > 0) \\ \psi(-u) = c\, e^{-i\frac{\pi}{2}(2-\alpha)}\, u^\alpha. \end{cases} \tag{25}$$

The corresponding stable variable X satisfies

$$\mathrm{E}\left(e^{-uX}\right) = e^{cu^\alpha} \qquad (u > 0), \tag{26}$$

and is not positive any more. Let us emphasize the condition

$$1 < \alpha < 2. \tag{27}$$

Until now we considered bricks (13) or (14) with $y > 0$ only. Combining bricks with $y < 0$ results in interchanging $\psi(u)$ and $\psi(-u)$. Combining bricks with real y's, using

$$\nu(dy) = A\, \frac{dy}{y^{1+\alpha}}\, 1_{\mathrm{I\!R}^+} + B\, \frac{dy}{|y|^{1+\alpha}}\, 1_{\mathrm{I\!R}^-}, \tag{28}$$

(1_{E} means the indicator function of the set E), we obtain any positive combination of the above $\psi(u)$ and $\psi(-u)$, that is

$$\begin{cases} \psi(u) = c\, e^{i\frac{\pi}{2}\beta}\, u^\alpha, & \text{for } u > 0 \\ \psi(-u) = \overline{\psi(u)}. \end{cases} \tag{29}$$

with $\alpha > 0$ and

$$\begin{cases} -\alpha \le \beta \le \alpha & \text{if } 0 < \alpha < 1, \\ \alpha - 2 \le \beta \le 2 - \alpha & \text{if } 1 < \alpha < 2. \end{cases} \tag{30}$$

In this way we obtain all stable laws of index different from one and two. "All" would need a proof; let us skip it. The symmetric stable laws correspond to $\beta = 0$.

When $\alpha = 2$ and $\alpha = 1$ the only stable laws are symmetric: the normal and the Cauchy laws. This ends the discussion of formula (9) and stable laws.

However, there are many other ways to combine our bricks (13) and (14). A most appealing way is to consider

$$\int_{|y|<1} (1 - e^{iuy} + iuy)\, \nu(dy) + \int_{|y|\ge 1} (1 - e^{iuy})\, \nu(dy), \tag{31}$$

where ν is a locally bounded measure on $\mathrm{I\!R} - \{0\}$ such that

$$\int_{|y|<1} y^2\, \nu(dy) + \int_{|y|\ge 1} \nu(dy) < \infty. \tag{32}$$

Let us remark that if we replace the domains of integration $|y| < 1$ and $|y| \ge 1$ by any open interval containing 0 and its complement with respect to $\mathrm{I\!R}$, we simply add a term of the form λiu (λ real). What is really important is ν, not the way we divide our blocks (13) and (14). For example, choosing ν as in (28), we obtain in this way all stable laws of index $\alpha \in]0, 2[$, including the Cauchy law corresponding to $\alpha = 1$ and $A = B$.

By adding the ψ-functions of a normal variable and a constant variable we obtain a most important class of ψ-functions, namely

$$\psi(u) = au^2 + biu + \int_{|y|<1} (1 - e^{iuy} + iuy)\nu(dy) + \int_{|y|\geq 1} (1 - e^{iuy})\nu(dy), \quad (33)$$

with $a \geq 0$, b real and ν as above.

Let us recall how we defined a ψ-function: $\psi(u)$ is a ψ-function if and only if $e^{-c\psi(u)}$ is a characteristic function whatever $c > 0$. Actually (33) describes *all* ψ-functions, and the corresponding laws are called *infinitely divisible*. Let us observe that an infinitely divisible law is the N-th convolution power of another law, for any N, and this statement can be taken as a definition. There are several other definitions. Most definitions and the above theorem (all ψ-functions can be written as (33)) are due to Lévy and Khintchin (see Lévy's book 1937). Note that ψ-functions appear also in a geometric context, as helix or screw functions (J.J. Schoenberg *Trans. Am. Math. Soc.* **44** (1938), 522–536; J. von Neumann and J.J. Schoenberg *Trans. Am. Math. Soc.* **50** (1941), 226–251).

Let us turn to the process aspect.

Can we find all processes X_t $(t \in \mathbb{R}^+)$, starting from zero, with stationary and independent increments? Starting from zero means $X_0 = 0$. Stationary increments means that $X_{t+h} - X_t$ has the same law as $X_h - X_0$ $(= X_h)$. Independent increments means that the increments on disjoints intervals $[a, b]$ and $[c, d]$, namely $X_b - X_a$ and $X_c - X_d$, are independent. Consequently,

$$
\begin{aligned}
E(e^{iuX_{t+s}}) &= E\left(e^{iuX_t} e^{iu(X_{t+s}-X_t)}\right) \\
&= E(e^{iuX_t}) E(e^{iuX_s}). \quad (34)
\end{aligned}
$$

This functional equation implies that $E(e^{iuX_t})$ never vanishes (because $E(e^{iuX_t}) = 0$ implies $E(e^{iuX_{t/2}}) = 0$, etc.) and that

$$E(e^{iuX_t}) = e^{-t\psi(u)}, \quad (35)$$

for all $t > 0$ and some function $\psi(u)$. According to our definition, $\psi(u)$ is a ψ-function, therefore (by Lévy-Khintchin) it is given by (33).

When $\psi(u) = au^2$ $(a > 0)$ we recognize Brownian motion. When $\psi(u) = biu$ we simply have a constant drift. When $a = b = 0$ $(a = 0$ is essential, $b \neq 0$ may be accepted) X_t is called a Lévy process and ν is called the associated Lévy measure.

Until now we derived a few consequences from the definition, but we did not establish the existence of a Lévy process. That is what we shall do now. Given a Lévy measure ν, with the only condition

$$\int_{|y|<1} y^2 \nu(dy) + \int_{|y|\geq 1} \nu(dy) < \infty, \quad (36)$$

how can we build a random function $X(t,\omega)$ ($t \in \mathbf{R}^+$, $\omega \in \Omega$, the probability space) which satisfies the given conditions? If we want to have a hint, we can try $\nu = \delta_1$, the Dirac measure at 1 (that gives the standard Poisson process), then $\nu = a\delta_b$ (b-multiple of a Poisson process, with a drift if $b < 1$). Let us go immediately to the general case.

Let λ be the Lebesgue measure on $\mathbb{R}^+$, and $\lambda \otimes \nu$ the tensor product of λ and ν :

$$\lambda \otimes \nu \, (dtdy) = \lambda(dt)\,\nu(dy). \tag{37}$$

The fundamental random set associated with $\lambda \otimes \nu$ is the Poisson point process of "intensity" $\lambda \otimes \nu$. Its defining property is that, given any box in the half plane $\mathbb{R}^+ \times \mathbb{R}$ (if you like Borel sets, you can say a Borel set instead of a box), the number of points of this random set in the box is a Poisson random variable whose parameter (mean number of points) is the $\lambda \otimes \nu$ measure of the box. The essential property is that the number of points in two disjoint boxes are independent random variables (actually, Poisson). The easiest construction is to divide $\mathbb{R}^+ \times \mathbb{R}$ into boxes B of finite $\lambda \otimes \nu$ measure ; then, for each B, choose n_B a random number according to the Poisson law of parameter $\lambda \otimes \nu\,(B)$, and choose randomly n_B points in B according to the $\lambda \otimes \nu$ measure ; and perform this construction in an independent way for different B's. A convenient approximate construction is to divide $\mathbb{R}^+ \times \mathbb{R}$ into small boxes B having a $\lambda \otimes \nu$ measure small compared to unity ; one then chooses $n_B = 1$ with probability $\lambda \otimes \nu\,(B)$ and $n_B = 0$ otherwise (events with more than one point may be neglected) ; small boxes look like points and the Poisson point process looks thus like the set of B's such that $n_B = 1$.

The random set obtained in the limit of very small boxes will be denoted by $E(\omega)$. It is important to remark that $E(\omega)$ can be constructed as a union of similar independent sets $E_j(\omega)$, corresponding to Lévy measures ν_j, whenever $\nu = \sum \nu_j$; actually, we shall use this remark in the most obvious case, when the measures ν_j's are carried by disjoint sets.

Almost surely, $E(\omega)$ contains the whole information that we need in order to construct the wanted process $X(t,\omega)$. Before proceeding to the construction, we observe that $E(\omega)$ allows us to recover the Lévy measure ν, because, given an interval I which does not contain 0, the density of the sequence of abscissas of the points belonging to $E(\omega)$ wih ordinates in I is precisely $\nu(I)$.

$E(\omega)$ describes the jumps of the process : if (s,y) belongs to $E(\omega)$, then $X(\cdot,\omega)$ has a jump y at time s. Therefore it is natural to try to define

$$X(t,\omega) = \sum_{(s,y)\in E(\omega),\, s<t} y. \tag{38}$$

This is a good definition in some, but not all cases. Let us begin with the good cases, when (38) works.

First case : ν is carried by $\mathbb{R}^+$ and

$$\int_{|y|<1} y\,\nu(dy) + \int_{|y|\geq1} \nu(dy) < \infty. \tag{39}$$

We may write $\psi(u)$ in the form

$$\psi(u) = biu + \int_{\mathbb{R}^+} (1 - e^{iuy})\,\nu(dy). \qquad (40)$$

From the formula

$$\mathrm{E}(e^{iuX_t}) = e^{-t\psi(u)}, \qquad (41)$$

we see that adding a constant drift bt to X_t results in subtracting biu from $\psi(u)$. Therefore we can suppose

$$\psi(u) = \int_{\mathbb{R}^+} (1 - e^{iuy})\,\nu(dy). \qquad (42)$$

The series in (38) converges a.s. for each t. It defines $X(t,\omega)$ as an increasing left-continuous function of t. The increments of $X(t,\omega)$ are stationary and independents on disjoint intervals and

$$\mathrm{E}(e^{iuX(t,\omega)}) = e^{-t\psi(u)}. \qquad (43)$$

Therefore (38) provides an explicit construction of the wanted process X_t. This gives all positive Lévy processes (that is, all increasing Lévy processes).

Second case: we drop the condition that ν is carried by $\mathbb{R}^+$ and keep the condition

$$\int_{|y|<1} |y|\,\nu(dy) + \int_{|y|\geq 1} \nu(dy) < \infty. \qquad (44)$$

Then $\nu = \nu_1 + \nu_2$, with ν_1 carried by $\mathbb{R}^+$ and ν_2 by $\mathbb{R}^-$. We define $X(t,\omega)$ as $X_1(t,\omega) + X_2(t,\omega)$, sum of two independent Lévy processes, the first being positive and the second negative, by considering separately the positive and negative terms in the series (38).

Third case: we assume that ν is symmetric and the general condition

$$\int_{|y|<1} y^2\,\nu(dy) + \int_{|y|\geq 1} \nu(dy) < \infty. \qquad (45)$$

Then the series in (38) is not necessarily absolutely convergent. However $\psi(u)$ can be written as

$$\psi(u) = \lim_{\epsilon \downarrow 0} \int_{\mathbb{R}^+ \setminus [-\epsilon,\,+\epsilon]} (1 - e^{iuy})\,\nu(dy), \qquad (46)$$

($A \setminus B$ designates points belonging to A and not to B) so that the series in (38) converges when ordered by decreasing $|y|$'s, and defines again a left-continuous version of the process X_t.

We now proceed to the general case. Then, ν is the sum of a measure carried by $|y| \geq 1$ and a measure carried by $|y| < 1$. The wanted process can be constructed as the sum of two independent processes; the first can be constructed as above (second case); only the second one needs a different approach. Therefore, the problem is to construct X_t when ν is a measure carried by the interval $(-1, 1)$ and satisfying

$$\int y^2 \nu(dy) < \infty. \tag{47}$$

In this case,

$$\psi(u) = \int (1 - e^{iuy} + iuy)\,\nu(dy). \tag{48}$$

We already observed that the Lévy process which satisfies

$$E(e^{iuX_t}) = e^{-ta(1-e^{iuy}+iuy)}, \tag{49}$$

is a centered Poisson process with intensity a, multiplied by y. The centered Poisson process is just the difference of the increasing Poisson process, whose jumps are on a point Poisson process, and its mean value, a constant drift. In order to obtain

$$E(e^{iuX_t}) = e^{-t\psi(u)}, \tag{50}$$

we have to combine such centered Poisson processes, while considering decreasing $|y|$'s for example. The resulting random function, $X(t,\omega)$, is no more a jump function as in (38). It is what Lévy calls a compensated jump function; for example, if ν is carried by $\mathbb{R}^+$ and $\int y\nu(dy) = \infty$, the sum of the jumps in (38) is infinite, but the sum of the drifts that we introduced is infinite too, so that we have compensation. We shall explain the situation on a deterministic example below.

Using the same image as in studying stable laws, we may say that the increasing Poisson processes are the bricks used in the construction of our random jump functions, and that the centered Poisson processes are the bricks for the construction of our random compensated jump functions. Anyway, we constructed $X(t,\omega)$ satisfying our requirements on X_t and a little more (almost sure left continuity, jumps described by the random set $E(\omega)$). The random function $X(t,\omega)$ is called a version of the process X_t.

Our construction gives immediately many almost sure properties of $X(t,\omega)$. For example, the jumps in a given interval of time, of length ℓ, constitute a point Poisson process with intensity $\ell\nu(dy)$. Since asymptotic properties of Poisson processes are well known, we have asymptotic properties of the sequence of jumps, which can be used in order to confirm or to falsify a modelisation of a natural phenomenon by a Lévy process.

The stable Lévy processes correspond to

$$\nu(dy) = A\,\frac{dy}{y^{1+\alpha}}\,1_{\mathbb{R}^+} + B\,\frac{dy}{|y|^{1+\alpha}}\,1_{\mathbb{R}^-}, \tag{51}$$

with $A > 0$, $B > 0$ and $0 < \alpha < 2$. The number α is called the index of the process. If we get rid of the drift, as is usually done, we saw that the corresponding ψ-function is of the form

$$\begin{cases} \psi(u) = c\, e^{i\frac{\pi}{2}\beta}\, u^{\alpha}, & \text{for } u > 0, \\ \psi(-u) = \overline{\psi(u)}. \end{cases} \tag{52}$$

with $|\beta| \leq \inf(\alpha, 2-\alpha)$ when $\alpha \neq 1$ and $\beta = 0$ when $\alpha = 1$ (Cauchy process). Choosing $c = 1$ as normalization, a stable Lévy process of index α is well defined by the condition of stable and independent increments and the condition

$$\mathrm{E}\left(e^{iuX_t}\right) = \exp(-te^{i\frac{\pi}{2}\beta} u^{\alpha}) \qquad (u > 0). \tag{53}$$

When $0 < \alpha < 1$ and $\beta = -\alpha$, we obtain the increasing stable Lévy processes, the "Lévy flights", also defined by

$$\mathrm{E}\left(e^{-uX_t}\right) = e^{-tu^{\alpha}} \qquad (u > 0). \tag{54}$$

The stability property expresses a simple scaling law: if we multiply the time by λ and the space coordinate by $\lambda^{1/\alpha}$, we obtain the same process.

Brownian motion correspond to the limit case $\alpha = 2$. It is just the missing stone in our construction of processes with stable and independent increments. Among them it enjoys two characteristic properties: (i) it is continuous (more pedantically, it has continuous versions $X(t, \omega)$), (ii) for each t it has a moment of order two (actually, it is a normal variable). Let us emphasize: contrary to Brownian motion, all Lévy processes are discontinuous, and they have no moments of order two. The increasing Lévy processes are pure jumps functions, and they have no moments of order one.

This ends the program indicated at the beginning of this paper. Let us add a few remarks.

1. In order to explain what a compensated jump function is Paul Lévy gives a deterministic example (1948, p. 151). Let $\varphi(t)$ be the 1-periodic function (meaning $\varphi(t + 1) = \varphi(t)$) such that $\varphi(t) = -t$ on $(-\frac{1}{2}, \frac{1}{2})$ (Paul Lévy leaves open the choice of $\varphi(t)$ for $t = \frac{1}{2}$ and we shall do the same). Then $\varphi(t) + t$ is a jump function, with unit jumps at all points $m + \frac{1}{2}$ ($m \in \mathbf{Z}$). Let

$$f(t) = \sum_{1}^{\infty} 2^{-j}\, \varphi(2^j t). \tag{55}$$

This function has jumps 2^{-j} at all points $2^j(m + \frac{1}{2})$; therefore the sum of all jumps over a finite interval is infinite. The divergence of the sum of jumps is just compensated by the divergence of the sum of drifts.

Actually, Lévy's example is not the first of this kind. In his thesis on trigonometric series, after defining his integral, Riemann gives an example of a nowhere continuous but integrable function, namely

$$g(t) = \sum_{1}^{\infty} \frac{1}{n^2}\, \varphi(nt), \tag{56}$$

(in Riemann's notation $\varphi(t) = -(t)$ and $\left(\frac{1}{2}\right) = 0$). Riemann anticipated many things.

2. Let S_t be an increasing Lévy process and $X(t,\omega)$ a Lévy or Wiener process (the Wiener process describes Brownian motion), independent from the process S_t. Then $X(S_t,\omega)$ is a Lévy process. A typical example is $S_t = $ "increasing stable Lévy process of index α", $0 < \alpha < 1$ and $X(t,\omega) = $ Wiener process. Then $X(S_t,\omega)$ is a symmetric stable Lévy process of index 2α. The increasing Lévy processes are called subordinators. They appear in a number of ways. In particular the reciprocal of local times of Lévy processes are subordinators; local times are increasing random functions whose variations over an interval measure the time spent by the processes in crossing the zero level. The stable subordinator of index $\frac{1}{2}$ corresponds to Brownian motion; therefore, its set of values (more precisely, the closure of its set of values) describes the zero-set of Brownian motion.

3. The closed range of a stable subordinator of index α has an interesting property in Fourier analysis. It carries a measure μ (namely, the image of the Lebesgue measure on a given time interval by the subordinator) which satisfies

$$\hat{\mu}(u) = 0\left(u^{-\frac{\alpha}{2}}\sqrt{\log u}\right) \qquad (u \to \infty), \tag{57}$$

($\hat{\mu}(.)$ means the Fourier transform of μ) and carries no non-zero measure μ such that, for some $\alpha' > \alpha$

$$\int_{\mathbb{R}} |\hat{\mu}(u)|^2 |u|^{1-\alpha'} du < \infty. \tag{58}$$

The last property means that it has a vanishing α'-capacity. Both properties imply that its Hausdorff dimension is α and moreover its Fourier dimension is α too. This is a simple way to obtain Salem sets (closed sets for which Hausdorff and Fourier dimensions are the same).

4. Lévy processes are Markov processes and therefore play a role in potential theory. As an example polar sets for α-stable Lévy processes (sets almost surely not hit) are exactly the sets of β-capacity zero, if $0 < \alpha < 1$ and $\alpha + \beta = 1$.

5. Since Lévy processes have no moments of the second order it may seem paradoxical to exponentiate them. However this is possible in the extreme situations when all jumps are negative or all jumps are positive. It is the beginning of the theory of Lévy multiplicative chaos.

6. All that we said can be generalized in several ways. In particular we can consider processes with values in $\mathbb{R}^d$ instead of real-valued processes. Then the Lévy measure has to be defined on $\mathbb{R}^d$, with the same condition as before

$$\int_{\|y\|\leq 1} \|y\|^2 \nu(dy) + \int_{\|y\|\geq 1} \nu(dy) < \infty, \tag{59}$$

and the jumps of the process are described by a random discrete set $E(\omega)$ in $\mathbb{R}^+ \times \mathbb{R}^n$, point Poisson process with intensity ν. We can consider also a multi-dimensional time, and that is relevant in the theory of Lévy multiplicative chaos.

7. Returning to subordinators, they also appear in relation with random Poisson coverings of $\mathbb{R}^+$. The topic was introduced and the first results obtained by B. Mandelbrot in 1972 (*Z. für Wahrsch.* **22**, 145–157).

Points 5 and 6 are developed in chapter 4 of the thesis of Fan Ai-hua (Orsay 1989) (see also *CRAS Paris* **308** (1989), 151–154, and a paper to appear in *Ann. Inst. H. Poincaré*). Point 4 can be found in many books; references appear in the review paper by S.J. Taylor "The measure theory of random fractals", *Math. Proc. Cambridge Math. Soc.* **100** (1986), 383–406. Point 3 is in a note by J.P. Kahane and B. Mandelbrot (*CRAS Paris* **261** (1965), 3931–3933) ($|\psi|$ should be replaced by $|\Re\,\psi|$ in the definition of $h(y)$). Point 2 is easy to see and classical (see, e.g., Lévy 1965). The reference to point 1 is B. Riemann, "Über die Darstellbarkeit einer Funktion durch eine trigonometrische Reihe" §6. The main part of this exposition is just taken from Paul Lévy.

References

Paul Lévy, Calcul des probabilités, Gauthier-Villars 1925.

Paul Lévy, Théorie de l'addition des variables aléatoires, Gauthier-Villars 1937 (also 1954).

Paul Lévy, Processus stochastiques et mouvement brownien, Gauthier-Villars 1948 (also 1965).

Introduction to Fractal Sums of Pulses

Benoit B. Mandelbrot
Mathematics Department, Yale University, New Haven, CT 06520-8283 USA

Abstract. In this paper, the classical Lévy flights are generalized, their jumps being replaced by more involved "pulses." This generates a wide family of self-affine random functions. Their versatility makes them useful in modeling. Their structure throws new conceptual light on the difficult issue of global statistical dependence, especially in the case of processes with infinite variance.

Keywords. Fractal sums of pulses. Global statistical dependence. Infinite variance. Lateral attractors. Lévy flights (generalized). Pulses.

A *Lévy flight* or *Lévy stable motion* (SLM) is well-known to be the sum of an infinity of step-functions of widely varying sizes. This paper introduces a generalized construction: the jumps are replaced by suitable affine reductions or dilations of "templates" represented by *kernel functions* K more general than a step. The result can be described as being an *affine convolution*. A template whose kernel is constant except on a (bounded) interval will be called *pulse*, and the resulting sums will be called *fractal sums of pulses* (FSP). A series of papers that started with [1], [2], [3] and [6] will describe the theory of self-affine FSP and (going beyond [4]) some of their concrete applications.

This paper concerns semi-random self-affine FSP constructed as follows. A pulse's height and location follow the same distribution as in a Lévy flight: the probability of the point $\{\lambda, t\}$ being found in an elementary rectangle of the (λ, t) address plane is $\propto \lambda^{-\delta-1} d\lambda dt$. In a Lévy flight the exponent δ is constrained to satisfy $0 < \delta < 2$. A major immediate difference is that in an FSP, the constraints are either $\delta > 0$ or $\delta > 1$, depending on the case. The pulse's width W (the length of the smallest interval in which the pulse varies) satisfies $W = \sigma \Lambda^\delta$, where $\sigma > 0$ is a scale constant, implying that $Pr\{W > w\} \propto w^{-1}$. The resulting FSP is called *semi-random* because Λ and W are functionally related (in fully random FSP, Λ and W are statistically independent). This paper is meant to show the great variety of distinct behaviors that can be found in an FSP, as we vary δ and three properties of the kernels:

a) Discontinuous versus very smoothly continuous;

b) Canceling (vanishing outside the interval in which they vary) versus non-canceling.

c) Atoms versus bursts. This is a useful but elusive distinction. When the pulse is made of a rise and fall followed later by another rise and fall, it can be decomposed into a burst of two indecomposable or atomic pulses.

The resulting templates are exemplified in Table I: cylinders (one discontinuous *rise* followed after the time W by one discontinuous *fall*), and multiple steps, cones (uniform rate rise followed by uniform rate fall), and uniform rate rises. Other templates are discussed in forthcoming papers.

In the standard study of attraction of a random function $X(t)$ to a limit such as Brownian motion, an early step is an affine rescaling of the form $T^{-H}X(T\rho)$. Being constructed to be self-affine, all FSP are invariant under this rescaling with the exponent $H = 1/\delta$. That is, using the physicists' language, each FSP defines its own "class of universality" with respect to a suitable affinity. But we shall introduce an alternative rescaling, to be called "lateral," and show that each semi-random FSP has an interesting "lateral attractor." The lateral attractor may be a Lévy flight (SLM), which might have been vaguely expected, because the construction starts with the Lévy measure. But the lateral attractor may also be a fractional Brownian motion (FBM), which is a surprise, and establishes deep new links between two independent theories that are known to have striking formal parallelism. The attractor may also be neither SLM nor FBM.

The increments of the FSP are globally dependent, but those of their SLM limits are independent. Thus, the FSP bring altogether new conceptual light on the probabilistic notion of global dependence and the related notion that a process is attracted by another process. Indeed, depending on the shape of K, global dependence is expressed in either of two ways: a) by a special exponent H familiar in such known contexts as FBM, or b) by a prefactor rather than a special exponent.

Separate papers investigate: A) semi-random FSP using additional classes of kernels and second differences of FSP; B) semi-random FSP with $W = \sigma\Lambda^\theta$, where $\theta \neq \delta$; they are not self-affine. Related papers investigate: C) Fully random FSP in which W is statistically independent of Λ with the measure $\propto w^{-\theta-1}dw$, where $0 < \theta < 1$; they are shown in [3] to be self-affine with $H = (1 - \delta)/\theta$; D) fractal sums of *micropulses* (FSM) which generate (FBM); see [1,2].

1. INTRODUCTION

The "normal" model of natural fluctuations is the Wiener Brownian motion process (WBM). By this standard, however, many natural fluctuations exhibit clear-cut "anomalies" which may be due to large discontinuities ("Noah Effect") and/or non-negligible global statistical dependence ("Joseph Effect"). I have long argued that the geometric features of surprisingly many of these anomalous aspects of nature are fractal. For example, for many large "Noah" discontinuities the tail probability distribution is "hyperbolic." That is, if it is large, a discontinuity U that exceeds the value u has a probability of the hyperbolic form $Pr(U > u) \sim u^{-\delta}$, with δ a positive constant. Second example: for large lags s, many globally correlated "Joseph" fluctuations have a correlation function of the form $C(s) \sim 2H(2H - 1)s^{2H-2}$, with $1/2 < H < 1$. [5] shows that one can model various instances of the Noah effect by the classical process of SLM, and various instances of the Joseph effect by the process of FBM.

SLM and FBM, however, are far from exhausting the anomalies found in nature; in particular, neither gives a satisfactory model of the shape of clouds, and many phenomena exhibit *both* the Noah and the Joseph effects and fail to be represented by either SLM or FBM. Hence, fractal modeling of nature demands "bridges," namely random functions (r.f.'s) that combine the infinite variance feature that is characteristic of SLM and the global dependence feature that is characteristic of FBM. One obvious bridge, fractional Lévy motion, is interesting mathematically, but has found no concrete use.

Furthermore, the mathematical theories of SLM and FBM exhibit striking parallels as well as discrepancies. One major discrepancy is in the allowable value of the exponent H which is defined by the condition that the distribution of $T^{-H}F(T\rho)$ is independent of T. For SLM, $1/2 < H < \infty$, while for FBM, $0 < H < 1$. This mismatch is a challenge. Being unexpected, the parallels are sometimes described as miraculous, but they have deep roots worth exploring.

2. DEFINITIONS

2.1 Stationarity and self-affinity

The function $F(t)$ is said to have *stationary increments* if the translated function

$$F(t_0 + t) - F(t_0)$$

has the same distributions for all values of t_0. A function $F(t)$ is said to be *self-affine of exponent $H > 0$* when the rescaled function

$$\rho^{-H}[F(t_0 + \rho t-)F(t_0)]$$

has the same distributions for all values of t_0 and $\rho > 0$. Some authors denote self-affinity by the improper term, self-similarity.

2.2 Pulse templates, pulses, affine convolutions, and fractal sums of pulses.

The graph of $K(t)$, a one-dimension function of a one-dimensional variable, will be called *generator)*, or (*pulse template)*, if $K(t)$ is constant outside an interval; we shall set the shortest such interval to be of length 1.

A *pulse* is a translated affine transform $K\left(\frac{t-t_n}{w_n}\right)$ of $K(t)$, where λ_n, t_n, and w_n are called the pulse's height, position, and width.

A *sum of pulses* is a function of the form

$$F(t) = \sum \lambda_n K\left(\frac{t - t_n}{w_n}\right).$$

Figure 1 is an example of a sum of these pulses. When $F(t)$ is a self-affine function, it will be called a *fractal sum of pulses*, FSP.

An *affine convolution* of the sequence $\{\lambda_n, t_n\}$ by the kernel $K(\cdot)$ is obtained when the pulse heights and widths are linked by a relation $w_n = \sigma \lambda_n^\delta$, where $\sigma > 0$ and $\delta > 0$ are prescribed. Thus, the semi-random FSP are affine convolutions of sequences $\{\lambda_n, t_n, w_n\}$.

2.3 The Lévy measure for the probability distribution of pulse height and position

The simplest pulse template is a step function redefined so that $w_n = 1$ for all n. The distribution of $\{\lambda_n, t_n\}$ that insures self-affinity in that case was discovered by Paul Lévy and is classical. The same distribution continues to be required in all FSP. Let the plane of coordinates t and λ be called *address plane*. Take a rectangle

$[\lambda, \lambda + d\lambda] \times [t, t + dt]$ in the address plane, such that $[\lambda, \lambda + d\lambda]$ does not contain $\lambda = 0$. Given an exponent $\delta > 0$ and two scale factors C' and C'', the Lévy measure is

$$C'\lambda^{-\delta-1}d\lambda dt \quad , \text{if} \quad \lambda > 0, \quad \text{and} \quad C''|\lambda|^{-\delta-1}d\lambda dt \quad , \text{if} \quad \lambda < 0.$$

Giving C' and C'' is, of course, equivalent to giving the overall scale $C' + C''$ and the skewness factor $C'/(C' + C'')$.

To define an FSP, the probability of finding an address point (λ, t) in the elementary rectangle is set equal to the Lévy measure. The number of address points in a domain $\mathcal{D}$ in the address space is taken to be a Poisson random variable whose expectation is the integral of the Lévy measure over $\mathcal{D}$. The total number of pulses is countably infinite.

2.4. Semi-random pulse templates

The simplest pulses are the step functions used by Paul Lévy to generate the SLM. The pulses examined in this paper and illustrated in Table 1 are *semi-random*: Λ is random with the Lévy measure, but the height Λ fully determines the width W. To insure that the FSP is self-affine, one must take

$$W = \sigma\Lambda^\delta, \quad \text{where} \quad \sigma > 0.$$

The resulting probability distribution of W is, independently of δ,

$$Pr\{W > w\} \propto w^{-1}.$$

The units in which Λ and W are measured are arbitrary and unrelated. If those units are identical, either $C' + C''$ or σ can be normalized to 1 by changing the unit. However, up to scale, the distribution of an FSP is determined by $(C' + C'')\sigma$ and the skewness $C'/(C' + C'')$. In the sequel, an important role is played by lateral limit theorems that are expressed most conveniently by fixing $C' + C''$ and allowing $\sigma \to \infty$. The consequences of $\sigma \to \infty$ are obvious when the pulses are "cylindrical:" a *rise* followed, after a span of $\Delta t = w$, by a *fall* of equal absolute value and opposite sign. Clearly, the contribution to $F(t + T) - F(t)$ from a pulse such that $w > T$ is not a pulse but an unattached rise or fall. Therefore, as $\sigma \to \infty$, each cylindrical pulse reduces to a rise or a fall, and the fact that a pulse has a bounded support becomes less and less significant.

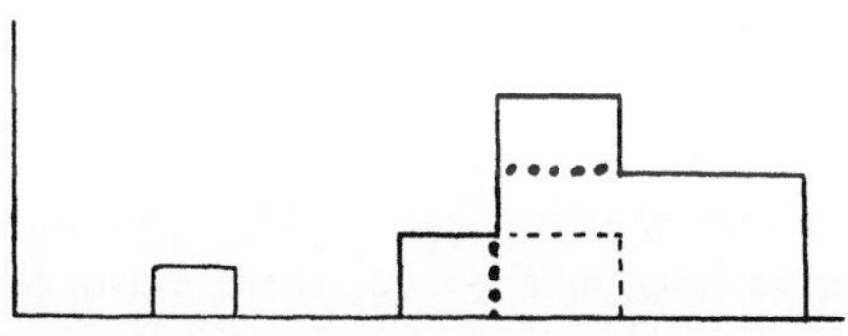

FIGURE 1: Schematic sum of pulses

3. SELF-AFFINITY AND THE EXPONENT $H = 1/\delta$; EXISTENCE OF GLOBAL DEPENDENCE

3.1. The self-affinity property of all FSP

For many combinations of a pulse shape and a value of δ, the semi-random FSP construction yields a well-defined random process, and δ is called *admissible* for these pulses. For other values of δ, the construction diverges and δ is called *excluded*. When the construction converges, it is easy to see that the resulting FSP are self-affine. All semi-random cases yield $H = 1/\delta$, just like in the Lévy case when the pulses are step functions. In one case to be described elsewhere, the construction of $F(t + T) - F(t)$ diverges, but that of the second difference $[F(t + T) - F(t)] - [F(t) - F(t - T)]$ converges and is self-affine.

3.2. First corollary of self-affinity: each FSP defines a special domain of attraction, hence the standard limit problem concerning random processes is degenerate

In the study of random functions, the standard next issue is whether or not there exists an exponent H such that, setting $F(0) = 0$,

$$\text{weak } \lim_{T \to \infty} T^{-H} F(\rho T)$$

is a non-degenerate function of ρ, called the "attractor" of F. The most familiar attractions are WBM and SLM in the case of independence and FBM in the case of dependence. Now suppose that $F(t)$ is a semi-random FSP. For it, the standard limit problem does not arise, since $T^{-H} F(\rho T)$ independent of T in distribution, each FSP defines its own domain of attraction of exponent $H = 1/\delta$.

Standard domains of attraction to an FSP. Given a self-affine attractor $X(t)$, the next challenge is to describe its domain of attraction, defined as the collection of r.f.'s $G(t)$ for which the rescaling (or renormalizing) function $A(T)$ can be selected so as to insure that

$$\text{weak } \lim_{T \to \infty} A^{-1}(T) G(\rho T) = X(t).$$

My study of the domain of attraction of a semi-random FSP has limited itself thus far to r.f.'s that are themselves sums of pulses, but involve a density other than Lévy's or a relation other than $w = \sigma \lambda^\delta$. I have not yet examined templates that depend on height.

Clearly, weak $\lim_{T \to \infty} T^{-H} F(\rho T)$ is unchanged if the distribution of pulse heights is changed over a bounded interval, for example if pulse height is restricted to $\lambda > \epsilon > 0$.

For the next obvious change, SLM suggests replacing the constants C' and C'' in the Lévy density by functions $C'(\lambda)$ and $C''(\lambda)$ that vary slowly for $\lambda \to \infty$. And FBM suggests replacing the relation $w = \sigma \lambda^\delta$ by $w = \sigma(\lambda) \lambda^\delta$, where $\sigma(\lambda)$ is also slowly varying. These changes lead to a nonself-affine generalized FSP. The questions is whether or not, as $T \to \infty$, there exists a rescaling $A(T)$ that makes $A^{-1}(T) G(T)$ converge weakly to a self-affine FSP. I have been content with verifying (see Section 7.6) that a sufficient condition is that $C'(\lambda)\sigma(\lambda) \to 1$; in that case, $T^{1/\delta} A(T)$ is the inverse function of $T = \sigma(\lambda) \lambda^\delta$, implying that $A(T)$ is slowly varying. The condition $C'(\lambda)\sigma(\lambda) \to 1$ is demanding, resulting in a narrow domain of attraction, a notion discussed in Section 3.4.

3.3. Second corollary of self-affinity: the global dependence property of all FSP

A corollary of Section 3.2 is that if F is a semi-random FSP, then $T^{-H}F(\rho T)$ fails to converge to a standard attractor relative to asymptotically independent increments, namely, either WBM or SLM. This implies that all semi-random FSP must fail to satisfy the usual criteria that express that dependence between increments is local.

In fact, they are uniformly globally dependent. For example, define for each t the following two functions

- the rescaled finite past $T^{-1/\delta}[F(t) - F(t - Tp)]$
- and the rescaled finite future $T^{-1/\delta}[F(t + Tp) - F(t)]$.

Because of pulses that contribute to both past and future, these random functions of p are not statistically independent. Because of self-affinity, their joint distributions are independent of T. This means that strong mixing is contradicted uniformly for all T.

3.4. Thoughts on the role of limit theorems suggested by the degeneracy of the standard limit problem in the case of FSP

To comment on the role of limit theorems in the light of Section 3.3, let us compare the "attractands" with their attractor. One wants the process of going to the limit to destroy the most idiosyncratic features of the attractand, while preserving features that have a degree of "universality." This is why the most important attractors continue to be: the nonrandom attractor for the laws of large numbers, the Gaussian r.v. for the central limit theorems, and WBM for the functional central limit theorems. These attractors' domains of attraction are very broad, being largely characterized by the absence of global independence and of significant probabilities for large values. By contrast, when a basin of attraction is narrow, the attractor yields specific information about the attractands. Thus, there is a sharp contrast between broad universality with little information and narrow universality with extensive information.

Down to specifics: for SLM, the dependence can be anything, as long as it is local, but the tails must be long and strictly constrained; for FBM, the tails can be anything, as long as they are short, but the dependence must be global and strictly constrained. Similarly, the sufficient condition $C'(\lambda)\sigma(\lambda) \to 1$ in Section 3.2 defines for each FSP a (partial) domain of attraction that is tightly constrained, with *the same exponent, for the tails and the global dependence.*

The resulting variety of forms creates a use for additional limit problems that would put order by destroying some of the FSP's overabundant specifications. That is, the finding in Section 3.3 must spur the search for alternative limit problems, for which the domains of attraction are broader, therefore reveal more "universal" properties of the FSP. Section 4 will advance one possibility.

4. THE CONCEPT OF LATERAL LIMIT PROBLEM AND THE EXPONENT α; UNISCALING ($\alpha = \delta = 1/H$) AND PLURISCALING ($\alpha = \min[2, \delta] \neq 1/H$) LATERAL ATTRACTORS

4.1. Background of the new "lateral" limit problem

Neither the random walk nor the Poisson process of finite density is self-affine, but both are attracted in the usual way to the Wiener-Brownian motion $B(t)$, which is self-affine with $H = 1/2$. That is, writing $F(0) = 0$, it is true in both cases that $T^{-1/2}F(\rho T)$ depends on T for $T < \infty$, but not in the limit $T \to \infty$; since the replacement of T by ρT transforms discrete time and F into real variables, rescaling time before taking a limit is necessary in the case of a random walk. But in the Poisson case one can rephrase the standard passage to the limit into a form that avoids rescaling time. One imagines N independent and identically distributed Poisson processes $F_n(t)$, then one forms

$$\tilde{F}_N(t) = \sum_{n=1}^{N} F_n(t),$$

and one finds that,

$$\text{weak limit}_{N \to \infty} N^{-1/\alpha} \tilde{F}_N(t) = B(t), \text{ with } 1/H = \alpha = 2.$$

This rephrasing of the passage to the limit is not important in the Poisson case, but it has the virtue that it continues to make sense in the case of FSP. The theory of the addition of independent identically distributed random variables tells us that, for fixed t, one must have $0 < \alpha \leq 2$ and the limit is a stable rv: Gaussian for $\alpha = 2$, and Lévy stable for $0 < \alpha < 2$. Table I lists the values of α for a selection of pulse shapes, and Section 7 gives an example of derivation.

4.2. The lateral limit problem as applied to FSP; the term "lateral"

Let us now observe that forming $\tilde{F}_N(t)$ for a semi-random FSP, and then letting $N \to \infty$ is equivalent to viewing $F(t)$ as a function of both t and $C' + C''$. Then we keep C'/C'' fixed and let $C' \to \infty$ and $C'' \to \infty$. One can think of the axis of $C' + C''$ as orthogonal to the axis of t, hence the term *lateral*.

For example, replacing C' by $C'_1 + C'_2$ and C'' by $C''_1 + C''_2$ can be interpreted as follows. The pulses corresponding to C'_1 and C''_1 can be called red and said to add up to $F_1(t)$, and the pulses corresponding to C'_2 and C''_2 can be called blue and said to add up to $F_2(t)$; all pulses together yield $\tilde{F}(2, t) = F_1(t) + F_2(t)$. In order to represent $F(N, t)$ graphically on a page, one must rescale it by the factor $N^{-1/\alpha}$.

4.3. An important corollary of the results in Table I: global dependence can be either "uniscaling" ($H = 1/\alpha$) or "pluriscaling" ($H \neq 1/\alpha$)

Table I uses α to denote the usual exponent of stability. A first glance shows familiar r.f.'s among the lateral limits. The SLM has independent increments and satisfies $0 < \alpha < 2$ and $H = 1/\alpha$. The FBM, except for $H = 1/2$, has globally dependent Gaussian increments, so that $\alpha = 2$, but satisfies $H \neq 1/\alpha = 1/2$.

For many years, my studies of global dependence concentrated on these two examples (broadened by fractional Lévy motion) and on the R/S statistic. This concentration made me think that global dependence can be tested and measured by a single exponent H. Given α, I thought that global dependence could be defined by $H \neq 1/\alpha$ and measured by $H - 1/\alpha$. It is a pleasure to note that I was prudent enough *not* to write of (in)dependence but rather of (R/S)-(in)dependence, but I confess that discrepancies were expected to belong to "mathematical pathology." I was thoroughly mistaken.

TABLE I

Pulse	$0 < \delta < 1$	$1 < \delta < 2$	$2 < \delta$
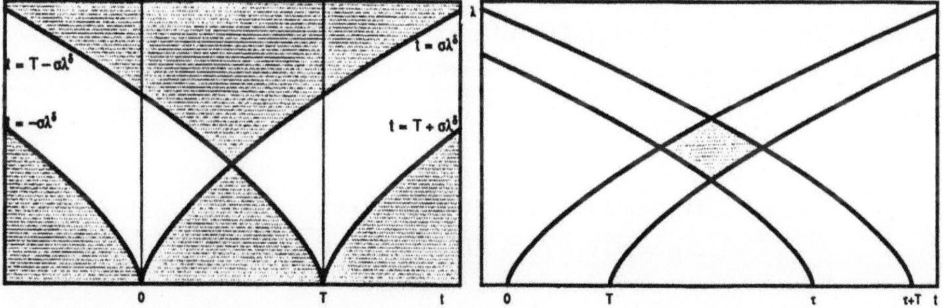	$\beta_0 = 0$	$\beta_0 = 0$	FBM: $\alpha = 2; \nu = \delta$
	SLM: $\alpha = \nu = \delta$		
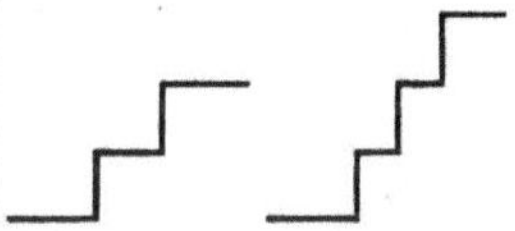	$\beta_0 = \beta_i$	$\beta_0 = \beta_i = 0$	Excluded: Divergence for low λs is non renormalizable
		$\nu = \infty$	$\nu = \delta$
		FBM: $\alpha = 2$	
		$\nu = \infty$	Excluded: Divergence for low λs is non renormalizable
	Excluded	SLM $\alpha = \delta$ $1/2 < H < 1$	FBM: $\alpha = 2; \nu = \delta$ $H < 1/2$

FIGURE 2   FIGURE 3

5. ADDRESS DIAGRAMS AND THE MECHANISM OF NON-LINEAR GLOBAL DEPENDENCE IN FSP

5.1. Address function and diagram; characteristic function of F

Once again, each semi-random pulse is represented by a point in the address space $\{t, \lambda\}$. Therefore, when a pulse template includes one or more time-intervals, one can define an address function and an address diagram, as seen momentarily. Examples are provided by this paper's illustrations.

When there is one time interval it will be denoted by $[0, T]$. The *address* function of $[0, T]$, denoted by $\varphi(t, \lambda \mid 0, T)$, will be the contribution of the pulse represented by $\{t, \lambda\}$ to $F(T)$, where we set $F(0) = 0$, and – as may be the case - to other increments of F as well.

Since $\{t, \lambda\}$ has a Poisson distribution, and the contributions of the address points to F are additive, the logarithmic characteristic function of F is simply the integral

$$\psi(\xi) = C(C', C'') \int (e^{i\xi\varphi(t,\lambda|0,T)} - 1)|\lambda|^{-\delta-1}d\lambda$$

carried over the address space. Here,

$$C(C', C'') = C' \quad \text{for} \quad \lambda > 0 \quad \text{and} \quad C(C', C'') = C'' \quad \text{for} \quad \lambda < 0.$$

When pulses are constant outside an interval, it follows that $\varphi = 0$ over a large *excluded* part of the address space. The analytic form of φ may depend on (λ, t), making it convenient to integrate $\psi(\xi)$ separately over a number of distinct domains. The boundaries of these domains will be said to form an *address diagram*, as exemplified in the case of cylinders by Figure 2.

5.2. Joint address diagrams and the nature of dependence in an FSP; its strongly non-linear character

The interdependence between the ΔF corresponding to two intervals $[0, T]$ and $[t, t + T]$, with $t \geq T$, is investigated by superposing their individual address diagrams. The result can be confusing, therefore it is important to look closely at the simplest case, when the pulse reduces to two steps, either cancelling (up then down or down then up) or noncancelling (up and up or down and down). We assume $0 < \delta < 1$ to postpone the need to face convergence problems.

For both of these pulses $\varphi(t, \lambda \mid 0, T)$ and $\varphi(t, \lambda \mid t, t + T)$ are either $= \lambda$ or $= 0$. Therefore, the domain of integration of φ splits into three parts: The first affects only $[0, T]$ and defines a r.v. Δ_{LL}. The second affects both $[0, T]$ and $[t, t + T]$ and defines a r.v. Δ_O. The third affects only $[t, t + T]$ and defines a r.v. Δ_{RR}.

That is,

$$\Delta_L = F(T) - F(0) = \Delta_{LL} + \Delta_O$$
$$\Delta_R = F(t + T) - F(t) = \pm\Delta_O + \Delta_{RR}.$$

Once again, Δ_{LL}, Δ_O and Δ_{RR} are independent and the symbol $\pm$ means $+$ when the pulse is noncancelling and $-$ when the pulse is cancelling.

The origin and the nature of the resulting dependence between Δ_{LL} and Δ_{RR} are clearest when t is several times larger than T (Figure 2), and easiest to follow if $C' = C''$ and their common value C is made to increase from 0.

The theory of Lévy stable variables suggests that, except for small C, the order of magnitude $\lambda(C)$ of the largest contribution to Δ_{LL} and Δ_{RR}, hence to Δ_L and Δ_R, satisfies $C[\lambda(C)]^\delta = 1$, i.e., $\lambda(C) \sim C^{1/\delta}$.

The underlying argument does not apply for small C, i.e., below the cross-over at $C \sim T$. For smaller C, Δ_L and Δ_R are both dominated by $\lambda \sim \pm T^{1/\delta}$. To the contrary, the addend Δ_O comes exclusively from tail values of λ, and is small. The dependence is therefore small, except when $t = T$, when we deal with neighboring intervals.

When C is very large, one finds that $\Delta_{LL} \propto C^{1/\delta}$ and $\Delta_{RR} \propto C^{1/\delta}$, while $\Delta_O \propto C^{1/2}$. Again, the dependence of Δ_L and Δ_R is small.

We are left with the midrange where $C \propto t$. There, both Δ_L and Δ_R are of the order of magnitude of the largest contributing λ. Once again, the dependence between Δ_L and Δ_R can be either of the order of 1 or small, according to whether or not it is the case that the same largest λ contributes to *both* Δ_L and Δ_R, through its contribution to Δ_O. Strong dependence has a probability of the order of 1, say 1/3.

As $C \to \infty$, the midrange $t \sim C$ also $\to \infty$, which shows that Δ_L and Δ_R become independent and do so in non-uniform fashion.

6. DISCUSSION OF TABLE I: EFFECTS OF PULSE SHAPE ON THE ADMISSIBLE δ, AND ON THE LATERAL ATTRACTOR

Changing the pulse shape greatly affects an FSP. First, it affects the domain of admissible δs. Next, it affects the attractor of $F(t)$ and in particular the dependence of α on δ. Several examples are summarized in Table I.

The left column of Table I illustrates a selection of pulse templates. Other templates are examined in follow-up papers. A striking immediate observation is that a discontinuous FSP can have continuous lateral attractors, and continuous FSP can have discontinuous lateral attractors.

The right column states that $\delta > 2$ is excluded when the pulse is non-cancelling but is admissible when the pulse is cancelling. In the latter case, the lateral attractor is FBM. In any event, $H < 1/2$ is incompatible with SLM.

The column second from the right shows that $1 < \delta < 2$ is admissible for all pulse templates and all values of δ, and that it yields varied and interesting results. Combining proven facts with inferences based on compelling heuristics, one is tempted to infer the following.

• For continuous pulses having left and right derivatives bounded away from 0 and ∞, the lateral attractor is FBM.

• For discontinuous or mixed pulses, the lateral attractor is SLM.

The column shows that the dependence of the limit on the template is far more complicated for $0 < \delta < 1$ than for $1 < \delta < 2$. When the pulse template is discontinuous, a formal extrapolation to $0 < \delta < 1$ of the results relative to SLM with $1 < \delta < 2$ is both meaningful and correct. When the pulse is continuous, the extrapolation to $0 < \delta < 1$ of the results relative to FBM with $1 < \delta < 2$ is meaningless, in fact, the construction diverges. As shown in a follow-up paper, one finds meaningful results by considering the second difference

$$\Delta\Delta F = [F(t+T) - F(t)] - [F(t) - F(t-T)].$$

The case $1 < \delta < 2$; sensitive dependence of the lateral attractor on the pulse templates; a bridge between FBM and SLM. Table I indicates that the attractor is FBM in the case of a smooth conical pulse, but it is SLM if the pulse has a discontinuity, however small, which occurs, for example, when the pulse is a stepped cone, namely a staircase made of many steps up followed by many steps down. "Intuition" suggests that, for finite C, these two pulse forms could not make much of a difference. This is confirmed by a comparative examination of the address diagrams. The diagram corresponding to the stepped cones approximates that of the smooth cone, except for large λ's. As the number of steps increase, so does the quality of the approximation, and it also spreads up to increasingly large λ's.

7. PROOFS OF THE CLAIMS IN TABLE I FOR THE CYLINDRICAL PULSES

The parameter σ is set to 1 in this Section. The address points with $\lambda > 0$ and with $\lambda < 0$ contribute to two independent parts of $F(T)$. The calculations leading to their characteristic functions are the same except for different values C' and C''. We restrict ourselves to the case $\lambda > 0$ and $C' = 1$.

7.1. The l.c.f. of $F(T)$

From Figure 1 it is clear that the address points that contribute to $F(T)$ fall into two domains described as left and right and denoted by $\mathcal{D}_L$ and $\mathcal{D}_R$. The strip $(\lambda, \lambda + d\lambda)$, where $\lambda > 0$, makes the following contribution to the l.c.f.

$$T[(e^{i\xi\lambda} - 1) + (e^{-i\xi\lambda} - 1)]\lambda^{-\delta-1}d\lambda \qquad \text{if} \quad \lambda > T^{1/\delta},$$

$$\lambda^\delta[(e^{i\xi\lambda} - 1) + (e^{-i\xi\lambda} - 1)]\lambda^{-\delta-1}d\lambda \qquad \text{if} \quad \lambda < T^{1/\delta}.$$

Integrating over λ and transforming to the rescaled variables $x = \lambda T^{-1/\delta}$ and $y = \xi T^{1/\delta}$ yields

$$\int_1^\infty [(e^{iyx} - 1) + (e^{-iyx} - 1)]x^{-\delta-1}dx + \int_0^1 [(e^{iyx} - 1) + (e^{-iyx} - 1)]x^{-1}dx.$$

This expression converges for all $\delta > 0$ and is the l.c.f. of a rescaled r.v. independent of T. Therefore, the rescaled increment $T^{-1/\delta}F(T)$ has a distribution independent of T. This is a property of self-affinity with $H = 1/\delta$. We know the mechanisms of self-affinity: in FBM, it is caused by global dependence without long-tailedness, in SLM it is caused by long-tailedness without global dependence, and in FSP it is caused by both long-tailedness and global dependence, acting together with the same value of H. The next issue is to separate the long-tailedness and dependence aspects.

7.2. The attractor in the case $\delta < 2$. Lateral attraction to symmetric Lévy stable increments with $\alpha = \delta$

Given N independent r.v.'s $F_n(T)$ with the above distribution, the behavior of $\Delta \tilde{F}_N = \sum_{n=1}^{N} F_n(T)$ depends sharply on the value of δ. When $\delta < 2$, the l.c.f. of $N^{-1/\delta} \Delta \tilde{F}_N$ can be written in the form

$$T \int_{(T/N)^{1/\delta}}^{\infty} (e^{i\xi u} + e^{-i\xi u} - 2)u^{-\delta-1}du$$

$$+ N \int_{0}^{T^{1/\delta}} (e^{i\xi \lambda N^{-1/\delta}} + e^{-i\xi \lambda N^{-1/\delta}} - 2)\lambda^{-1}d\lambda.$$

When $N \to \infty$, the first term converges to the well-known l.c.f. of a symmetric Lévy stable r.v. with the stability parameter $\alpha = \delta$ and the scale parameter proportional to $T^{1/\delta}$. The second term is of order $N^{1-2/\delta}$, therefore converges to zero. Hence, for fixed T, $F(T)$ belongs to the symmetric δ-stable domain of attraction.

7.3. The dependence structure of $F(T)$ in the case $\delta < 2$. Lateral attraction to a Lévy stable r.f.'s with $\alpha = \delta$ and independent increments

We proceed to the multidimensional structure of the FSP. We show that the multidimensional distributions of the FSP are attracted to that of a symmetric SLM. To prove it we need to find the limit in distribution of a linear combination

$$\sum_{i=1}^{k} \xi_i N^{-1/\delta} \sum_{n=1}^{N} F_n(T_i),$$

where $F_n(T_i), i = 1, 2, \ldots, k$, are nonoverlapping increments of an FSP copy F_n, over (possibly different) time spans T_i. The limit should be the corresponding linear combination of independent δ-stable variables with scale parameters proportional to the respective $T_i^{1/\delta}$. Here, we consider increments over two disjoint time spans T_1 and T_2, i.e., $k = 2$. The general case is not mathematically more involved but requires an overload of notation. Our assertion follows from the same reasoning as in the one-dimensional case, if we show that the expression

$$N \int_{\mathcal{D}} [(e^{i(\xi_1-\xi_2)\lambda N^{-1/\delta}} - 1) - (e^{i\xi_1 \lambda N^{-1/\delta}} + e^{-i\xi_2 \lambda N^{-1/\delta}} - 2)]\lambda^{-\delta-1}dtd\lambda$$

converges to zero, where $\mathcal{D}$ is the dotted region depicted in Figure 3. But this expression is again of order $N^{1-2/\delta}$, which establishes the result.

7.4. The attractor in the case $\delta > 2$. Lateral attraction to Gaussian increments

This case is very different, since the variance $EF^2(T)$ is finite, therefore the rescaled sum $N^{1/2}[\tilde{F}_N - E\tilde{F}_N]$ is asymptotically Gaussian, i.e. stable with $\alpha = 2$.

7.5. The dependence structure of $F(T)$ in the case $\delta > 2$. Lateral attraction to a fractional Brownian r.f.'s with $H = 1/\delta < 1/2$

Examination of the characteristic function of a linear combination of nonoverlapping increments of the FSP shows that it has the second derivative at 0, i.e. every linear combination has the second moment, and multidimensional distributions of the FSP are attracted to multidimensional Gaussian distributions. Note that this Gaussian process in the limit must have stationary increments and be self-affine with the constant $H = 1/\delta$ since these properties are preserved under convolution and convergence in distribution. It is well-known that FBM with $H = 1/\delta$ is the unique Gaussian process satisfying these requirements.

7.6. Some semi-random FSP belonging to the domain of standard attraction of a semi-random self-affine FSP

We replace the constants C' and σ with the slowly varying (at ∞) functions $\gamma(\lambda)$ and $\sigma(\lambda)$ such that the function $w = \sigma(\lambda)\lambda^\delta$ is monotonically increasing. Writing the inverse function of $w(\lambda)$ as $\lambda = w^{1/\delta}L(w)$ yields the identity

$$L(w) = \sigma^{-1/\delta}(w^{1/\delta}L(w)),$$

which is an implicit equation for $L(w)$ and will be used momentarily without having to be solved.

When $\lambda > 0$, the strip $(\lambda, \lambda + d\lambda)$ makes the following contribution to the l.c.f.

$$T[(e^{i\xi\lambda} - 1) + (e^{-i\xi\lambda} - 1)]\gamma(\lambda)\lambda^{-\delta-1}d\lambda \qquad \text{if} \quad \lambda > T^{1/\delta}L(T),$$

$$\sigma(\lambda)\lambda^\delta[(e^{i\xi\lambda} - 1) + (e^{-i\xi\lambda} - 1)]\gamma(\lambda)\lambda^{-\delta-1}d\lambda \qquad \text{if} \quad \lambda < T^{1/\delta}L(T).$$

Integrating over λ and transforming to the rescaled variables $x = \lambda T^{-1/\delta}L^{-1}(T)$ and $y = \xi T^{1/\delta}L(T)$ yields

$$\int_1^\infty [(e^{iyx} - 1) + (e^{-iyx} - 1)]x^{-\delta-1}\{\gamma[xT^{1/\delta}L(T)]L^{-\delta}(T)\}dx$$

$$+ \int_0^1 [(e^{iyx} - 1) + (e^{-iyx} - 1)]x^{-1}\{\gamma[xT^{1/\delta}L(T)]\sigma[xT^{1/\delta}L(T)]\}dx.$$

The integral over $(1, \infty)$ converges for all $\delta > 0$. Assuming that the functions γ and σ are such that also the second integral is finite (e.g. γ and σ are both bounded in the neighborhood of zero), the above expression gives the l.c.f. of a rescaled r.v. $T^{-1/\delta}L^{-1}(T)F$.

This l.c.f. may depend on T. If so, the following question arises: under what conditions on $\gamma(\lambda)$ and $\sigma(\lambda)$, and hence on $L(\lambda)$, does this l.c.f. converge to that prevailing in the FSP case $\gamma(\lambda)\sigma(\lambda) = C'\sigma$? (We know that the product $C'\sigma$

determines the type of an FSP.) Because of the identity that links $\sigma(\lambda)$ and $L(w)$ the two factors written between braces are identical (asymptotically, when $T \to \infty$); therefore the two halves of the l.c.f. yield the same condition on convergence. It is

$$\lim_{\lambda \to \infty} \gamma(\lambda)\sigma(\lambda) = C'\sigma.$$

In other words, the functions $1/\gamma(\lambda)$ and $\sigma(\lambda)$ must vary slowly with λ and asymptotically proportionally to each other.

The question concerning whether or not these conditions are also necessary has not been addressed yet.

Acknowledgments. In 1977-78, I studied semi-random FSP with a one-dimensional t and a multi-dimensional F: early simulations for the second row of Table I were performed by M.R. Laff, and I conjectured (then J. Hawkes proved, that the closure of the set of values of $F(t)$ remains of Hausdorff-Besicovitch dimension $D = \delta$. In the mid-1980s, I studied an application of semi-random FSP with a multi-dimensional t and one-dimensional F: early simulations performed by S. Lovejoy are reported in [4]. I also made conjectures concerning the random FSP; in due time, they were proved in [1] and [3]. This paper was discussed at length with R. Cioczek-Georges. Diagrams were drawn by H. Kaufman.

References

[1] R. Cioczek-Georges and B.B. Mandelbrot, *A class of micropulses and antipersistent fractional Brownian motion*, to appear.

[2] R. Cioczek-Georges and B.B. Mandelbrot, *Alternative micropulses and fractional Brownian motion*, to appear.

[3] R. Cioczek-Georges, B.B. Mandelbrot, G. Samorodnitsky and M. Taqqu, *A class of cylindrical pulse processes*, to appear in *Bernoulli*.

[4] S. Lovejoy and B.B. Mandelbrot, *Fractal properties of rain and a fractal model*, Tellus 37A (1985), 209–232.

[5] B.B. Mandelbrot, *The Fractal Geometry of Nature*, W.H. Freeman, New York, 1982.

[6] B.B. Mandelbrot, *Fractal sums of pulses I: Affine convolution, global dependence, and a versatile family of self-affine random functions*, to appear.

Time Scales in Noisy Conservative Systems

Albert Fannjiang

Department of Mathematics, UCLA, Los Angeles, CA 90024-1555, USA.

1 Introduction

A dynamical system, continuous or discrete in time, is *conservative* if it preserves an invariant measure with smooth density. The invariant measure is not unique unless the dynamical system is ergodic. For concreteness, we present the discussions and results mostly for the *classical systems* [1] for which the state space is a periodic domain $\mathcal{M}$ in R^d and the continuous time system

$$\frac{d\boldsymbol{x}(t)}{dt} = \boldsymbol{u}(\boldsymbol{x}(t)) \tag{1.1}$$

preserves the Lebesgue measure, namely

$$\nabla \cdot \boldsymbol{u}(\boldsymbol{x}) = 0. \tag{1.2}$$

The vector field $\boldsymbol{u}(\boldsymbol{x})$ is assumed to be bounded. We consider small white noise perturbation

$$d\boldsymbol{x}_\varepsilon(t) = \boldsymbol{u}(\boldsymbol{x}_\varepsilon(t))dt + \sqrt{\varepsilon}d\boldsymbol{w}(t) \tag{1.3}$$

where $\sqrt{\varepsilon}$ is the noise magnitude and $w(t)$ is the $d-$dimensional Brownian motion. The random perturbation is chosen as to preserve the same invariant measure of the dynamical system.

There are clearly many possible choices of noises depending on the actual physical contexts from which the dynamical system and noise arise. To fix the idea, we think of (1.1) as the Lagrangian description of the incompressible fluid flow $u(x)$ and ε as the molecular diffusivity of some passive scalar particles. In this context, it is of considerable interest to consider the *annulus*, the *strip* or the *cube* as the state space with *reflecting boundaries*(to preserve the total number of particles) since most fluid dynamical experiments have such set-ups (see, e.g., [10], [19]).

In another situation of solid state physics where particles (e.g. electrons) move in a potential force field (e.g. electric field), the Hamiltonian flow $u(x)$ on the phase space preserves the invariant Lebesgue measure (Liouville theorem) which is not finite in view of the fact that the momentum variables are effectively unbounded. In that case, the more reasonable random processes of modeling the microscopic collisions are those possessing some finite invariant measures. One such candidate is the Ornstein-Ulenbeck process which leaves invariant the Maxwellian distribution. The motion of a particle in a potential force field under the small perturbation of the Ornstein-Ulenbeck process will be addressed in a separate paper.

In general, different noises give rise to different results on the problems addressed in this paper. But we expect those results are stable within each type of noise according to the correlation properties, invariant measures,.... etc and are not sensitive to the details of the random noises.

In the mathematical apparatus developed in [7], the perturbation ε plays the role of parametrization of time in the following sense. The unperturbed skew-symmetric operator $u \cdot \nabla$ generates an unitary group of transformations $e^{tu \cdot \nabla}$ so that the L^p norm is unchanged with time , for all $1 \leq p \leq \infty$,

$$||e^{tu \cdot \nabla} - \langle \cdot \rangle||_{p \to p} = 1, \quad \forall 1 \leq p \leq \infty. \tag{1.4}$$

126

Here $\langle \cdot \rangle$ is the average w.r.t. the invariant measure and $\|\cdot\|_{p\to q}$, for $1 \le p, q \le \infty$, is the operator norm from L^p to L^q. (1.4) reflects the reversibility of the unperturbed system (1.1) and the progression in time is not observable in the sense of operator norm. Of course, if a weaker notion of measuring the evolution is used, then the time progression may be observable. The perturbed generator

$$\mathcal{L}^\epsilon = \frac{\epsilon}{2}\Delta + u(x)\cdot\nabla \tag{1.5}$$

gives rise to a semi-group of contractions $\mathcal{P}_t^\epsilon = e^{t\mathcal{L}^\epsilon}$ with a positive decay rate $N_\epsilon(t)$ such that

$$\|\mathcal{P}_t^\epsilon - \langle\cdot\rangle\|_{2\to 2} = e^{-N_\epsilon(t)}. \tag{1.6}$$

The semi-group property of $\mathcal{P}_t^\epsilon$ implies immediately that $N_\epsilon(t)$ is a super-additive function of t

$$N_\epsilon(t+s) \ge N_\epsilon(t) + N_\epsilon(s), \quad \forall t, s \ge 0 \tag{1.7}$$

with $N_\epsilon(0) = 0$, $\forall \epsilon > 0$. In particular, $N_\epsilon(t)$ is a increasing function of t. We could consider more general $\|\cdot\|_{p\to p}$ norm in (1.6) too. But $\|\cdot\|_{2\to 2}$ seems most natural in view of its relation with the usual definition of dissipation rate

$$\frac{d}{dt}\langle(\mathcal{P}_t^\epsilon f - \langle f\rangle)^2\rangle = -\epsilon\langle(\nabla\mathcal{P}_t^\epsilon f)^2\rangle. \tag{1.8}$$

From (1.6) and (1.8), we get

$$e^{-N_\epsilon(t)} = 1 - \sup_f \left\{ \int_0^t ds\, \frac{\epsilon\langle(\nabla\mathcal{P}_s^\epsilon f)^2\rangle}{\langle(f - \langle f\rangle)^2\rangle} \right\}. \tag{1.9}$$

We are going to characterize a number of time scales in Section 2 in terms of the asymptotics of $N_\epsilon(t)$ as $\epsilon \to 0$ coupled with $t \to \infty$. One of the time scales discussed in Section 2 is the *dissipation time t_{diss}* which is the least such that

$$N_\epsilon(t) \to \infty, \quad \forall t \gg t_{diss}, \tag{1.10}$$

as $\epsilon \to 0$. Namely, it is the time scale on which the evolution $\mathcal{P}_t^\epsilon$ stablizes. Because the stability is measured in the operator norm, the asymptotic behaviours of (1.3) beyond t_{diss} are essentially independent of initial positions. t_{diss} can be substantially smaller than $\frac{1}{\varepsilon}$ and when this happens, it is called the phenomenon of convection enhanced dissipation. The idea is that the unperturbed system (1.1) may create small spatial scales in relatively short time on top of which the small diffusion ε can act effectively and smooth out the initial distributions rather quickly. Examples of this kind are given in Section 3.

Another important time scale t_m, called the *martingale time* in Section 4, is one beyond which the Markov process (1.3) behaves like martingale when looked at from suitable reference frames. The perturbed process (1.3) is conviently decomposed into a martingale and a fluctuation with the former capturing the final state of diffusion. The martingale time is defined by the ratio of the fluctuation to the per unit time variance of the martingale.

The noise contaminates the system completely on the *diffusion time* which is roughly the martingale time independent of initial points. This approach to the final state of diffusion is stated in Section 5 as the limit theorem for the perturbed system (1.3) on the diffusion time scale. This is quite different from the usual central limit theorem in that the *convection enhance diffusion* generally occurs which alters the scaling drastically. Examples of this kind are given in Section 6 and more examples can be found in [8] and [9].

The dissipation time characterizes the mixing mechanism and the martingale time characterizes the vanishing of the non-transporting fluctuation. Together with the diffusion time scale, they and their relations constitute the major part of this paper.

Finally, we comment on the transient crossover times. Contrary to traditional belief, the transient dispersion may take place well beyond the *dissipation time* t_{diss}, if $t_{diss} \ll t_m$, and consequently the transient dispersion in this case is initial point independent.

This paper is a brief account of the full blown version [7].

Before we entering next section, let us set up some notations. Given two sequences of positive numbers $a_\varepsilon, b_\varepsilon$, we say that

$$a_\varepsilon \overset{<}{\sim} b_\varepsilon \ \ \text{if} \ \ \overline{\lim_{\varepsilon \to 0}} \frac{a_\varepsilon}{b_\varepsilon} < \infty; \tag{1.11}$$

$$a_\varepsilon \gtrsim b_\varepsilon \quad \text{if } \varliminf_{\varepsilon \to 0} \frac{a_\varepsilon}{b_\varepsilon} > 0; \tag{1.12}$$

and $a_\varepsilon \sim b_\varepsilon$ if both (1.11) and (1.12) hold. Also

$$a_\varepsilon \ll b_\varepsilon \quad \text{if } \varlimsup_{\varepsilon \to 0} \frac{a_\varepsilon}{b_\varepsilon} = 0; \tag{1.13}$$

$$a_\varepsilon \gg b_\varepsilon \quad \text{if } \varliminf_{\varepsilon \to 0} \frac{a_\varepsilon}{b_\varepsilon} = \infty. \tag{1.14}$$

2 Decay Rate and Dissipative Time

First of all, the following theorem provides general upper and lower bounds for $N_\varepsilon(t)$

Theorem 2.1

$$C\varepsilon \le \frac{N_\varepsilon(t)}{t} \le -Re(\mu_\varepsilon), \quad \textit{uniformly in } t \in [0, \infty), \quad \textit{as } \varepsilon \to 0, \tag{2.1}$$

where μ_ε (possibly not unique) are the nonzero eigenvalues with the largest real part of $\mathcal{L}^\varepsilon$.

The lower bound is proved by the Trotter product formula and the upper bound follows immediately from the theorem of spectral radius [20].

Moreover, these bounds are realizable on the initial and final time scales in the following sense.

Theorem 2.2 *There exists a time $t_{sp} \ge 0$(probably divergent as $\varepsilon \to 0$) such that*

$$N_\varepsilon(t) \sim -Re(\mu_\varepsilon)t, \quad \forall t \gg t_{sp} \tag{2.2}$$

and

$$N_\varepsilon(t) \sim \varepsilon t, \quad \textit{uniformly in } t \in [0, T], \forall T < \infty \tag{2.3}$$

if u is continuously differentiable.

(2.2) is again part of the spectral radius theorem [20]. For bounded time interval, (2.3) is a regular perturbation problem and is proved by a theorem of Blagoveshchenskii and Freidlin [7].

Now, if we also know that

$$-\operatorname{Re}(\mu_\varepsilon) \overset{<}{\sim} \varepsilon \tag{2.4}$$

there would be no problems with time scales: t_{sp} can be set to zero and

$$N_\varepsilon(t) \sim \varepsilon t, \quad \text{uniformly in } t \in [0, \infty). \tag{2.5}$$

But in general, Theorem 2.2 fails for time scales tending to infinity when

$$-\operatorname{Re}(\mu_\varepsilon) \gg \varepsilon \tag{2.6}$$

i.e. when the interaction of convection and diffusion is so significant that $\varepsilon\Delta$ is no longer a regular perturbation to $u \cdot \nabla$. It becomes a singular perturbation problem of the secondary eigenvalues and results in a nontrivial dependence of t_{sp} on ε. The coexistence of multiple time scales and asymptotics prohibits one from extracting directly global time information from the local ones such as one can for symmetric Markov processes via the logarithmic Sobolev inequalities. Here we want to emphasize that the eigenvalue μ_ε is only an indicator of the split of different time scales, but knowing μ_ε alone would not help much in understanding the dissipation mechanism, unless t_{sp} is also known, and even that may not tell us the important quantities such as the time scale beyond which the noise takes over, for t_{sp} may be too large such that $t_{sp} \gg \frac{-1}{\operatorname{Re}(\mu_\varepsilon)}$ and $N_\varepsilon(t) \to \infty$ long before t_{sp}.

Let us define the *dissipation time scale* t_{diss} to be the one beyond which the perturbed semigroup stablizes, i.e.

$$N_\varepsilon(t) \to \infty, \quad \text{as } \varepsilon \to 0, \ \forall t \gg t_{diss}. \tag{2.7}$$

We take t_{diss} to be the least of those times for which (2.7) holds.

Since the decay rates of the initial and final phases are ε and $-\operatorname{Re}(\mu_\varepsilon)$ respectively, we have

$$-\frac{1}{\operatorname{Re}(\mu_\varepsilon)} \leq t_{diss} \leq \frac{1}{\varepsilon}. \tag{2.8}$$

We call it the phenomenon of *convection enhanced dissipation* when the dissipation time is much shorter than $\frac{1}{\varepsilon}$,

$$t_{diss} \ll \frac{1}{\varepsilon}. \tag{2.9}$$

Convection enhanced dissipation (2.9) is related to "convection enlarged spectral gap" (2.6) at least in the following way

Proposition 2.1 *If the convection enhanced dissipation (2.9) takes place, then*

$$- Re(\mu_\varepsilon) \gg \varepsilon. \tag{2.10}$$

Thus convection enhanced dissipation (2.9) is an precursor of (2.6) and the coexistence of multiple time scales for the rate function $N_\varepsilon(t)$. But the converse of Proposition 2.1 may not be true unless one can also show

$$t_{sp} \sim t_{diss} \tag{2.11}$$

, and if this is the case, then the natural guess would be

$$t_{diss} \sim -\frac{1}{\text{Re}(\mu_\varepsilon)}. \tag{2.12}$$

But, in general, neither (2.11) nor (2.12) is true in view of the recent studies of the evolution processes of nonnormal operators such as $\mathcal{L}^\varepsilon$ (see [16] and the references therein). The evolution $\mathcal{P}_t^\varepsilon$ typically has various transient behaviors which sometimes show up on the most interesting time scales and the eigenvalues such as μ_ε do not necessarily provide the relevent information. What (2.6) does tell us is the existence of initial-data sensitive, transient behaviours taking place before and up to the dissipation time t_{diss}.

The summary of this section is that on the order one time scale the rate function behaves like εt and on the longest time scale it is like $-\text{Re}(\mu_\varepsilon)t$. In between, there is the dissipation time scale t_{diss} that tells us when the evolution stablizes.

3 Examples of Convection Enhanced Dissipation

Let us begin with a negative result of when (2.9) does not occur.

Theorem 3.1 *If the unperturbed system (1.1) is not ergodic and it has a non-constant invariant measure with twice continously differentiable density, then*

$$t_{diss} \sim \frac{1}{\varepsilon}. \tag{3.1}$$

We do not know any continous time systems for which (2.9) does occur. The problem simply is that nontrivial ergodic flows are hard to write down. Note that ergodicity alone is not sufficient to have (2.9). For instance, take $u(x) = (u_1, u_2, ..., u_d)$ in d-dimensional torus, where u_i are rationally independent constant. In this example, nothing intersting is going on; one has

$$- \operatorname{Re}(\mu_\varepsilon) \sim \varepsilon \tag{3.2}$$

(cf. [13]) and thus

$$t_{diss} \sim \frac{1}{\varepsilon}. \tag{3.3}$$

Some notion of mixing or hyperbolicity seem relevent as indicated in the next example (Theorem 3.2).

But for discrete time systems such as the iterated maps perturbed by convolutions, (2.9) can manifest in a rather dramatic manner for the alternating baker's maps [1]. Let F, F' denote the usual baker's map and its $\frac{\pi}{2}$−rotated version acting on the period cell $[0, 1]^2$:

$$F(x, y) = \begin{cases} (2x, 1/2y) & \text{mod } 1, \quad \text{if } 0 \le x < 1/2 \\ (2x, 1/2(y+1)) & \text{mod } 1, \quad \text{if } 1/2 \le x < 1 \end{cases} \tag{3.4}$$

$$F'(x, y) = \begin{cases} (1/2x, 2y) & \text{mod } 1, \quad \text{if } 0 \le y < 1/2 \\ (1/2(x+1), 2y) & \text{mod } 1, \quad \text{if } 1/2 \le y < 1 \end{cases} \tag{3.5}$$

Define the unperturbed system $\mathcal{P}_t^0$ as

$$\mathcal{P}_t^0 = (\mathcal{P}_1^0)^t, \quad \text{with } \mathcal{P}_1^0 = F \circ F'. \tag{3.6}$$

Let (3.6) be perturbed by convolution with heat kernel with variance ε

$$\mathcal{P}_t^\varepsilon = (\mathcal{P}_1^\varepsilon)^t, \quad \text{with } \mathcal{P}_1^\varepsilon = H_\varepsilon * (F \circ F') \tag{3.7}$$

where $H_\varepsilon(x) = \frac{1}{2\pi\varepsilon} e^{-|x|^2/2\varepsilon}$. We have

Theorem 3.2 *For (3.7),*

$$t_{diss} \lesssim \log \frac{1}{\varepsilon}. \tag{3.8}$$

Faced with the two extreme situations of Theorem 3.1 and 3.2, one may wonder where the generic cases lie. Is Theorem 3.1 or Theorem 3.2 exceptional? The word *generic* or *exeptional* becomes meaningful by putting a topology or a measure over the group of transformations. Without entering the discussion in depth, let us simply remark that the ergodicity is generic and the mixing is exceptional in the weak topology for the general abstract dynamical systems (namely, not restricted to the classical systems such as (1.1))[12].

4   Martingale Time and Diffusive Time

As demonstrated in [15] and [14], it is very useful to decompose the sample path $x_\varepsilon(t)$ into a martingale $\rho^\varepsilon(x_\varepsilon(t))$ and a fluctuation $\chi^\varepsilon(x_\varepsilon(t))$

$$x_\varepsilon(t) = \rho^\varepsilon(x_\varepsilon(t)) - \chi^\varepsilon(x_\varepsilon(t)) \tag{4.1}$$

where $\chi^\varepsilon(x) = (\chi_1^\varepsilon(x), \ldots, \chi_d^\varepsilon(x))$ is the zero mean, periodic solution of

$$\mathcal{L}^\varepsilon \chi^\varepsilon = u - \langle u \rangle. \tag{4.2}$$

Here $\langle \cdot \rangle$ is the average w.r.t the invariant measure. χ^ε is called the corrector in homogenization theory whose significance can be seen from the identity (cf. [2],[8])

$$\begin{aligned}
\sigma_\varepsilon(e_i) &= \varepsilon + \varepsilon \frac{1}{|\mathcal{M}|} \int_{\mathcal{M}} dx (\nabla \chi_i^\varepsilon)^2 \\
&= \varepsilon \frac{1}{|\mathcal{M}|} \int_{\mathcal{M}} dx (\nabla \rho_i^\varepsilon)^2, \quad \forall i
\end{aligned} \tag{4.3}$$

where $\sigma_\varepsilon(e_i)$ is the *effective diffusivity* in the direction e_i which is supposed to be the variance per unit time of the sample path $x_\varepsilon(t)$ on the longest time scale. For simplicity, let us assume, from now on, that $\{e_i\}$ is a complete set of orthornormal eigenvectors of the effective diffusivity. This time scale can be legitimately called the *diffusion time*. The asymptotic behavior of (4.3) as $\varepsilon \to 0$ has been studied in detail in [8] for periodic flows and in [9] for random flows. But actually how long the diffusion time is, the homogenization theory does not tell us.

It is easy to check from (4.1) and (4.2) that ρ^ε is harmonic w.r.t. $\mathcal{L}^\varepsilon$ so that $\rho^\varepsilon(x_\varepsilon(t))$ is a martingale. Moreover, it is a stochastic integral

$$\rho^\varepsilon(x_\varepsilon(t)) = \sqrt{\varepsilon} \int_0^t \nabla\rho^\varepsilon(x_\varepsilon(s))dw_s + \rho^\varepsilon(x_\varepsilon(0)). \tag{4.4}$$

so that the variance

$$E_x\left\{(\rho^\varepsilon)^2(x_\varepsilon(t))\right\} = (\rho^\varepsilon)^2(x_\varepsilon(0)) + \varepsilon \int_0^t ds\, E_x\left\{(\nabla\rho^\varepsilon)^2(x_\varepsilon(s))\right\}. \tag{4.5}$$

Before the sample path $x_\varepsilon(t)$ reaches its final state of diffusion signified by the effective diffusivity σ_ε, it has to attain the state of martingale represented by $\rho^\varepsilon(x_\varepsilon(t))$ first and the condition for this is that the non-martingale fluctuation is small relative to the martingale component in a uniform manner,

$$\lim_{\varepsilon\to 0} \frac{E_x\left[(\chi_i^\varepsilon)^2(x_\varepsilon(t\tau)\right]}{\varepsilon \int_0^t ds\, E_x\left\{(\nabla\rho^\varepsilon)^2(x_\varepsilon(s))\right\}} = 0,, \quad \forall t \gg t_m(x, e_i) \tag{4.6}$$

uniformly in $\tau \in [\tau_0, T]$, $\forall 0 < \tau_0 < T < \infty$, $\forall i$, in view of (4.5). Here E_x denotes the average w.r.t. the Wiener process.

Condition (4.6) defines the initial-position-x sensitive *martingale time* scale $t_m(x, e_i)$ which also depends on the direction e_i. To seek an initial-position independent *martingale time* $t_m(e_i)$, one has to look beyond the *dissipation time* t_{diss} so that the condition becomes

$$\frac{1}{t}\sigma_\varepsilon^{-1}(e_i)\langle(\chi_i^\varepsilon)^2\rangle \to 0, \quad \forall t \gg t_m(e_i), \tag{4.7}$$

in view of (4.3). Thus it seems most natural to define the *martingale time* through the formula

$$t_m(e_i) = \max\left\{\sigma_\varepsilon^{-1}(e_i)\langle(\chi_i^\varepsilon)^2\rangle,\, t_{diss}\right\}, \tag{4.8}$$

beyond which the perturbed process (1.3) behaves like a martingale independent of the initial point x. It is easy to see that

$$t_m(e_i) \lesssim \frac{1}{\varepsilon} \tag{4.9}$$

since $t_{diss} \lesssim \frac{1}{\varepsilon}$ and

$$\sigma_\varepsilon(e_i) > \varepsilon\frac{1}{|\mathcal{M}|}\int_\mathcal{M} dx(\nabla\chi_i^\varepsilon)^2 \geq C\varepsilon\frac{1}{|\mathcal{M}|}\int_\mathcal{M} dx(\chi_i^\varepsilon)^2 \tag{4.10}$$

which is an application of the Poincare inequality.

5 Limit Theorem

In the previous section, the diffusion time t_{diff} is loosely defined as the time scale beyond which the perturbed process (1.3) approaches the final state of diffusion. In the following theorem, we give a precise estimate on how large d_{diff} has to be. In order to state the result in its simplest form, we assume the manifold $\mathcal{M}$ to be the $d-$dimensional torus and we take the variables x in (1.1) and (1.3)to be the "extended" coordinates of the covering space R^d.

Theorem 5.1 *The family of processes $\frac{1}{\lambda_\epsilon}\sqrt{\sigma_\epsilon^{-1}}\left\{x_\epsilon(\lambda_\epsilon^2 t) - \lambda_\epsilon^2\langle u\rangle t\right\}\cdot e_i$ converge in law to the one dimensional Brownian motion in $[t_0, T], \forall 0 < t_0 < T < \infty$, if $\lambda_\epsilon^2 = \lambda_\epsilon^2(e_i) \gg t_m(e_i)$ and*

$$\lim_{\epsilon \to 0}\left\{ N_\epsilon(\lambda_\epsilon^2) + (\frac{d}{2}+1)\log(\lambda_\epsilon^2\epsilon)\right\} = \infty. \tag{5.1}$$

The convergence holds in measure with respect to the initial positions $x \in \mathcal{M}$.

Several remarks concern the above theorem.

For experimental set-ups such as those of [10] and [19], the covering space is the infinite strip and has only one "extended" direction. So the limit theorem for this case is also limited to the extended coordinate.

The tightness of $\frac{1}{\lambda_\epsilon}\sqrt{\sigma_\epsilon^{-1}}\left\{x_\epsilon(\lambda_\epsilon^2 t) - \lambda_\epsilon^2\langle u\rangle t\right\}$ requies the following stronger (than (1.6)) ultracontractivity estimate

Theorem 5.2

$$\|\mathcal{P}_t^\epsilon - \langle\cdot\rangle\|_{1\to\infty} \leq \frac{C_\delta}{(\epsilon t)^{d/2}}e^{-N_\epsilon(\delta t)}, \quad \forall 0 < \delta < 1, \forall t > 0 \tag{5.2}$$

for some positive constant C_δ.

The estimate of (5.2) is obtained in [7] by the duality of $\mathcal{P}_t^\epsilon$ and its adjoint $\mathcal{P}_t^{\epsilon*}$ whose generator is, because of the incompressibility (1.2),

$$\mathcal{L}^{\epsilon*} = \frac{\epsilon}{2}\Delta - u(x)\cdot\nabla. \tag{5.3}$$

Thus, in addition to Theorem 5.2, one also has

Theorem 5.3

$$\|\mathcal{P}_t^{\varepsilon*} - \langle\cdot\rangle\|_{1\to\infty} \leq \frac{C_\delta}{(\varepsilon t)^{d/2}} e^{-N_\varepsilon(\delta t)}, \quad \forall 0 < \delta < 1, \forall t > 0 \tag{5.4}$$

for some positive constant C_δ.

The ultracontractivity estimates are clearly singular at $t = 0$ but the ε dependence in the denominators of (5.2) and (5.4) may not be optimal. And because of it, a logarithmic term is needed in the assumption (5.1) and so is the cut-off t_0.

In any case, if

$$\lambda_\varepsilon^2(e_i) \gg \frac{1}{\varepsilon} \tag{5.5}$$

then (5.1) is automatically satisfied due to (2.8). The diffusion time and the martingale time are probably of the same order of magnitude. If this is the case, (5.1) at worst overestimates $t_{diff}(e_i)$ by a logarithm of $\frac{1}{\varepsilon}$.

6 Examples of Convection Enhanced Diffusion

We note that the scaling in Theorem 5.1 is not the usual one of the martingale central limit theorem (cf. [6]) because the flow sensitive, effective diffusivity σ_ε generally satisfies the power law

$$\sigma_\varepsilon(e_i) \sim \varepsilon^\alpha, \quad \text{as } \varepsilon \to 0 \tag{6.1}$$

with

$$-1 \leq \alpha \leq 1, \tag{6.2}$$

or possibly with a logarithm of $\frac{1}{\varepsilon}$ (cf. (6.10) and [8]).

When $\alpha < 1$, (6.1) is called the *convection enhanced diffusion* and is studied in detail in [8] and [9] using variational methods. We begin with two examples of laminar Rayleigh-Benard convection: one is the square cellular flow in two dimension; the other, the hexagonal flow in three dimension, both of which are completely integrable (hence non-ergodic), so by Theorem 3.1,

$$t_{diss} \sim \frac{1}{\varepsilon}. \tag{6.3}$$

The cellular flow is given by the periodic stream function

$$\psi(x, y) = \sin x \sin y \tag{6.4}$$

such that $u(x, y) = \nabla^\perp \psi(x, y) = (-\psi_y, \psi_x)$ and the trajectories of u are exactly the level sets of ψ. For (6.4), it was calculated in [3], [17], [18] and [8] that

$$\sigma_\varepsilon(e_i) \sim \sqrt{\varepsilon}, \ \forall i \tag{6.5}$$

and

$$\langle (\chi_i^\varepsilon)^2 \rangle \sim 1, \ \forall i \tag{6.6}$$

so

$$t_{diff}(e_i) \sim t_m(e_i) \sim \frac{1}{\varepsilon} \tag{6.7}$$

by (6.4), (4.8) and Theorem 5.1.

The hexagonal flow is given by

$$u_3(x) = \sin z \sum_i \cos (q_i \cdot x) \tag{6.8}$$

$$u_\perp(x) = \frac{1}{|q|^2} \nabla^\perp \frac{\partial u_3}{\partial z} \tag{6.9}$$

where $u_\perp$ denotes the velocity in the $x - y$ plane and the constant vectors q_i form an equilateral triangle in the $x - y$ plane. It was calculated in [5] that

$$\sigma_\varepsilon(e_i) \sim \varepsilon \log \frac{1}{\varepsilon}, \ \forall i \tag{6.10}$$

and it can be shown that

$$\langle (\chi_i^\varepsilon)^2 \rangle \sim 1, \ \forall i \tag{6.11}$$

hence (6.7).

As a contrast, let us consider the convection diffusion by the perturbed baker's map (3.7). It was calculated numerically in [4] for the randomly perturbed single baker's map (without the alternation) that

$$\langle (\chi_1^\varepsilon)^2 \rangle \sim 1, \tag{6.12}$$

and

$$\sigma_\varepsilon(e_1) \sim \sqrt{\varepsilon} \tag{6.13}$$

where e_1 is the direction of compressing and folding. With the combination of the baker's map and its rotated version, we expect that

$$\langle (\chi_i^\varepsilon)^2 \rangle \sim 1, \ \forall i \tag{6.14}$$

and

$$\sigma_\varepsilon(e_i) \sim \sqrt{\varepsilon}, \ \forall i. \tag{6.15}$$

Thus the martingale time is much larger than the dissipation time

$$t_m \sim \frac{1}{\sqrt{\varepsilon}} \gg t_{diss} \tag{6.16}$$

by Theorem 3.2.

7 Transient Dispersion

The first crossover time before the final state of diffusion is reached is the martingale time $t_m(e_i)$ which may be substantially larger than the dissiptive time t_{diss} as we have seen in the example of alternating baker's map.

When $t_m(e_i)$ is indeed equal to t_{diss}, the crossover may happen before and up to t_{diss} and the transient dispersion is generally initial point sensitive. This is the case, for instance, when the fluid flow u is not ergodic and consequently $t_{diss} \sim \frac{1}{\varepsilon}$. If, furthermore, the flow is laminar convective cells such as (6.4) and (6.8)-(6.9), then it was proposed that the ε independent anomalous diffusion takes place

$$E_x\{x_\varepsilon^2(t)\} \sim t^{1/2} \tag{7.1}$$

for

$$1 \ll t \ll \frac{1}{\varepsilon} \tag{7.2}$$

and initial point x sufficiently close to the boundary of the convective cells (cf. [11],[21] and [5]). Here 1 in (7.2) represents the order of the turnover time.

But if $t_m(e_i) \gg t_{diss}$, then the first crossover time is instead

$$\sigma^{-1}(e_i)\langle (\chi_i^\varepsilon)^2 \rangle \tag{7.3}$$

and an *initial point x independent* transient dispersion may take place between t_{diss} and $t_m(e_i)$. In this case, there may be multiple transient regimes: some depend on initial points; others do not.

We plan to address this issue in a more systematic way in a separate paper.

References

[1] V. I. Arnold and A. Avez: *"Ergodic Problems of Classical Mechanics"*, Redwood City, Calif. : Addison-Wesley, 1989.

[2] A. Bensoussan, J. L. Lions and G. C. Papanicolaou: *"Asymptotic Analysis for Periodic Structures"*, North-Holland, Amsterdam (1978).

[3] S. Childress: *Alpha-effect in Flux Ropes and Sheets*, Phys. Earth Planet Inter. **20**, 172-180(1979).

[4] S. Childress and I. Klapper:*"On Some Transport Properties of Baker's Maps"*, J. Stat. Phys. **63**(no.5-6): 897-914(1991).

[5] A. M. Dykhne, M. B. Isichenko and W. Horton:*"Diffusion in Laminar Rayleigh-Benard Convection: boundary layers versus boundary tubes"*, Physics of Fluids **6:7**, 2345-51(1994).

[6] S. Ethier and T. G. Kurtz:*"Markov Processes-Characterization and Convergence"* , John Wiley & Sons, Inc., New York, 1986.

[7] A. Fannjiang:*"Random Perturbations of Conservative Dynamical Systems"*, to appear.

[8] A. Fannjiang and G. Papanicolaou:*"Convection Enhanced Diffusion in Periodic Flows"*, SIAM J. Appl. Math. Vol. 54, No. 2, pp.333-408, 1994.

[9] A. Fannjiang and G. Papanicolaou: *"Convection Enhanced Diffusion in Random Flows"*, to appear.

[10] J. P. Gollub and T. H. Solomon:*"Complex Particle Trajectories and Transport in Stationary and Periodic Convective Flows"*, Physica Scripta **40** 1989, 430-435

[11] E. Guyon, Y. Pomeau, J. P. Hulin and C. Baudet:*"Dispersion in the Presence of Recirculation Zones"* Nuclear Physics B (Proc. Suppl.)2, 271-280(1987).

[12] P. R. Halmos: *"Lectures on Ergodic Theory"*, Chelsea, New York (1958).

[13] Yu. I. Kifer:*"The Spectrum of Small Random Perturbations of Dynamical Systems"*, in *Multicomponent Random Systems*, ed. R. L. Dobrushin and Ya. G. Sinai, Advances in Probability and Related Topics G, Marcel Dekker, New York, 423-450(1980).

[14] S. M. Kozlov:*"The Method of Averaging and Walks in Inhomogeneous Enviroments"*, Russian Math. Surveys 40:2 (1985), 73-145.

[15] G. Papanicolaou, D. W. Stroock and S. R. S. Varadhan :*"Martingale Approach to Some Limit Theorems"* Conference on Statistical Mechanics, Dynamical Sustems, and Turbulence, Duke University, M. Reed ed., Duke Univ. Math. Series **3**.

[16] S. C. Reddy and L. N. Trefethen :*"Pseudospectra of the Convection-Diffusion Operator"*, SIAM J. Appl. Math., to appear.

[17] M. N. Rosenbluth, H. L. Berk, I. Doxas and W. Horton: *"Effective Diffusion in Laminar Convective Flows"*, Phys. Fluids **30**, 2636-2647(1987)

[18] B. Shraiman: *"Diffusive Transport in a Rayleigh-Benard Convection Cell"*, Phys. Rev. A **36**, 261(1987).

[19] T. H. Solomon, E. R. Weeks and H. L. Swinney:*"Chaotic Advection in a Two-dimensional Flow: Levy Flights and Anomalous Diffusion"*, preprint 1994.

[20] K. Yosida:*"Functional Analysis"*, Springer-Verlag, New York, 1974.

[21] W. Young, A. Pumir and Y. Pomeau:*"Anomalous Diffusion of Tracer in Convection Rolls"* Physics of Fluids A (Fluid Dynamics)1:3, 462-9 (1989).

Geometric constructions in multifractality formalism

Michael Blank

C.N.R.S., Observatoire de Nice, BP 229, 06304 Nice Cedex 4, France
e-mail: blank@obs-nice

Abstract. We consider two geometrical approaches for the investigation of multifractal properties of both sets and measures, based on the idea, that a multifractal object is a union of monofractal Cantor-like ones. The first "naive" approach uses a notion of a local dimension of a set or a measure, while the second one uses statistics of empty and full elements of partitions of the phase space for different scales.

Keywords. Fractal, multifractal, Cantor-like set, measure.

1 Introduction

One of the most widespread objects in the modern physical literature is the investigation of multifractal properties of measures and functions (see, for example [1,3,7] and references therein). However there were practically no attempts to investigate multifractal properties of sets. The author knows only, on the one hand, numerical investigation of this problem for the set of prime numbers [5], where actually something like a structure function for a set was calculated, but no definitions of the multifractality were given; and, on the other hand, some examples of so called "multi-scale" fractals [3,4], consisting of several Cantor sets with different dimensions. The main aim of the paper is to discuss connections between these two objects: fractals and multifractals and to propose the geometric version of multifractal formalism, which could be applied for the both of these objects. Therefore we give the main attention to the explanation of our approach, but not to details of proofs and possible generalizations (for example, for the sake of simplicity we shall consider here only the one-dimensional case (D=1) and sometimes even zero-dimensional sets).

Even from the philological point of view it is clear, that fractals and multifractals are relative one to another. However, the first of these notions is applied only for sets, while the second for measures and functions. So it is of interest to elaborate a unified approach for both sets and measures. On the other hand, our aim is to find something like a geometrical description of the notion of multifractality.

On leave from Russian Academy of Sciences, Inst. for Information Transmission Problems, Ermolovoy Str. 19, 101447, Moscow, Russia

Let us remind some definitions. Due to the idea, introduced by Parisi G. and Frisch U. ([1], 1985) the measure μ is called multifractal, if there exists a function $d : [0,1] \to I\!\!R^+$ such that $\mu(B_\epsilon(x)) \sim |B_\epsilon(x)|^{d(x)}$. In the other terms this relation could be expressed in terms of so called dimension spectrum:

$$d(\mu, \alpha) = \frac{1}{1-\alpha} \lim_{\epsilon \to 0} \frac{\log \sum_k \mu(\Delta_k)^\alpha}{\log 1/\epsilon}$$

where $\{\Delta_k\}$ is a partition of the unit interval into equal nonintersecting subintervals of length ϵ. Now, if this function is not a constant we say that the measure is multifractal. This is a thermodynamic description of multifractality.

Now let us answer a question, what does a multifractal set mean. From a geometrical point of view, one could think about such set as a union of monofractal ones, where by a monofractal set we mean a Cantor-like set. For example, we can consider a usual Cantor set, when we are throughing out the middle 1/3 interval, and then to unite it with a new Cantor set with different dimension, constructed in these throughed out middle intervals.

2 "Naive" multifractality construction

Let us define the notion of the multifractal set rigorously. At first, we consider a "naive" definition. We define the notion of a *local* dimension in the following way:

$$\dim(\mathrm{A}, \mathrm{x}) = \overline{\lim_{\epsilon \to 0}} \dim(\mathrm{A} \cap \mathrm{B}_\epsilon(\mathrm{x}))$$

where $\dim(A)$ is a fractional dimension of the set A. Now if $\dim(\mathrm{A}, \mathrm{x}) = \mathrm{const}$ for all $x \in A$ we say that the set is monofractal and multifractal otherwise. Remark, that we could consider here different dimensions (Hausdorff, box etc) and will obtain in the general case different local dimensions.

It would be worthwhile to remind definitions of these dimensions. Consider a partition of the interval $[0,1]$ into equal intervals of length ϵ and let $N_\epsilon(A)$ be a number of intervals Δ_i such that $\Delta_i \cap A \neq \emptyset$. Then the B-dimension (box dimension) of a set A is

$$\dim_{\mathrm{B}}(\mathrm{A}) = \overline{\lim_{\epsilon \to 0}} \frac{\log \mathrm{N}_\epsilon(\mathrm{A})}{\log 1/\epsilon}$$

To obtain the H-dimension (Hausdorff dimension) $\dim_{\mathrm{H}}(\mathrm{A})$ of a set A we need a covering (instead of a partition) Δ_i of the set A. Then $\dim_{\mathrm{H}}(\mathrm{A})$ is defined as follows:

$$\dim_{\mathrm{H}}(\mathrm{A}) = \sup\{\alpha : \mathrm{S}_\alpha(\mathrm{A}) = \infty\} = \inf\{\alpha : \mathrm{S}_\alpha(\mathrm{A}) = 0\}$$

where

$$S_\alpha(A) = \inf_\epsilon \inf_{\Delta : \mathrm{diam}(\Delta_i) < \epsilon} \sum_i (\mathrm{diam}(\Delta_i))^{D\alpha}$$

In the 1-dimensional case one could consider also another statistics to define the dimension of a closed set A. Note that in this case the complement of the set is a union of intervals l_i. In [6] Ya.G.Sinai proposed to consider the statistics $N_n^{(S)}(A)$ of the number of intervals l_i such that their lengths $|l_i|$ satisfy the inequality

$$2^{-n} \leq |l_i| < 2^{-n+1}$$

Then one could define the dimension of the set A as follows:

$$\dim_S(A) = \lim_{n \to \infty} \frac{\log N_n^{(S)}(A)}{\log 2^n}$$

It was proved [6] under some technical conditions that this dimension coincides with the usual Hausdorff dimension. For example, this is true for the set of zeros of one-dimensional Brownian motion. Unfortunately, it is not clear how to generalize this definition for sets with topological dimension larger than 1. In this sense the two previous definitions are better, because they admit straightforward multidimensional generalizations.

The counterpart of the local dimension definition for measures could be described in the same way. At first, we define the notion of the dimension of a measure. Once more, we could do it in different ways. The simplest way is to use a box-like definition. Consider a partition of the interval $[0,1]$ into equal subintervals of length ϵ and let $N_\epsilon(\mu)$ be a number of intervals Δ_i such that $\mu(\Delta_i) \geq \epsilon$. Then by the B-dimension (box dimension) of the measure me we shall mean

$$\dim_B(\mu) = \overline{\lim_{\epsilon \to 0}} \frac{\log N_\epsilon(\mu)}{\log 1/\epsilon}$$

Now just as in the previous case we define a local dimension

$$\dim_B(\mu, x) = \overline{\lim_{\epsilon \to 0}} \dim_B(\mu | B_\epsilon(x))$$

where $\mu | A$ is a restriction of the measure μ to the set A, i.e. for any measurable subset $S \subset [0,1]$ we have $(\mu|A)(B) = \mu(A \cap B)/\mu(A)$.

We also could define the dimension of a measure in a Hausdorff-like manner. To do it consider for a fixed value $\epsilon > 0$ the set of all subsets $A_\epsilon \subset [0,1]$ such that $\mu(A_\epsilon) \geq 1 - \epsilon$. Then by the H-dimension (Hausdorff dimension) of the measure μ we shall mean

$$\dim_H(\mu) = \overline{\lim_{\epsilon \to 0}} \sup_{A_\epsilon} \dim_H(A_\epsilon)$$

Note that our definition of the dimension of a measure differs from the usual one, corresponding to the dimension of the support of a measure:

$$\dim_H(\mu) = \inf\{ \dim_H(A) : \mu(A) = 1 \}$$

because this definition doesn't give a possibility to investigate local properties, and the local dimension is different from the well known pointwise dimension, i.e. from

$$\dim_{\mathrm{p}}(\mu, \mathrm{x}) = \varlimsup_{\epsilon \to 0} \frac{\log \mu(\mathrm{B}_\epsilon(\mathrm{x}))}{\log \epsilon}$$

3 Geometric multifractal spectrum

Even this naive approach gives us a possibility to investigate simple examples of multifractality of measures and sets. However, especially in homogeneous cases, this description gives us not too much information about the multifractal properties, compare to the dimension spectrum described above. Therefore we also introduce such notion in our second approach to the problem, but in a different way. To have a simple picture in mind, let us start with a relatively new for a such consideration object - "discrete" sets - subsets of the set of integers $I\!N$, which is zero-dimensional, considered as a subset of $I\!R^1$. This object deserves also an independent interest due to the fact in computer simulations for dimension estimates we have dealings only with sets on discrete lattices, which could be considered as subsets of integers. Moreover, properties of subsets of integers are universal from the point of view of our geometrical approach to multifractality. Indeed, we shall show that any multifractal object has its counterpart as a subset of integers. Note, that the definition of the dimension for a fractal subset of the set of integers was given first by Bedford T. and Fisher A.M. [2] and slightly differs from ours.

It is natural that a fractal "discrete" set is a set with zero density, i.e. $N_n(A)/n \to 0$ as $n \to \infty$, where

$$N_n(A) = \#\{x \in A : x \le n\}$$

is a number of the elements in the set A that are less or equal to n. Now the most natural candidate to be a monofractal (selfsimilar) set is a power-type sequence $\{k^m\}_k$ for some integer m. Sets of this type are the replacement of usual Cantor sets here. To define the notion of the dimension we as usual apply coverings or partition into boxes. In this case there is no difference between this approaches. So, let us consider a partition of the integers between 1 and n into consecutive groups of equal lengths (boxes). The smallest possible length of a such box here is equal to 1, therefore the number of nonempty boxes is equal to $N_n(A)$. Therefore the dimension could be defined in the same manner as before:

$$\dim(\mathrm{A}) = \varlimsup_{\mathrm{n} \to \infty} \frac{\log \mathrm{N}_\mathrm{n}(\mathrm{A})}{\log \mathrm{n}}$$

Examples of more complex subsets of integers that we want to consider are the following:

(1) $\{[k^\alpha]\}$ - ($\dim = 1/\alpha$);

(2) $\{\sum_{k=1}^{n} \alpha_k 3^k\}$, $\alpha_k \in \{0, 2\}$ - integer Cantor set (dim $= \frac{\log 2}{\log 3}$);

(3) the set of prime numbers (dim $= 1$, $N(A, n) \sim n/\log n$);

(4) the set of zeros of a simple symmetrical random walk (dim $= 1/2$);

Here $[x]$ means the closest integer to the number $x \in \mathbb{R}$. It is of interest, that the dimension could be larger than 1. If we set $0 < \alpha < 1$ in the first example, we shall obtain this. The explanation is that in the considered sequence there would be a lot of equal elements, which will give us the dimension larger than 1.

A question arises if is it possible to find an inclusion of the set of integers into some set with a more rich topology, for example into $[0, 1]$, to investigate the fractal properties in the usual way. From the first sight, it seems that the natural way to do it is to consider an inclusion into the space of binary sequences, which is isomorphic to $[0, 1]$. Indeed, let us suppose that in the set A there are no equal elements and consider a sequence l_A whose k-th element is equal to 1 if $k \in A$ and equal to 0 otherwise. Then, defining by L_A both the closure of all shifts to the left of the sequence l_A and the corresponding subset of the unit interval, we can investigate the fractal properties of the set L_A, instead of the genuine set A. However, it appears that for all interesting examples, when the density of the subset of integers is equal to zero, the Hausdorff dimension of the set L_A is also zero. Therefore this approach is not very fruitful. There is also another inclusion into the unit interval, given by the map $1/n$. It could be proved that the entropy of the sequence, obtained in this way, is equal to the dimension of the subset of integers, described above.

Now following the idea, that the multifractal set is a union of monofractal ones let us solve an inverse problem: how to find this representation for an arbitrary subset of integers. Considering this from the statistical point of view, we come to the representation of the number of "full" boxes in the following series:

$$N_n(A) = \sum_k R_k(n)\, n^{d_k}$$

where d_k are dimensions of the monofractals and $R_k(n)$ are corresponding multiplicities (roughly speaking, "weights" of these monofractals). To be well defined the coefficients $R_k(n)$ could grow (decay) not faster than some power of logarithm n. This brings us to the definition of the asymptotic series for $N_n(A)$ as follows:

$$\varlimsup_{n \to \infty} \frac{\log \left| N_n(A) - \sum_{k=1}^{l} R_k(n)\, n^{d_k} \right|}{\log n} < d_l$$

for any $l \geq 2$, where the first exponent d_1 is set to be equal to the dimension of the set A:

$$d_1 = \varlimsup_{n \to \infty} \frac{\log N_n(A)}{\log n}$$

Then to estimate the multiplicity $R_1(n)$ one has to consider the largest integer value p such that

$$\varlimsup_{n \to \infty} \frac{\log N_n(A)\, n^{-d_1}}{\log^{(p)}(n)} = C \neq 0$$

where $\log^{(p)}(n)$ is the p-multiple logarithm of n. The case $p = 0$ corresponds to a finite multiplicity. Then we have to subtract the term $C \log^{(p)}(n) n^{d_1}$ from $N_n(A)$ and repeat the procedure of finding of values of p, because it is possible that there are lesser terms in $R_1(n)$. The last step continues up to the moment, when for any p the above limit equals zero. Then using the residual term instead of $N_n(A)$ we continue the first step of the procedure to calculate d_2, etc.

Theorem 1. The representation above is well defined.

Proof. Let us suppose that we have two different representations for $N_n(A)$:

$$N_n(A) = \sum_k R_k(n)\, n^{d_k} = \sum_k R'_k(n)\, n^{d'_k}$$

satisfying the inequality above for any $l \geq 2$ and such that the functions R_k and R'_k are monotonous. According to definition $d_1 = d'_1$. Then subtracting the second series from the first we have:

$$0 = \sum_k (R_k(n)\, n^{d_k} - R'_k(n)\, n^{d'_k})$$

$$\leq |R_1(n) - R'_1(n)|\, n^{d_1} + \sum_{k>1} |R_k(n)\, n^{d_k} - R'_k(n)\, n^{d'_k}|$$

From where it follows that $|R_1(n) - R'_1(n)| \to 0$ as $n \to \infty$. Subtracting the first term and continuing this procedure for $k > 1$ we obtain the statement of Theorem. **Q.E.D.**

According to our scheme the sequence $\{d_k\}$ corresponds to the dimensions of the monofractal components. However, we see also a new object - coefficients R_k, which we call multiplicities. Indeed, let us consider the case, when $p_k = 0$ for all k, but we have several nonzero $R_k(n) = C_k$ with integer values. Then we could interpret these values as a number of repetitions of the monofractals in the representation.

Considering examples above from the point of view of this geometric description of multifractality we have the following.

Theorem 2. The sets in the examples (1), (2) and (4) are monofractals with multiplicities equal to 1. Under the Riemann hypothesis the set in the example (3) is also monofractal.

It's of interest, that the set of prime numbers has zero density, but its dimension is equal to 1 with zero multiplicity. This shows that we couldn't define a fractal zero-density set by the property that its dimension is not an integer.

Proof. The first two examples are very simple, so we consider only the third and the fourth. For the set of prime numbers, using the Riemannian Hypothesis we have an estimate for the remainder term of type: $O(n^{1-\epsilon})$ that gives us the monofractality of this set. However till now only an estimate $O(n * e^{-C(\log n)^{0.6}})$ is proven which is insufficient for us to rule out the multifractality, stated in some numerical works (see, for example, [5]).

Let us prove the monofractality of the set in the example (4). This is a random set and we shall prove only that if we consider mathematical expectations of $E[N_n(A)]$ in the definition of multifractality rather than the random values $N_n(A)$, then our asymptotic series consists of only one term. The mathematical expectation of a number of returns of a symmetrical random walk during the first $2n$ steps is equal to

$$E[N(2n)] = \sum_{k=1}^{n} \mathcal{P}\{\mathcal{S}_{2k} = 0\}$$

where $\mathcal{S}_k$ is a sum of independent random values $+1$ or -1 with probability $1/2$. It is well known that

$$\mathcal{P}\{\mathcal{S}_{2k} = 0\} = 1/\sqrt{\pi k} * R(k)$$

where as usual in the estimates using Stirling's formulae, the remainder term satisfies the inequality

$$e^{-1/(3n)} \leq R(k) \leq e^{1/(3n)}$$

Thus

$$E[N(2n)] = \sum_{k=1}^{n} 1/\sqrt{\pi k} * R(k)$$

estimating the main term (the remainder term we may drop because it is exponentially small) we have:

$$\sum_{k=1}^{n} 1/\sqrt{k} = \int_{1}^{n} \sqrt{x}\,dx + Q(n) = 2\sqrt{n} + \tilde{Q}(n)$$

where $\tilde{Q}(n)$ coincides up to a constant with an integral

$$\int_{1}^{n} (\sqrt{[x]} - \sqrt{x})\,dx < \sum_{k=1}^{n} (1/\sqrt{k} - 1/\sqrt{k+1})$$

$$< \sum_{k=1}^{n} 1/(2k\sqrt{k}) < \text{const} < \infty$$

because $1/\sqrt{k} - 1/\sqrt{k+1} < 1/(2k\sqrt{k})$. Therefore

$$|E[N(2n)] - 2\sqrt{n}/\sqrt{\pi}| < \text{const} < \infty$$

Hence the dimension spectrum has only one nonzero value $d_1 = 1/2$ with the multiplicity $C_1 = 2/\sqrt{\pi}$. **Q.E.D.**

Remark. Our examples show that the values d_k of the spectrum could be larger than the space dimension (see example (1) with $\alpha < 1$) and the multiplicities could be negative. The latter means that from the geometrical point of view we have not to add the corresponding Cantor-like set, but to subtract it.

Returning from the "discrete" set of integers to the "continuous" set $[0, 1]$, we see that we can define the notion of geometrical multifractality of a subset $A \subset [0, 1]$ in the same way as above, considering the asymptotic series (as $\epsilon \to 0$) for the number $N_\epsilon(A)$ of "full" boxes of size ϵ

$$N_\epsilon(A) = \sum_k R_k(\epsilon)\,\epsilon^{-d_k}$$

instead of the asymptotic series for $N_n(A)$ as $n \to \infty$. Here we shall consider multifractality from the point of view of box-type definitions.

Theorem 3. For any C, q positive d and nonnegative integer q there exists a "continuous" monofractal set with dimension d and multiplicity $R(\epsilon)$, growing with less than power rate.

Proof. We shall follow the construction of the usual Cantor set with the exception that on the n-th step we shall throw out not the middle $1/3$ of the interval, as usual, but the middle interval of relative length $\varphi(n)$, where the integer argument function $0 < \varphi(n) < 1$. Then on the n-th step we shall have 2^n "full" intervals of length

$$\epsilon = 2^{-n} \prod_{k=1}^{n} (1 - \varphi(n))$$

Therefore choosing a function

$$\varphi(n) \to \varphi_\infty \in (0, 1/2) \text{ as } n \to \infty$$

we could obtain arbitrary dimensions and multiplicities. **Q.E.D.**

To clarify this construction, note that $\varphi(n) \equiv 1/3$ gives a usual Cantor set.

As we mentioned above, there is a correspondence between "continuous" and "discrete" sets from the point of view of geometrical multifractality.

Theorem 4. For any "continuous" monofractal set $A_c \subset [0, 1]$ there exists a "discrete" monofractal set $A_d \subset I\!N$ with the same dimension and multiplicity.

To prove this statement it suffices to construct a "discrete" set with prescribed dimension and multiplicity. It could be done in the same way as in the "continuous" case. Let us show how to obtain a constant nonzero multiplicity. We consider a sequence $\{(n/C)^\alpha\}$. Then $N_n = Cn^{1/\alpha}$.

There is also another way to construct "discrete" sets with prescribed dimension and multiplicity. Let us start from the set of the type $\{n^{\alpha(n)}\}$. Then after the n-th point we put new $\psi(n)$ points. If we suppose, that $\psi(n)/n^\alpha \to 0$

as $n \to \infty$, then we can also obtain the nontrivial multiplicity. We do not know the counterpart of this construction for "continuous" sets.

4 Geometric multifractal spectrum for measures

Consider now how the proposed approach works for the case of measures. In the same manner as for sets we can use the statistics of boxes $N_\epsilon(\mu)$ whose μ-measure is not less than ϵ instead of N_ϵ-statistics, which we have used for sets. Indeed, assume that the geometrically monofractal measure is a Cantor-like one. The latter means that it could be constructed in the same way as a Cantor set and during the each step of the construction we define a measure uniformly distributed on the remained intervals. Then by the Cantor-like measure we mean the limit in the weak topology of measures of the measures constructed above. Then by the multifractal measure it is natural to understand a sum of such Cantor-like measures of different dimensions, and to define its dimension spectrum by the representation of $N_\epsilon(\mu)$ in a series of the type, considered in Section 3.

It is of interest co compare our multifractal spectrum for measures with the usual dimension spectrum. Of course, for the Cantor-like measures the both spectra will have only one term. However it is far from clear that from the fact that one of these spectra is monofractal it follows that the other one is also monofractal.

5 Applications

The paper represents only the first attempt of mathematical description of the multifractality problem for sets. Therefore we could derive rigorously the spectrum $\{d_k\}$ only for the simplest examples. However we applied this technique numerically for the problem of multifractality of the set of shocks of one-dimensional Burgers equation with Brownian initial data. Statistical properties of solutions of this type were considered in detail in [6] and [7]. For our analysis we used data obtained by a program written by A.Gordon and A.Troussov for 1000 realizations with 10^6 initial data of Brownian type in each for different moments of time (from 100 to 3000 in conventional time units). Analysis of the statistics of "full" boxes N_n shows that our approach here gives only the first element of spectrum d_1 and all others are equal to zero. Therefore the set of shocks is monofractal (at least up to the obtained precision). However the graph of the function $\log(N_n)$ over $\log(n)$ shows also another peculiar behaviour. The graph consists of two pronouncely linear pieces - one for small n ($100 < n < 700$) and another one for large n ($1000 < n < 3000$) with different slopes. This shows that in spite the monofractal features of the set it could have quite different statistics for large and small scales. Such statistics could be explained if we suppose that the set under consideration is constructed as follows. For the first n steps we follow the usual Cantor set construction, but then from the $n + 1$-th step we change it to the construction of a Cantor set with a different dimension. Then the graph of statistics N_n will have the described shape.

The author is grateful to Ya.Sinai and U.Frisch for very helpful discussions. The work was supported by the French Ministry of Higher Education.

References

[1] Parisi G., Frisch U. in "Turbulence and predictability in geophysical fluid dynamics", Ghil M., Benzi R., Parisi G., eds. (North-Holland, Amsterdam, 1985), p.84.

[2] Bedford T., Fisher A. M. Analogues of the Lebesgue density theorem for fractal sets of reals and integers. *Proc. London Math. Soc.*, **64** (1992), 95-124.

[3] Evertsz C. J. G., Mandelbroit B. B. Multifractal measures. in "Chaos and fractals", Peitgen H.-O., Jurgens H., Saupe D.. eds. (Springer Verlag, New York, 1992), 849-881.

[4] Falconer K. L. The geometry of fractal sets, Cambridge University Press, 1985.

[5] Wolf M. Multifractality of prime numbers. *Physica* A **160** : 1 (1989), 24-42.

[6] Sinai Ya. G. Statistics of shocks in solutions of inviscid Burgers equation, *Commun. Math. Phys.* **148** (1992), 601-622.

[7] Vergassola M., Dubrulle B., Frisch U., Noullez A. Burgers' equation, devil's staircases and the mass distribution for large-scale structures, *Astron. & Astrophys.* **289** (1994), 325-356.

PART 3:

LÉVY FLIGHTS IN DYNAMICAL SYSTEMS

Lévy Walks in Chaotic Systems:
Useful Formulas and Recent Applications

T. Geisel

Institut für Theoretische Physik und SFB Nichtlineare Dynamik
J.W.Goethe Universität Frankfurt
D – 60054 Frankfurt am Main, Germany

Abstract: The article starts with a review of early work on statistical descriptions of Lévy walks and compiles some useful equations that are not easily extractable from the original literature. A general renewal formalism gives accurate results not only for the mean-square displacement, but also for the power spectral density where other methods were deficient. The latter is important for comparison with experiments, in which diffusive properties are often accessible by means of the velocity power spectrum. The formalism is applied to dissipative as well as Hamiltonian chaotic systems where nonlinear dynamics supplies a generic mechanism for enhanced anomalous diffusion and $1/f$-noise. In a physical application to semiconductor nanostructures (antidot lattices) chaotic Lévy walks are responsible for a *negative* Hall resistance, which was measured in transport experiments.

1 Scope

A decade ago the first examples of a class of random walks with enhanced anomalous diffusion were discovered [1] which subsequently made their appearance in many other chaotic systems and soon were christened Lévy walks [2,3]. They play a role in dissipative and conservative dynamical systems as well as in fluid mechanics. As the mean-square displacement diverges faster than linearly in time, the diffusion coefficient is *infinite*. We first called these processes "accelerated diffusion" [1] to distinguish them from "enhanced" diffusion, which in those times was used for processes where the diffusion coefficient *exists*, but strongly increases and eventually diverges when certain bifurcations are approached. Meanwhile, as the latter processes have almost been forgotten, "enhanced diffusion" is used by many authors to denote random walks with *infinite* diffusion coefficients. The nomenclature "Lévy walks" was chosen [2,3] because in individual steps the average jump length is infinite like in the well-known Lévy flights. In contrast to the latter, however, Lévy walks assume that a certain time is needed to complete each jump depending on its length. This aspect makes them more physical than Lévy flights and is the main reason for their widespread applicability.

The first investigated examples were chaotic systems in the form of one-dimensional maps [1] where a trajectory (or particle) exhibits free paths of nearly constant velocity and finite duration T. Due to some type of intermittency the distribution of free paths $\Psi(T)$ follows an inverse power law. After this model we studied the extended Sinai billiard, a particle moving in a periodic array of reflecting disks in the plane. Here the existence of arbitrarily long free paths is associated with an algebraic free path distribution $\Psi(T) \sim T^{-3}$ as well, but the mean-square displacement $\sigma^2(t)$ only exhibits weakly enhanced diffusion $\sigma^2(t) \sim t \ln t$ [4]. As the average path length is finite, this is not a real Lévy walk and we will not discuss it in detail here. On the other hand, another type of Lévy walk showed up in *conservative* chaotic systems for the first time, when we modified this model to a softer version in which a smooth two-dimensional periodic potential replaces the reflecting disks [5]. Now many of the peculiar phenomena of the theory of nonintegrable Hamiltonian systems could come into play and indeed our analysis showed that the physical mechanisms underlying this Lévy walk are entirely different from those leading to anomalous diffusion in the Sinai billiard. The persistence in free paths is caused here by a trapping in a self-similar hierarchy of cylindrical cantori, which act like tubes in phase space. The trapping in turn can be described in terms of a statistical theory by Meiss and Ott [6]. These Lévy walks in conservative systems were then found in many other physical situations in particular in fluid dynamical and plasma physical models which admit a Hamiltonian formalism. They also are responsible for a negative Hall resistivity observed in certain semiconductor microstructures.

After a review of a general formalism for an analytic description of Lévy walks in Sect. 2, we will apply it to one-dimensional maps in Sect. 3 and to conservative chaotic systems in Sect. 4. Experiments and theories on Lévy walks in semiconductor microstructurs are described in Sect. 5. We conclude with some remarks on similar quantum mechanical phenomena.

2 Analytic Description of Lévy walks

In our early paper [1] we already gave a full analytic description of Lévy walks in terms of a renewal process. For some reason many later authors did not use this formalism but rather adopted a particular continuous-time random walk (CTRW) description, which in some cases leads to wrong results. Although our formalism is already a decade old, it may thus be worthwhile to review its essentials and to compile some useful formulas.

We want to describe a random walker which successively performs statistically independent free paths of constant velocity v_0 whose durations T (or lengths L) are distributed according to an inverse power law

$$\Psi(T) \sim T^{-\nu-1} \tag{1}$$

where $\Psi(T)$ is the probability density function. The time spent in residence or in localized motion between free paths is assumed to be negligible with respect to the latter and a free path in the opposite direction is assumed to have equal probability. In many cases Eq. (1) only holds in a certain regime up to a cutoff time $T^* = 1/\varepsilon$ beyond wich the decay is exponential. For technical reasons it is essential that we calculate with such a cutoff and let $\varepsilon \to o$ in the end. The following can also be generalized easily to cases where the free path velocity is assumed to depend on the length of a free path $v_0 = v_0(T)$.

As we had been able to treat a previous model dealing with dispersive (suppressed) anomalous diffusion [7] in a CTRW-formalism, our first guess was that it should be possible to describe also the enhanced anomalous diffusion by some CTRW-formalism. We first modeled the motion by a CTRW with a coupled space-time memory [8] of the type

$$\varphi(\mathbf{L}, T) = \Psi(T)\delta(L - v_0 T). \tag{2}$$

In this approximation, the continuous motion in a free path of length L and duration T is replaced by instantaneous displacements by L (jumps) after waiting on a site up to time T. Many of the asymptotic results came out correctly, but it is perhaps not too surprising that such a rough approximation failed in some cases. In particular, for $\nu \leq 1$ the velocity power spectrum $S(\omega)$ came out wrong, e.g. as $S(\omega) \sim \omega^{\nu}$ instead of $S(\omega) \sim \omega^{-2+\nu}$ for $\nu < 1$. (A leading $\delta(\omega)$ term comes in addition). Since the important case of $1/f$-noise was also concerned (for $\nu = 1$), we had to find another description which captures details of the motion more accurately, i.e. the renewal formalism. After this formalism was published in Ref. [1], the possibility of a CTRW-description with a coupled space-time memory was also pointed out in a comment by Shlesinger and Klafter [2]. The CTRW-description of Lévy walks has maintained the deficiencies mentioned above until recently, when it was modified in a so-called "velocity" model as opposed to the "jump" model by Zumofen and Klafter [9,10].

The description in terms of a renewal process [1] as described below avoids these problems and yields accurate results for the asymptotic behavior of all statistical quantities of interest. It is based on the velocity autocorrelation function

$$C(t) = < \mathbf{v}(t) \cdot \mathbf{v}(0) > = 2 < v_x(t)v_x(0) >, \tag{3}$$

where the last equality requires square symmetry in the plane. Picking an arbitrary time $t' = 0$ on the time axis, the probability $w_0(T)dT$ that the particle is in a free path of duration T is proportional to $\Psi(T)$ and also proportional to T

$$w_0(T)dT = \frac{T}{<T>}\Psi(T)dT. \tag{4}$$

The denominator ensures the correct normalization and is the average duration of a free path

$$<T> = \int_0^\infty T\Psi(T)dT. \tag{5}$$

The probability that this path persists until time t is the fraction $(T-t)/T$. The probability $w_{0,t}(T)dT$ that the particle is in a free path of duration T between time 0 and t is thus

$$w_{0,t}(T)dT = \frac{T-t}{<T>}\Psi(T)dT. \tag{6}$$

Using the longitudinal velocity $v_0 = L/T$ mentioned above, we obtain the velocity autocorrelation function $C(t)$ from the integral of Eq. (6) over all possible durations

$$C(t) = \frac{v_0^2}{<T>}\int_t^\infty (T-t)\Psi(T)dT. \tag{7}$$

This equation only includes events consisting of a single path between times 0 and t. Sequences consisting of two or more free paths (independent events) between 0 and t have equal probability to end with a positive or negative velocity ($\pm v_0$) and thus do not contribute to Eq. (7). We have shown this in more detail for the Sinai-billiard [4]. Denoting the Laplace transforms of $C(t)$ and $\Psi(T)$ by $\tilde{C}(s)$ and $\tilde{\Psi}(s)$, Eq. (7) turns into

$$\tilde{C}(s) \propto \frac{1}{s} + \frac{1}{<T>}\frac{\tilde{\Psi}(s)-1}{s^2}. \tag{8}$$

From Eq. (8) we obtain all statistical quantities of interest, e.g. the power spectral density as

$$S(\omega) = \int_{-\infty}^{\infty} <\mathbf{v}(t)\cdot\mathbf{v}(0)> e^{-i\omega t}dt = \lim_{\delta\to 0}[\tilde{C}(\delta+i\omega) + \tilde{C}(\delta-i\omega)]. \tag{9}$$

For the mean-square displacement of the particle

$$\sigma^2(t) = <[\mathbf{r}(t) - \mathbf{r}(0)]^2> \tag{10}$$

we use the identity

$$\sigma^2(t) = 2\int_0^t (t-\tau)C(\tau)d\tau \tag{11}$$

whose Laplace transform is $\widetilde{\sigma^2}(s) = 2s^{-2}\tilde{C}(s)$. Thus from Eq. (8) we obtain

$$\sigma^2(t) = 2\mathbf{L}^{-1}\left\{s^{-3} + <T>^{-1} s^{-4}[\tilde{\Psi}(s) - 1]\right\} \tag{12}$$

where $\mathbf{L}^{-1}$ denotes the inverse Laplace transform. Finally, the diffusion coefficient, where it exists is defined as $D = \lim \sigma^2(t)/2t$, and from Eq. (10) follows as

$$D = \int_0^\infty <\mathbf{v}(t)\cdot\mathbf{v}(0)> dt \tag{13}$$

or

$$D = (1/2)S(\omega = 0). \tag{14}$$

The above expressions are general. We now make use of the particular form of $\Psi(T)$ as an inverse power law Eq. (1). Detailed analytic expressions for the

Laplace transforms $\tilde{\Psi}(s)$ can be found in Ref. [7]. It is useful to consider their limiting behaviour for small s and without cutoff ($\varepsilon = 0$)

$$\tilde{\Psi}(s) \sim \begin{cases} -b\, s^\nu & (0 < \nu < 1) \\ c\, s\, \ln s & (\nu = 1) \\ - < T > s & (\nu > 1) \end{cases} \tag{15}$$

where b and c are constants. For $\nu > 1$ the next term is $-\Gamma(1-\nu)[(\nu-1)< T >s]^\nu$ as long as $\nu < 2$.

For $\nu > 1$ we can use this result in Eqs. (8) and (12) to obtain $S(\omega)$ and $\sigma^2(t)$. In cases where $< T >$ diverges ($0 < \nu \leq 1$) we must consider a repesentation of $\Psi(\mathrm{T})$ with a finite cutoff (small ε) whereby the Laplace transform becomes much more complicated [1]. For $\nu = 1$ the correlation function $C(t)$ can be obtained by direct integration. The asymptotic behaviour of the mean-square displacement in the long-time limit is then calculated by making use of Tauberian theorems and of Karamata's theorem

$$\sigma^2(t) \sim \begin{cases} t^2 & (0 < \nu \leq 1) \\ t^{3-\nu} & (1 < \nu < 2) \\ t\ln t & (\nu = 2) \\ t & (\nu > 2) \end{cases} \tag{16}$$

These results pertain to Eq. (1) in the limit $\varepsilon \to 0$, i.e. where the cutoff $T^* \to \infty$. It should be stressed that in cases where it is essential ($0 < \nu \leq 1$) it is understood that the limits are considered in the order $t \to \infty$ before $\varepsilon \to 0$. If this is not respected and $\varepsilon = 0$ is assumed in the very beginning (Eq. (1)), the calculations are considerably simpler as the simple Laplace transforms Eq. (15) can be used, but instead of t^2 lead to the form $t^2/\ln t$ for $\nu = 1$ [11].

The power spectral density is calculated from Eqs. (8) and (9) or by directly Fourier transforming $C(t)$ in the case $\nu = 1$. In the low-frequency regime one obtains [1]

$$S(\omega) \sim \begin{cases} \omega^{-1}(\varepsilon/\omega)^{1-\nu} & (0 < \nu < 1) \\ -(1/\ln\varepsilon)\omega^{-1} & (\nu = 1) \\ \omega^{\nu-2} & (1 < \nu < 2) \\ |\ln\omega| & (\nu = 2) \\ const. & (\nu > 2) \end{cases} \tag{17}$$

Here the low-frequency regime is meant as the regime $\varepsilon \ll \omega \ll \pi$, i.e. where ω is large compared with the cutoff frequency $\varepsilon = 1/T^*$. The cutoff allows to maintain the normalizability of $S(\omega)$ for $\nu \leq 1$, i.e. for cases of $1/f$-noise. The divergence of the integral $\int \omega^{\nu-2} d\omega$ otherwise would imply a divergence of the velocity variance $< v^2 > = C(t = 0) \propto \int S(\omega) d\omega$, which is not realistic in physical contexts.

The function $S(\omega)$ is of major importance as it characterizes the diffusion process (e.g. Eq. (14)) and in experiments (e.g. in fluid dynamics) is often measurable more directly than $\sigma^2(t)$. This is particularly important where $1/f$-noise is found ($\nu \simeq 1$). Moreover, in cases where a Kubo formula holds, it is closely related to the frequency-dependent conductivity $\sigma(\omega)$. It is more sensitive to details of the motion than $\sigma^2(t)$. E.g. in a CTRW-approximation (Eq. (2)) where the continuous motion in free paths is replaced by a discontinuous motion as mentioned above, one obtains $S(\omega) \sim \omega^{-\nu}$ [2] instead of $S(\omega) \sim \omega^{\nu-2}$ as it should be for $\nu < 1$ (besides a $\delta(\omega)$ term).

3 Anomalous diffusion in dissipative chaotic systems

Dissipative chaotic systems can exhibit diffusive motions, which in many cases may be described by one-dimensional maps [12]. We pointed out that anomalous diffusion may occur as well, if the maps allow for intermittent behaviour [7]. If the laminar phases consist in a persistent sticking near a marginally stable fixed point, the diffusion coefficient vanishes (dispersive anomalous diffusion). We knew that we could find the opposite case i.e. of enhanced anomalous diffusion with infinite diffusion coefficient for maps in which the laminar phases consisted in persistent free paths. Our interest in working out a theory for this case, however, was stimulated mainly, when an experimental observation of $1/f$-noise in a Josephson analog was reported [13], where such a theory seemed to be relevant [1].

For sufficient damping the chaotic dynamics of Josephson junctions and other systems with discrete translational symmetry can be described approximately by circle maps [12]

$$x_{t+1} = f(x_t), \qquad f(x + n) = f(x) + n. \tag{18}$$

The actual form of the map, for which we assume reflection symmetry $f(-x) = -f(x)$ depends on the particular physical problem and its parameters. As an example we have studied a map which near $x \simeq m$ takes the form

$$f(x) = (1 + \varepsilon)(x - m) + a(x - m)^z + m - 1 \quad (m \lesssim x < m + \tfrac{1}{2}) \tag{19}$$

as illustrated in Fig. 1. As usual the exponent z distinguishes universality classes, $z = 2$ is the generic case. Most studies of intermittency assume maps with $\varepsilon = 0$. An invariant measure does not exist in this case [14], but is needed nevertheless to define the injection probability into the laminar regions and in turn the distribution of free paths $\Psi(T)$. To avoid this problem we introduced the parameter ε leading to a cutoff time $T^* = 1/\varepsilon$ and let $\varepsilon \to 0$ in the end. The calculations thereby become much more complicated (see e.g. Eq. (20) vs. Eq. (1)), but the existence of the distribution $\Psi(T)$, the central quantity is guaranteed.

When an orbit x_t reaches the vicinity of a point $x = m$ in Fig. 1 it is transfered to an almost equivalent position in the neigbouring cell, i.e. near $x = m + 1$. This transfer is repeated successively resulting in a laminar motion of correlated jumps (free paths) over many cells. The probability density $\Psi(T)$ of the duration of free paths is calculated in a continuous-time approximation [1]

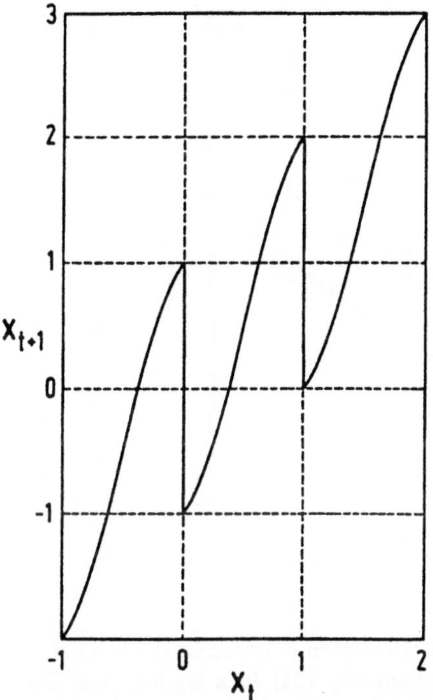

Fig. 1. Example of a map satisfying Eqs. (18,19) shown for 3 unit cells of x_t with successors x_{t+1} spreading over 5 unit cells. The reduced map (x_{t+1} mod 1 in the unit cell $0 \leq x_t \leq 1$) has laminar regions in the vicinity of fixed points at $x = 0$ and $x = 1$.

$$
\begin{aligned}
\Psi(T) = {} & \frac{2\varepsilon^{\nu+1}}{a^\nu} \left[\left(1 + \frac{2^{z-1}\varepsilon}{a} \right) e^{\varepsilon(z-1)T} - 1 \right]^{-\nu} \\
& + \frac{2\varepsilon^{\nu z}}{a^{\nu z - 1}} \left[\left(1 + \frac{2^{z-1}\varepsilon}{a} \right) e^{\varepsilon(z-1)T} - 1 \right]^{-\nu z}
\end{aligned}
\tag{20}
$$

where $\nu = 1/(z-1)$. For $z = 2$ one can obtain the correlation function $C(t)$ by direct integration of Eq. (7)

$$
C(t) = - \left[\ln \left(1 + \frac{a}{2\varepsilon} \right) \right]^{-1} \ln \left[1 - \frac{a}{a + 2\varepsilon} e^{-\varepsilon t} \right].
\tag{21}
$$

In the limit $\varepsilon \to 0$ and $s/\varepsilon \gg 1$ its Laplace transform becomes

$$
\tilde{C}(s) = \left(\ln \frac{a}{2\varepsilon} \right)^{-1} \left(\frac{\gamma}{s} + \frac{1}{s} \ln \frac{s}{\varepsilon} \right)
\tag{22}
$$

where γ is Euler's constant.

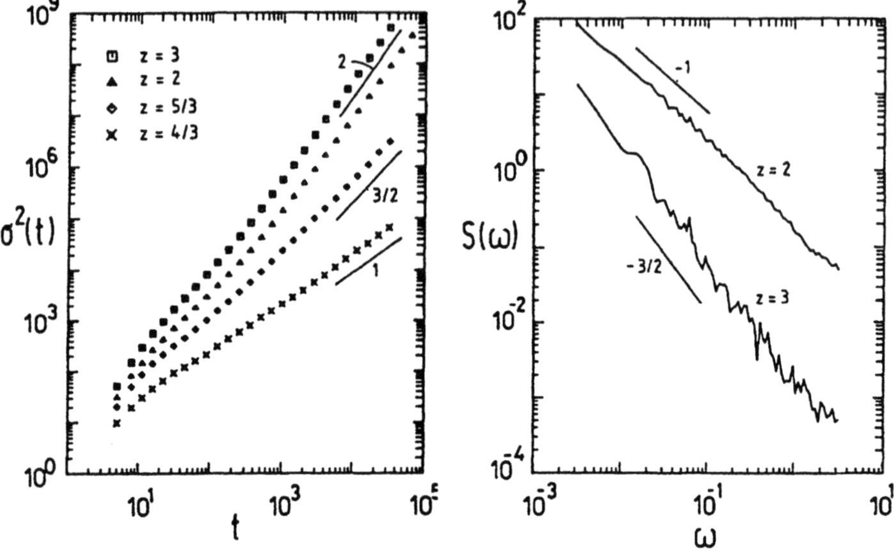

Fig. 2. (a) Computer simulation illustrating the anomalous asymptotic growth of the mean-square displacement Eq. (24) in a log-log plot. Theoretically predicted slopes are indicated by straight lines. The anomalous growth (for $z \geq 3/2$) is accelerated as compared with normal diffusion ($z = 4/3$).
(b) Velocity power spectrum $S(\omega)$ exhibiting $1/f$-noise in a computer simulation for $z = 2$ and $z = 3$. Straight lines indicate the analytic result Eq. (25). While this equation correctly yields $S(\omega) \sim \omega^{-3/2}$ for $z = 3$, the CTRW-description mentioned in the preceding section would lead to the power law $S(\omega) \sim \omega^{-1/2}$.

Equation (20) gives representations of the algebraic distributions. Eq. (1) with a cutoff time $T^* = 1/\varepsilon$, if we identify

$$\nu = \frac{1}{z-1}.\tag{23}$$

The formulas given in the previous section for $\sigma^2(t)$ and $S(\omega)$ thus apply to the chaotic diffusion process in the present maps with the following specifications [1]

$$\sigma^2(t) \sim \begin{cases} t^2 & (z \geq 2) \\ t^{3-1/(z-1)} & (3/2 < z < 2) \\ t \ln t & (z = 3/2) \\ t & (1 < z < 3/2) \end{cases}\tag{24}$$

$$S(\omega) \sim \begin{cases} \omega^{-1}(\varepsilon/\omega)^{1-1/(z-1)} & (z > 2) \\ -(1/\ln\varepsilon)\omega^{-1} & (z = 2) \\ \omega^{-2+1/(z-1)} & (3/2 < z < 2) \\ |\ln\omega| & (z = 3/2) \\ const. & (1 < z < 3/2) \end{cases} \qquad (25)$$

where $\varepsilon \ll \omega \ll \pi$ and $\varepsilon \to 0$. These results are illustrated by computer simulations in Figs. 2a and b. The mean-square displacement $\sigma^2(t)$ determined by averaging over 2000 orbits shows anomalously enhanced diffusion for $z > 3/2$. The power spectral density $S(\omega)$ (Fig. 2b) shows $1/f$-noise. Note that for $z = 3$ one clearly observes the power law $S(\omega) \sim \omega^{-3/2}$ in agreement with Eq. (25), whereas the continuous-time random walk description (CTRW) mentioned in the preceding section would lead to the wrong power law $S(\omega) \sim \omega^{-1/2}$. For $z \geq 2$ the power spectra are also markedly different from those found in other intermittent systems [15,16]. In an experiment on a Josephson analog $1/f$-noise like in the case $z = 2$ was observed [13]. As discussed in Ref. [1] this is an unambiguous indication of a chaotic Lévy walk with enhanced anomalous diffusion $\sigma^2(t) \sim t^2$.

4 Anomalous Diffusion in Hamiltonian Chaotic Systems

Dissipative chaotic systems and Hamiltonian chaotic systems exhibit quite different phenomenologies. The mechanisms leading to diffusion and anomalous diffusion thus may be very different, although the statistical description of Lévy walks as outlined in Sect. 2 turns out to apply similarly in both cases. In particular, while the mechanism of Lévy walks in *dissipative* systems (Sect. 3) is nongeneric and only arises under special circumstances, we have demonstrated that *Hamiltonian* systems offer a new generic mechanism for Lévy walks and $1/f$-noise [5]. It was thus predictable that Lévy walks would soon be observed in many different Hamiltonian systems. Sect. 5 will be devoted to the description of one particular physical application, superlattices on semiconductor heterojunctions. In the present section we want to outline the general mechanism as reported in the early publications [5]. It is based on a typical self-similar hierarchy of barriers in phase-space (cantori) in combination with geometrical contraints forcing trajectories into long persistent free paths. The hierarchy of barriers can be described in a Markov tree model [6], but it should be emphasized that this usually leads to an inhibition of diffusion (due to trapping in local barriers) and not to Lévy walks. In the cases which we consider here, the cantorus barriers have the shape of cylinders, within which trajectories propagate axially at nearly constant velocity.

As an example we treat the conservative motion of a particle in an analytic two-dimensional potential $V(x,y)$. More generally one may consider Hamiltonian chaotic systems with periodic symmetry. In the physical application of Sect. 5 the particle represents a ballistic electron of effective mass m and energy $E = E_F$ (Fermi energy) moving in the superlattice potential $V(x,y)$. Depending on the preparation technique, the potential can have various shapes. One may describe it in terms of a general 2-d Fourier expansion of a periodic potential

$$V(x,y) = V(\mathbf{r}) = \sum_{\mathbf{G}} V_{\mathbf{G}}\, e^{i\mathbf{G}\cdot\mathbf{r}} \tag{26}$$

where $\mathbf{G}$ is a reciprocal lattice vector. Here we will consider the three lowest order terms

$$V(x,y) = A + B(\cos x + \cos y) + C\cos x \cos y \tag{27}$$

where the third term is needed for the generic situation of a nonintegrable Hamiltonian. For numerical calculations we mainly use the parameters $A = 2.5$, $B = 1.5$, and $C = 0.5$. This potential has the form of an egg-carton as is shown in Fig. 3. It has potential minima in the centers of the cells, potential maxima at the four corners, and saddle points at the midpoints of the edges.

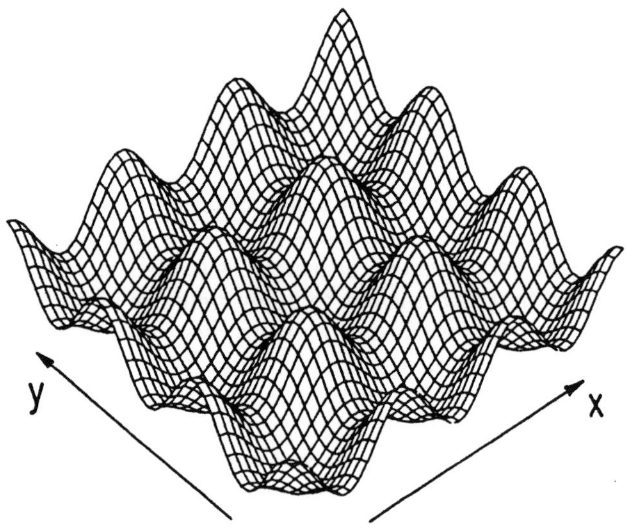

Fig. 3. Periodic Potential $V(x,y)$ with $A = 2.5$, $B = 1.5$, and $C = 0.5$.

With the above choice of parameters the potential minima are $V_{\min} = 0$, the maxima $V_{\max} = 6$, and the saddles have the potential $V_S = 2$. For total energies $E \leq V_S$ the particle is confined to a single cell. For energies $E > V_S$ the particle can make transitions to adjacent cells. Depending on the initial conditions we found that the particle is drifting or shows diffusive motion. With increasing energies $E > V_{\max}$ an integrable (free particle) situation is approached. We investigate the most interesting case $V_S < E < V_{\max}$ where the diffusive motion has a persistent character with long free paths, which are interrupted by episodes where the particle is trapped for a while in a cell. This is reminiscent of the low-friction limit of Brownian motion models [17], where a particle that is excited to a momentary energy above the potential barrier thermalizes so slowly to lower energies that it can perform a long free path.

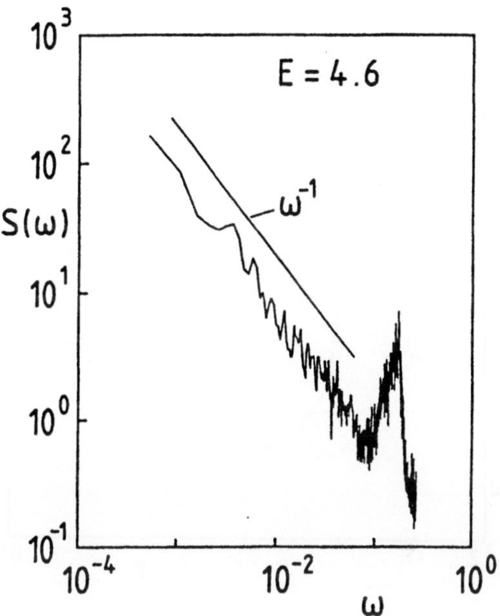

Fig. 4. Velocity power spectral density of diffusive motions. In a broad energy range above the saddle point energy V_S one finds $1/f$-noise $S(\omega) \sim \omega^{-\alpha}$ as exemplified for $E = 4.6$.

Assuming mass $= 1$, the potential (27) leads to the equations of motion

$$\frac{dv_x}{dt} = (B + C \cos y) \sin x$$
$$\frac{dv_y}{dt} = (B + C \cos x) \sin y \tag{28}$$

For a characterization of the diffusion process we have performed a power spectral analysis of the velocity. Fig. 4 shows an example of $S(\omega)$ for diffusive motion at the energy $E = 4.0$ in a log-log plot where $S(\omega)$ behaves like ω^{-1}. We have carried out more detailed numerical studies which have shown that $S(\omega)$ diverges like $\omega^{-\alpha}$ $(0.7 \leq \alpha \leq 1.1)$ in a range of energies E extending from the saddle-point energy $V_S = 2.0$ to $E = 4.6$. Considering that $S(\omega)$ is closely related to the mean-square displacement (Eq. (11)), and in particular that $S(\omega = 0) = 2D$ (where D is the diffusion coefficient), we can conclude that the diffusion is anomalously accelerated for $2.0 < E \leq 4.6$. For $5.0 \leq E \leq 6.0$ there is a normal diffusion process characterized by a finite diffusion coefficient.

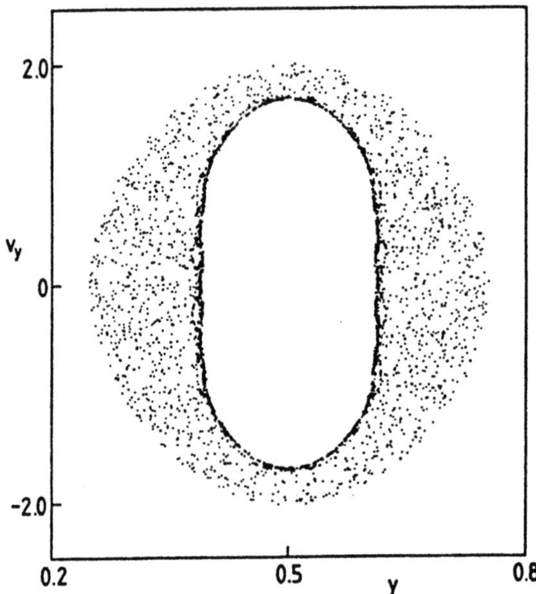

Fig. 5. Poincaré surface of section at the boundaries of the cells ($x = 2\pi n$) for energy $E = 4.0$. The points represent particle trajectories leaving the cell in the perpendicular direction. Position is measured in units of 2π.

In order to understand the origin of the observed anomalous diffusion and $1/f$-noise we have determined Poincaré surfaces of section at the boundaries of the cells at $x = 2\pi n$ (Fig. 5). Whenever the particle left a cell in $\pm x$-direction we have recorded its y-coordinate ($mod\, 2\pi$) and its v_y-coordinate. Every point in Fig. 5 thus represents a motion of the particle leaving the cell in the perpendicular $\pm x$-direction, and localized motions within a well cannot show up. As the distinction will be important, I will reserve the term *orbit* only for the discrete dynamics within the Poincaré surface of section and the term *trajectory* only for the particle motion in the perpendicular direction.

For diffusive motions one typically finds surfaces of section as shown in Fig. 5 for the energy $E = 4.0$. In the island there are periodic and quasiperiodic orbits, which are not shown. They pertain to unlimited free paths (drift motions), as the particle consecutively crosses the edges of the cells. On the other hand, orbits in the chaotic sea surrounding the island remain there only a *finite* time.

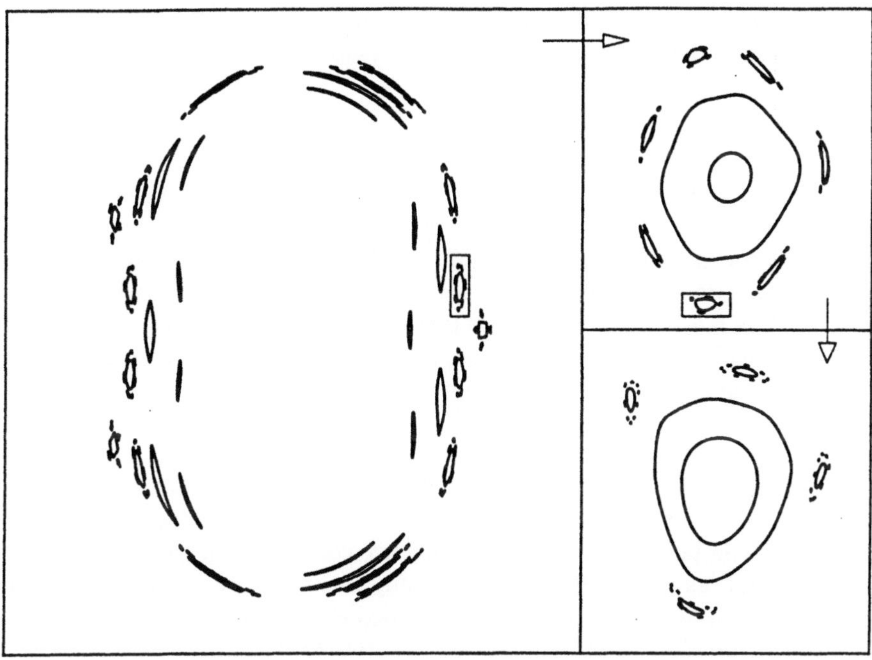

Fig. 6. a) Isolation of island chains near the boundary of the central island of Fig. 5.
b) Magnification of the box shown in Fig. 6a above, displaying the self-similar hierarchy
of daughter islands within the chaotic sea.

(Hereby we mean that the particle trajectory turns into a local motion within a
well, before it can reach the surface of section again.) When they reach its outer
boundary, the energy condition to cross the saddle is no longer fulfilled. The free
path of the particle thus persists only a finite time and gives rise to diffusion.

Near the inner boundaries of the chaotic sea in Fig. 5 the orbits seem to have
a higher density, which we attribute to the finite observation time. This fact also
points to the origin of $1/f$-noise. The orbit can be seen to stick near daughter
islands surrounding the central islands in Fig. 5. To illustrate this in more detail for $E = 4.0$ Fig. 6a shows three island chains indicated by representative
quasiperiodic orbits. They were isolated by selecting special initial conditions.
The magnification in Fig. 6b reveals four levels in a hierarchy of daughter islands
around daughter islands. We know that generically this hierarchy continues ad
infinitum [18]. Every island in the chaotic sea is encircled by cantori [19], partial barriers which the orbit can penetrate. The deeper the orbit enters into the
hierarchy of nested cantori, the longer it remains trapped before it can leave the
chaotic sea (hereby I mean that the particle trajectory turns into a local motion

within a well, before it can reach the Poincaré section again). This hierarchy of time scales is the origin of the observed $1/f$-noise.

We now give a statistical formulation of the mechanism outlined above. The free paths of duration T with a probability density $\Psi(T)$ can be treated as a renewal process as in Sect. 2. Eqs. (7) and (11) give us the correlation function and mean-square displacement. It would be desirable to obtain $\Psi(T)$ analytically for the mechanism of cantorus trapping. Too little, however, is known with mathematical rigour about the transport across cantori. In this situation we are forced to make a number of assumptions, which reduce the dynamics of our model to a simpler statistical model. We do know rigorously that every island in the chaotic sea is encircled by a set of cantori belonging to a continuum of rotation numbers [19]. Those with sufficiently irrational rotation number (e.g. in terms of a continued fraction expansion) have a small flux across and act as barriers [20]. The latter cantori form a sequence converging to the boundary circle of the island, as is illustrated schematically in Fig. 7a. The boundary circle is believed to always be a critical KAM-torus, i.e. a KAM-torus at the transition to destruction. It was shown that critical tori exhibit scaling properties [21], which also imply scaling for the fluxes across the encircling low-flux cantori [20]. Embedded between these cantori are island chains (daughter islands) where the scheme repeats on a finer scale (Fig. 7a). Thus it is confirmed that there are sequences of partial barriers nested hierarchically within each other, and that the hierarchical trapping mechanism must be present. The question is how the transport across cantori occurs in detail, and whether the dynamics might be complicated by additional phenomena.

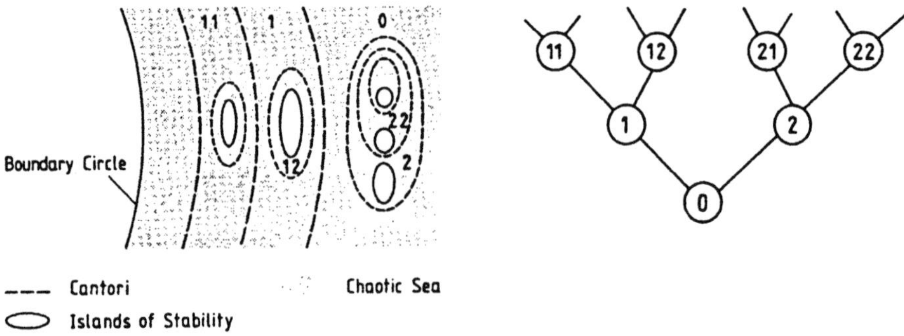

Fig. 7. a) Schematic representation of the nested hierarchy of cantori. The numbers indicate Markov states, b) In a Markov approximation, transitions across low-flux cantori correspond to a random walk on a hierarchical lattice as is illustrated here by a binary tree.

The hierarchical nesting of cantori implies that the accessible regions between the cantori are organized on a tree. This is illustrated in Fig. 7b by the example of a binary tree. The numbered circles, i.e. the sites of the lattice correspond to the accessible regions, whereas the lines represent crossing of a cantorus. In order to enter the hierarchy from the outer region of the chaotic sea (stem of the tree), an orbit can cross one of several possible cantori in each step, i.e. move towards the center of an island, or cross a cantorus of a daughter island. One may assume for simplicity that successive crossings of low-flux cantori can be treated as a Markov process. The chaotic dynamics thus becomes statistically equivalent to a random walk on a hierarchical lattice or Markov tree [6], as will be described below. Let us assume a constant branching ratio m and number the accessible sites (regions) by a sequence of integers $\underline{\ell} = \ell_1 \ell_2 \cdots \ell_n$ as illustrated in Fig. 7. The stem of the tree is the site $\underline{\ell} = \ell_1 = 0$ and denotes the outer region of the chaotic sea. With each branching (cantorus crossing) an integer ℓ_{n+1} is added to the sequence, where $\ell_{n+1} = 1$ stands for crossing of a cantorus of the same island, and $\ell_{n+1} = 2, \cdots, m$ refers to daughter islands.

Let $p_t(\underline{\ell}|\underline{\ell}')$ denote the conditional probability that the first transition to $\underline{\ell}'$ occurs at time t, if at time 0 there was a transition to $\underline{\ell}$. The analogous probability for direct transitions (without intermediate transitions to other sites) is denoted by $d_t(\underline{\ell}|\underline{\ell}')$ and is easily determined from the transition rates. The conditional probability p_t must satisfy a set of coupled integral equations [6]

$$p_t(k|0) = d_t(k|0) + \sum_{j=1}^{m} \int_0^t d_\tau(k|kj)p_{t-\tau}(kj|0)d\tau. \tag{29}$$

This equation expresses the fact that the walker can jump from k to 0 either directly, or by jumping to a higher site kj at some intermediate time τ and descending to the stem 0 in the remaining time. The latter transitions fulfill a similar integral equation

$$p_t(kj|0) = \int_0^t p_\tau(kj|k)p_{t-\tau}(k|0)d\tau \tag{30}$$

i.e. in order to go from kj to 0, the walker must hop to the connecting site k at some intermediate time τ. The system of equations is solved by Laplace transform and a scaling ansatz assuming scaling of the transition rates [6,5].

From the solution $p_t(k|0)$ we can determine the distribution $\Psi(T)$ of the durations of free paths. Recall that a free path of the particle persists as long as the discrete orbit is trapped in the hierarchy of cantori. It ends, when the orbit reaches the outer region of the chaotic sea. In the simplified model of the random walker, the duration of the free path is then determined by the time interval between a transition from 0 to a trapped state k and the first transition back to 0. Since within this model the transitions are taken as independent events, the previous and following history of the walker is irrelevant, and the probability for those transitions is proportional to $\sum_k p_\tau(k|0)$. Therefore the distribution of free times is

$$\Psi(T) \propto \sum_{k=1}^{m} p_T(k|0) \tag{31}$$

The Laplace transform of $\Psi(T)$ is thus obtained as [5]

$$\tilde{\Psi}(s) = 1 + cs^{\mu} + \cdots \tag{32}$$

where c is a constant and the exponent μ is determined implicitly by the scaling of transition rates. This is associated with an asymptotic algebraic decay of $\Psi(T)$

$$\Psi(T) \sim T^{-1-\mu}. \tag{33}$$

For $\mu = 1$, we obtain instead

$$\tilde{\Psi}(s) = 1 + bs(\gamma + \ln s) + \cdots \tag{34}$$

with the same asymptotic decay as Eq. (33). By identification with Eq. (1) we can now directly apply the results of Sect. 2.

The low frequency behaviour of $S(\omega)$ results as

$$S(\omega) \sim \omega^{\mu-2} \tag{35}$$

and the asymptotic divergence of the mean-square displacement results as

$$\sigma^2(t) \sim \begin{cases} t^{3-\mu} & \text{for } \mu \geq 1 \\ t^2 & \text{for } \mu \leq 1 \end{cases} \tag{36}$$

The mechanism treated in this section is thus able to qualitatively explain and recover the observed $1/f$-spectrum by Eq. (35). The actual value of μ, however, remains an open problem. Identifying $2 - \mu$ with the exponent α of Fig. 4 gives a value $\mu = 1$. More generally we found values of μ fluctuating between 0.9 and 1.3.

Above we have described the first case where a Lévy walk was observed in a chaotic Hamiltonian system. Similar Lévy walks were found and analyzed for the classical analog of Bloch electrons in a magnetic field [22] and for the motion of guiding centers in electrostatic plasma waves [23]. Our results were confirmed by further studies of the particle dynamics in periodic and quasiperiodic potentials [24]. The above statistical formalism can also be applied to many other examples that were discovered in the meantime in particular for transport in fluids which admit a Hamiltonian description. The first observation of such a Lévy walk in a chaotic fluid dynamical system was made by Pasmanter in numerical simulations of transport in shallow tidal seas [25] to describe an anomalous time-dependent spread of pollutants that had been measured in various seas. The system can be modeled by a Hamiltonian with two coordinates and a periodic (tidal) driving. Further examples were then reported in other simulations [26,27] and experiments [28] of passive scalar transport in fluids, transport in the presence of convection cells [29] and capillary waves [30].

5 Application to semiconductor microstructures: Channeling and negative Hall resistance in antidot lattices

Semiconductor physics and technology have reached a stage, where the artificial creation of diverse microstructures can be realized. High-purity AlGaAs/GaAs heterojunctions permit a ballistic motion of charge carriers in two dimensions with elastic mean free paths of the order of 10 μm. In addition, lateral structures such as quantum dots, antidots, quantum wires, or lateral surface superlattices are imposed in search of novel electronic properties for future devices. E.g. the lateral surface superlattices (LSSL) serve to break the lateral free-particle behaviour of the electron and to produce minigaps in the band structure [31]. These lateral structures represent a two-dimensional potential for the charge carriers and thus are realizations of the model studied in Sect. 4. At present the spatial scales of the structures are still larger ($\geq$ a factor of 10 for LSSLs) than the Fermi wavelength. The particle dynamics can thus be described by classical approximations, where the chaotic behaviour shows up. On the other hand, as the spatial scales are reduced further, these systems will also become interesting objects and a testing field for studies in quantum chaos.

A particular type of LSSLs consists of periodic arrays of forbidden circular regions, so-called antidots. In a sense they are electronic realizations of a (non-ideal) Sinai billiard. In these systems, magnetrotransport measurements have reveiled a sequence of resistivity peaks at moderate magnetic fields [32] which can be explained by KAM-islands and cantorus hierarchies [33]. Furthermore, for small magnetic fields a surprising negative Hall resistance was measured, which as will be discussed can be related to the existence of Lévy walks in the form of chaotic channeling trajectories [34].

The classical approximation for the dynamics of an electron wavepacket in a two-dimensional potential $V(x,y)$ and a perpendicular magnetic field B is described by the Hamiltonian

$$H = (p - eA)^2/2m + V(x,y), \tag{37}$$

where A is the vector potential and m the effective mass of the electron [22]. To model antidot arrays we will use the potential (see Fig. 8a)

$$V(x,y) = V_0 \left[\cos(2\pi x)\cos(2\pi y)\right]^\beta, \tag{38}$$

where β controls the steepness of the antidots and the prefactor V_0 of the potential is chosen such that the ratio of the dot diameter at the Fermi energy to the distance of adjacent dots is one third, similar as in the experiments [32]. For certain magnetic field regimes we find trajectories enclosing 1, 2, 4, 9, 21, and even more antidots for a steep potential ($\beta = 64$) and orbits enclosing 1 and 4 antidots for a smooth potential ($\beta = 4$) as well as chaotic orbits, as shown in Fig. 8b.

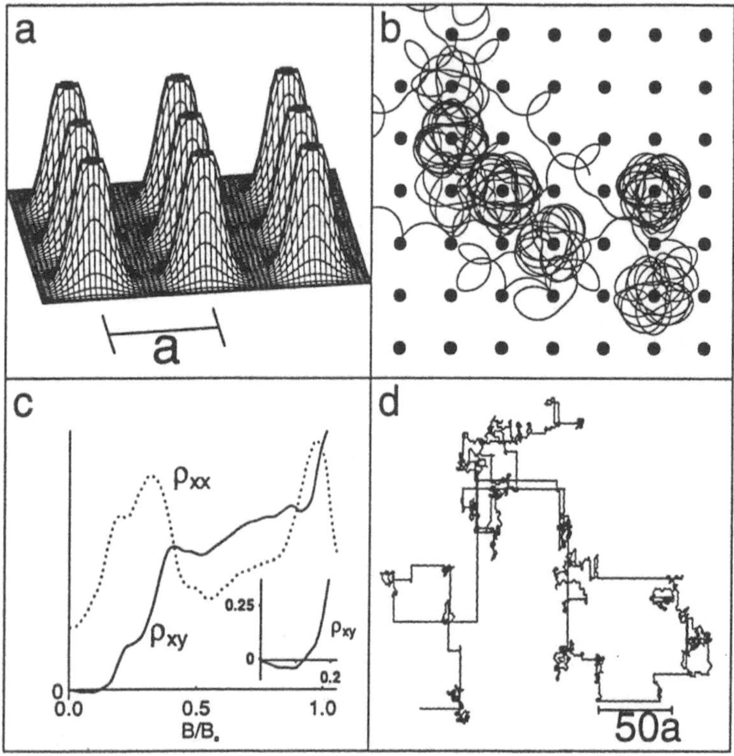

Fig. 8. (a) The antidot potential Eq. (38) for $\beta = 4$. (b) Chaotic electron trajectory that is trapped several times around an antidot. (c) Magnetoresistance ρ_{xx} (dotted line) showing peaks at magnetic fields where chaotic electrons are often trapped around 1 and 4 antidots and Hall resistivity ρ_{xy} (solid line) showing a negative value for small magnetic fields (see inset). (d) Chaotic electron trajectory for small magnetic fields ($B/B_0 = 0.1$) where the negative Hall effect occurs (note the much larger scale than in (b)).

With this model and using linear response theory we were able to closely reproduce the magnetoresistance peaks [33] (dotted line in Fig. 8c) as measured in the experiments by Weiss et al. [32]. The peaks are caused by chaotic trajectories like the one in Fig. 8b that are trapped around 1 or 4 antidots for long times due to cantorus hierarchies around what we call cyclotron tori.

A surprising feature is the negative Hall-resistance measured [32] and reproduced theoretically [34] for small magnetic fields ($B/B_0 < 0.12$) where the free cyclotron radius would be larger than 5 unit cells. This can be seen in the inset of Fig. 8c. In this magnetic field regime chaotic electron trajectories tend to stay within the channels between the antidots along the axes for 3 to 1000 or even more unit cells. They perform a Lévy walk in a magnetic field, which we predicted and investigated for a similar potential previously [22]. It shows enhanced anomalous diffusion and is caused again by trapping in a hierarchy of cantori around cylindrical KAM-surfaces in phase space as described in Sect. 4. The Hall conductivity is determined by the correlation function $\langle v_x(t)v_y(0)\rangle$ which exhibits a negative long-time tail (Fig. 9) reminiscent of Eq. (33). Its negative value is related to the way they eventually leave the channel (Fig. 9, inset). Due to the magnetic field the channeling trajectories tend to move on one side of the channel and thus they have an increased probability to hit one of the antidots on this side such that they leave the channel to the opposite side and thereby turn against the Lorentz force. The combination of the nonlinear dynamics that causes chaotic orbits to remain in the channel and the above mentioned geometrical effect when leaving the channel cause the correlation function and the Hall resistivity (Fig. 8c) to be negative. This phenomenon thus is a direct manifestation of a Lévy walk in the system.

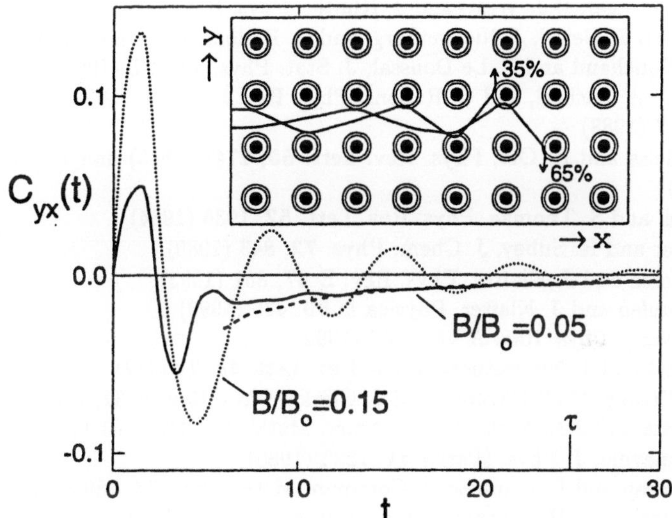

Fig. 9. The velocity correlation function $C_{xy}(t)$ for $B/B_0 = 0.05$ (solid line) exhibits a large negative peak and a negative algebraic tail. Both are caused by channeled chaotic trajectories (inset), which tend to leave the channel in a direction opposite to the Lorentz force $ev \times \mathbf{B}$ and thus give a negative contribution to the correlation. Above $B/B_0 = 0.15$ other chaotic trajectories dominate, which revolve around 1 or 4 antidots and thereby produce oscillations in $C_{xy}(t)$ (dotted line for $B/B_0 = 0.15$).

It is worthwhile mentioning that an analog of enhanced anomalous diffusion can also be found in quantum mechanics if the spectrum is a fractal [35]. Cantor spectra show up e.g. in quantum mechanical descriptions of the above systems, i.e. for Bloch electrons in magnetic fields [36]. Here the width of wavepackets grows diffusively. More generally the variance $\sigma^2(t)$ can be related to the Hausdorff dimension D_0 of the spectrum, $\sigma^2(t) \sim t^{2D_0}$ [35]. Rigorous arguments yield $2D_0$ as an upper bound for the exponent. Continuously varying exponents $2D_0$ between 1 and 2 can be found e.g. in Fibonacci chains. The decay of correlations also exhibits long-time tails. If the generalized dimension D_2 of the initial spectral measure is $0 < D_2 < 1$, it follows that the correlation function decays as a power law t^{-D_2} irrespective of a singular continuous or an absolutely continuous spectrum [37].

References

1 T. Geisel, J. Nierwetberg, and A. Zacherl, Phys. Rev. Lett. **54**, 616 (1985)

2 M.F. Shlesinger and J. Klafter, Phys. Rev. Lett. **54**, 2551 (1985), comment on Ref. [1]

3 M.F. Shlesinger, B.J. West, and J. Klafter, Phys. Rev. Lett. **58**, 1100 (1987)

4 A. Zacherl, T. Geisel, J. Nierwetberg, and G. Radons, Phys. Lett. A **114**, 317 (1986); J. P. Bouchaud and P. Le Doussal, J. Stat. Phys. **41**, 225 (1985)

5 T. Geisel, A. Zacherl, and G. Radons, Phys. Rev. Lett. **59**, 2503 (1987) and Z. Phys. **71**, 117 (1988)

6 J. D. Meiss and E. Ott, Phys. Rev. Lett. **55**, 2741 (1985) and Physica D **20**, 387 (1986)

7 T. Geisel and S. Thomae, Phys. Rev. Lett. **52**, 1936 (1984)

8 J. Klafter and R. Silbey, J. Chem. Phys. **72**, 843 (1980)

9 G. Zumofen and J. Klafter, Phys. Rev. E **47**, 851 (1993)

10 G. Zumofen and J. Klafter, Physica D **69**, 436 (1993)

11 X.-J. Wang, Phys. Rev. A **45**, 8407 (1992)

12 T. Geisel and J. Nierwetberg, Phys. Rev. Lett. **48**, 7 (1982)

13 R.F. Miracky, M.H. Devoret, and J. Clarke, Phys. Rev. A **31**, 2509 (1985)

14 A. Lasota and J.A. Yorke, Trans. Amer. Math. Soc. **186**, 481 (1973)

15 P. Manneville, J. Phys. (Paris) **41**, 1235 (1980); Y. Pomeau and P. Manneville, Commun. Math. Phys. **74**, 189 (1980)

16 I. Procaccia and H. Schuster, Phys. Rev. A **28**, 1210 (1983)

17 e.g. T. Geisel, in *Physics of Superionic Conductors*, ed. M.B. Salamon, Topics in Current Physics, Vol. 15, p. 201. Berlin, Springer 1979

18 see e.g. R.S. MacKay, Lect. Notes in Phys. **247**, 390 (1986)

19 I. C. Percival, AIP Conference Proceedings **57**, 1179 (1980); Aubry, S., in *Solitons and Condensed Matter Physics*, eds. A.R. Bishop and T. Schneider, p. 264. Berlin, Springer 1978

20 R. S. MacKay, J. D. Meiss, and I. C. Percival, Phys. Rev. Lett. **52**, 697 (1984) and Physica **13** D, 55 (1984); D. Bensimon and L.P. Kadanoff, Physica D **13**, 82 (1984)

21 L. P. Kadanoff, Phys. Rev. Lett. **47**, 1641 (1981); S. J. Shenker and L.P. Kadanoff, J. Stat. Phys. **27**, 631 (1982); R. S. MacKay, Physica **7** D, 283 (1983); J. M. Greene, R. S. MacKay, and J. Stark, Physica D **21**, 267 (1986)

22 T. Geisel, J. Wagenhuber, P. Niebauer, and G. Obermair, Phys. Rev. Lett. **64**, 1581 (1990); and Phys. Rev. B. **45**, 4372 (1992)

23 G. Petschel and T. Geisel, Phys. Rev. A **44**, 7959, (1991)

24 G.M. Zaslavsky, R.Z. Sagdeev, D.K. Chaikovsky, and A.A. Chernikov, Sov. Phys. JETP **68**, 995 (1989); and D.K. Chaikovsky and G.M. Zaslavsky, Chaos **1**, 463 (1991)

25 R.A. Pasmanter, Fluid Dyn. Res. **3**, 320 (1988)

26 A.A. Chernikov, B.A. Petrovichev, A.V. Rogalsky, R.Z. Sagdeev, and G.M. Zaslavsky, Phys. Lett. A **144**, 127, (1990), and B.A. Petrovichev, A.V. Rogalsky, R.Z. Sagdeev, and G.M. Zaslavsky, Phys. Lett. A **150**, 391 (1990)

27 T.M. Antonsen and E. Ott, Phys. Rev. A **44**, 851 (1991)

28 T.H. Solomon, E.R. Weeks, and H.L Swinney, Phys. Rev. Lett. **71**, 3975 (1993)

29 P. Tabeling and O. Cardoso, Bull. Soc. Franc. Phys. **73**, 13 (1989)

30 R. Ramshankar, D. Berlin, and J.P. Gollub, Physics Fluids A2, 1955 (1990)

31 H. Sakaki, K. Wagatsuma, J. Hamasaki, and S. Saito, Thin Solid Films **36**, 497, (1976) R. T. Bate, Bull. Am. Phys. Soc. **22**, 407 (1977); A.C. Warren, D.A. Antoniadis, H.I. Smith, and J. Melgnailis, IEEE Electron. Device Lett. **6**, 294 (1985)

32 D. Weiss, M.L. Roukes, A. Menschig, P. Grambow, K.v. Klitzing, and G. Weimann, Phys. Rev. Lett. **66**, 2790 (1991); and R. Schuster, K. Ensslin, J.P. Kotthaus, M. Holland, and S.P. Beaumont, Superlattices and Microstructures **12**, 93 (1992)

33 R. Fleischmann, T. Geisel, and R. Ketzmerick, Phys. Rev. Lett. **68**, 1367 (1992)

34 R. Fleischmann, T. Geisel, and R. Ketzmerick, Europhys. Lett. **25**, 219 (1994)

35 T. Geisel, R. Ketzmerick, and G. Petschel, Phys. Rev. Lett. **66**, 1651, (1991)

36 G. Petschel and T. Geisel, Phys. Rev. Lett. **71**, 239 (1993)

37 R. Ketzmerick, G. Petschel, and T. Geisel, Phys. Rev. Lett. **69**, 695 (1992)

Transport and large scale stochasticity for a nonperiodic generalisation of the standard map

Sadruddin Benkadda, Brigitte R. Ragot and Yves Elskens

Equipe Turbulence Plasma de l'URA 773 CNRS–Université de Provence,
Institut méditerranéen de technologie, Château-Gombert,
F-13451 Marseille cedex 20, France

Abstract. In the standard map, accelerator modes generate ballistic transport, which may dominate over anomalous transport. We investigate Cohen's symplectic map of the cylinder, which has generally no accelerator modes.

1 Introduction

Our traditional model of area-preserving maps is the Chirikov-Taylor standard map

$$I_{n+1} = I_n + K \sin \theta_n$$
$$\theta_{n+1} = \theta_n + \lambda I_{n+1} \tag{1.1}$$

which is doubly periodic :
- with respect to θ : if $(I_n{}', \theta_n{}') = (I_n, \theta_n + 2\pi)$ then $(I_{n+1}{}', \theta_{n+1}{}') = (I_{n+1}, \theta_{n+1}+2\pi)$; thus the map (1.1) on the plane $\mathbb{R}^2$ defines a map on the cylinder $\mathbb{R}\mathrm{x}\mathbb{T}$ (with $\mathbb{T}=\mathbb{R}/2\pi$) ;
- with respect to I : if $(I_n{}', \theta_n{}') = (I_n +2\pi/\lambda, \theta_n)$ then $(I_{n+1}{}', \theta_{n+1}{}') = (I_{n+1}+2\pi/\lambda, \theta_{n+1}+2\pi)$; thus (1.1) also projects to a map for $(\lambda I, \theta)$ on the torus $\mathbb{T}^2$.

Parameter λ is usually eliminated by rescaling I : the dynamics of (1.1) depends only on $K' = K\lambda$.

The fact that a shift by $2\pi/\lambda$ in action I_n implies a change in the angle θ_{n+1} by 2π permits the existence of so-called accelerator modes : fixed points of the torus map may correspond to orbits of (1.1) in the plane for which I_n diverges like n and θ_n like n^2. For some intervals of values of parameter K', (1.1) admits stable (elliptic) and unstable (hyperbolic) accelerator modes. For such orbits, the limit $\lim_{n\to\infty} <(I_n -I_0)^2>/n$ does not exist ; thus they prevent us from defining a "diffusion coefficient" for such values of K'.

Apart from this, we note that, while periodicity with respect to one coordinate (the "angle") is reasonable for hamiltonian dynamics, periodicity with respect to both conjugate variables I and θ is unusual. This invites us to study a simply-periodic map of the plane. Our model was introduced by Cohen [5] to describe the motion of a charged particle in asymmetric magnetic traps ; this map $(I_n, \theta_n) \to (I_{n+1}, \theta_{n+1})$ combines two steps like (1.1) :

$$J_n = I_n + K \sin \theta_n$$
$$\psi_n = \theta_n + \lambda_1 J_n$$
$$I_{n+1} = J_n + K \sin \psi_n \qquad (1.2)$$
$$\theta_{n+1} = \psi_n + \lambda_2 I_{n+1}$$

with parameters $K \geq 0$, $\lambda_2 \geq \lambda_1 \geq 0$ (one can eliminate one of the parameters by rescaling I). The mapping (1.2) reduces to a standard map (1.1) with parameter K' in two cases :

- $\lambda_1 = 0$: then $K' = 2\lambda_2 K$;
- $\lambda_1 = \lambda_2$: then $K' = \lambda_1 K$ (and (1.2) is the iterate of (1.1)).

  The phase space structure of (1.2) depends on the ratio

$$\varphi = \frac{\lambda_1}{\lambda_1 + \lambda_2} \qquad (1.3)$$

- For commensurate (λ_1, λ_2), $\varphi = p/q \in Q$ and Cohen's map (1.2) induces a map on the torus with period $2\pi q/(\lambda_1 + \lambda_2)$ for the action : in this case, the map (1.2) admits accelerator modes for appropriate values of K.
- For incommensurate (λ_1, λ_2), $\varphi \notin Q$ and Cohen's map (1.2) induces only a map of the cylinder, with action $I \in \mathbb{R}$; there are no accelerator modes.

We compare two basic aspects of the dynamics of Cohen's map (1.2) with the corresponding aspects of the standard map (1.1) :
- the large scale stochasticity threshold K_c, i.e. the value of K (for given λ_1, λ_2) at which the last (KAM) rotational invariant circle on the cylinder is destroyed [9] ;
- one basic transport property [10] : the usual "diffusion coefficient" $<(I_n - I_0)^2>/2n$.

Other non-periodic maps of the cylinder were also investigated in various contexts [4]. Cohen's map has the advantage of being simple and quasi-periodic under action translations.

2 Large scale stochasticity threshold

To estimate the threshold K_c for the breakdown of the last invariant rotational circle of map (1.2), it is convenient to consider this map as a Poincaré return map for the periodic time-dependent formal hamiltonian

$$H(I, \theta, \tau) = \frac{I^2}{2} + K \left(\delta_{\lambda_1 + \lambda_2}(\tau) + \delta_{\lambda_1 + \lambda_2}(\tau - \lambda_1) \right) \cos \theta \qquad (2.1)$$

where $\delta_\alpha(\tau)$ is the α-periodic Dirac comb distribution. With the rescaling

$$t = \tau \frac{2\pi}{\lambda_1 + \lambda_2}$$

$$J = I \frac{\lambda_1 + \lambda_2}{2\pi} \tag{2.2}$$

it is rewritten to describe the motion of a charged particle in the field of an infinity of waves with integer velocities :

$$H'(J,\theta,t) = \frac{J^2}{2} + M \sum_{m \in \mathbf{Z}} A_m \cos(\theta - mt - \alpha_m) \tag{2.3}$$

where

$$M = K \frac{\lambda_1 + \lambda_2}{4\pi^2} \tag{2.4}$$

$$A_m \, e^{i\alpha_m} = 1 + e^{-2\pi im\varphi}$$

The form (2.3) is suited to the determination of the large scale stochasticity threshold using the resonance overlap picture [3,6,7,9]. Note that, if $\varphi = p/q \in \mathbf{Q}$, the waves' parameters (A_m, α_m) are q-periodic and phase space is periodic with respect to the action : then a countable infinity of invariant circles break down at the same value K_c.

As the half width of the resonance associated with wave (A_m, α_m) is $2\sqrt{A_m M}$, the resonance overlap parameter for resonances m and r is

$$s_{m,r} = 2 \frac{\sqrt{A_m} + \sqrt{A_r}}{|m-r|} \sqrt{M} \tag{2.5}$$

Hence a first estimate for K_c would follow from the criterion

$$\inf_m s_{m,m+1} = s_{crit} \tag{2.6}$$

with $s_{crit}=1$ for the primary resonance overlap estimate [3].

As the waves (A_m, α_m) do not all have the same amplitude, some will lead to resonance overlap for smaller K than others. Thus some ranges of action values in phase space will exhibit "large" scale chaos while others (with $I \approx m$ with small A_m) form "gaps" between them [5]. This is seen on Fig. 1 for K=0.5, $\lambda_1 = 1$, $\lambda_2 = \pi$: while the second step of (1.2) is a standard map with $K' = \lambda_2 K = \pi/2$ (which exhibits large scale chaos), Cohen's map still exhibits invariant (KAM) rotational circles, separating large ($\Delta I \approx 5 > 2\pi/\lambda_2$) chaotic domains.

In particular, if $\varphi = p/q \in \mathbf{Q}$, with q=2r, the amplitude A_r vanishes ; in this case, the resonance overlap criterion must involve $s_{r-1,r+1} = 4\sqrt{M}\sqrt{A_{r-1}} = 4\sqrt{M}\sqrt{A_{r+1}} = 4\sqrt{M}\sqrt{2}\sin(\pi\varphi)$. Similarly, for irrational φ, A_m may be arbitrarily small and one has $\inf_m s_{m,m+1} = 4\sqrt{M}\sqrt{2}\sin(\pi\varphi)$. For $\varphi = p/q$ with q=2r+1, no amplitude A_m vanishes, and $\inf_m s_{m,m+1} \geq 4\sqrt{M}\sqrt{2}\sin(\pi\varphi)$.

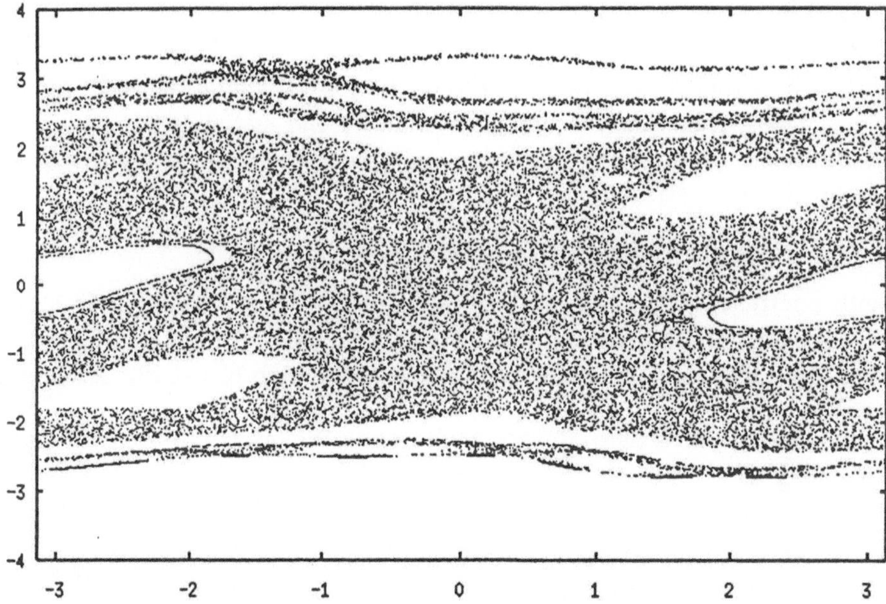

Fig. 1. Phase portrait of Cohen's map for K=0.5, $\lambda_1 = 1$, $\lambda_2 = \pi$.

The estimate of K_c provided by (2.6) with $s_{crit}=1$ amounts to approximating dynamics induced by (2.3) by that of nearest-neighbour resonances only. This picture is satisfactory for the standard map dynamics, where all primary resonances have the same amplitude. But it is not appropriate for Cohen's map in regions where the primary resonances have small amplitudes and where "further" resonances significantly affect the dynamics too. For instance, for $\lambda_1=1$ and $\lambda_2=72$, Cohen finds numerically $K_c{}^{num} \approx 0.1$ while the primary criterion (2.6) yields $K_c{}^{(1)} \approx 0.4$. To explain this discrepancy for $\varphi \ll 1$, he proposes a heuristic approximation of (1.2) by a standard map with a renormalised (action-dependent) parameter, and he obtains a more satisfactory estimate $K_c{}^{est} \approx \lambda_1{}^{-1/2}(\lambda_1+\lambda_2)^{-1/2}$.

To estimate the large scale stochasticity threshold more accurately, the general procedure [6] is to change variables (by a Kolmogorov transformation) from (I,θ) to a couple (I',θ') for which all primary resonances are eliminated ; then one applies the criterion inf $s = s_{crit}$ to the "secondary" resonance overlap parameters of the hamiltonian for (I', θ').

As this procedure involves numerical evaluation of several constants, it has been tested [9] for specific values of (λ_1,λ_2) and specific choice of relevant resonances ; for $\lambda_1=1$, $\lambda_2=72$, and secondary resonances surrounding a noble torus, one finds $K_c{}^{(2)}=0.09$, in reasonable agreement with $K_c{}^{num}$.

Cohen's original argument [5] can be considered as a crude version of the renormalisation ; the present argument [9] brings it to a rigorous form in the appropriate framework.

3 Transport

For $K > K_c$, trajectories on the cylinder may be unbounded and one may characterize their large scale behaviour by the statistics of $(I_n - I_0)^2 / 2n$. With the random-phase approximation, one finds the so-called quasi-linear diffusion estimate

$$\frac{<(I_n - I_0)^2>}{2n} \approx D_{QL} \qquad (3.1)$$

with coefficient $D_{QL} = K^2 / 2$ if $\lambda_1 > 0$ ($D_{QL} = K^2$ if $\lambda_1 = 0$). In (3.1), one averages over initial data (I_0, θ_0) with Lebesgue measure[1]. However, (3.1) cannot be accurate : it is always violated if the map has stable accelerator modes.

For the standard map, Rechester, Rosenbluth and White [8,11,12] found an asymptotic expansion (independent of (I_0, θ_0)) of $(I_n - I_0)^2 / 2n$ as $K \to \infty$ and $n \to \infty$ by Fourier transforming the distributions of $(I_n - I_0, \theta_n - \theta_0)$:

$$\frac{D_{as}}{D_{QL}} = 1 + \frac{4}{\sqrt{2\pi K\lambda}} \cos\left(K\lambda - \frac{\pi}{4}\right) + \frac{4}{\pi K\lambda} \cos^2\left(K\lambda - \frac{\pi}{4}\right) + o(K^{-1}) \quad (3.2)$$

Figure 2 displays this well-known result. The divergences near integer values of $K/2\pi$ are due to accelerator modes ; their contributions are not powerlike functions of K^{-1} and cannot be obtained from (3.2). The validity of the derivation of (3.2) is discussed in [10].

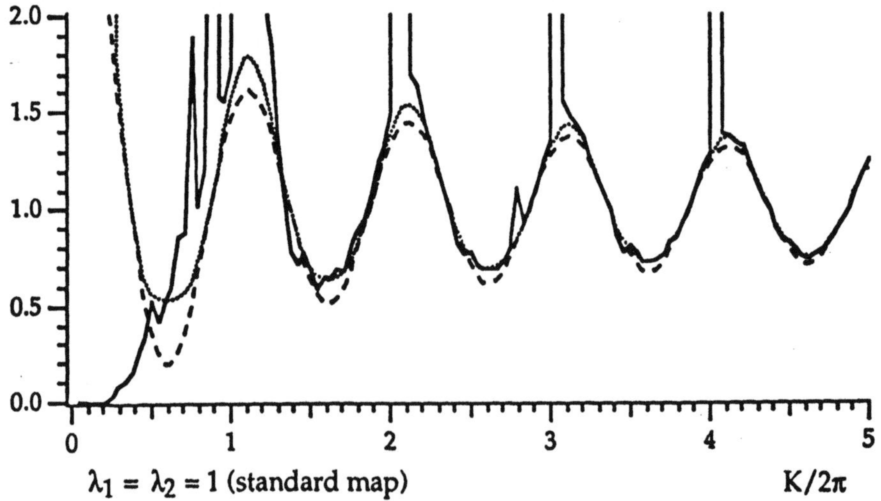

Fig. 2. Solid line : D_n / D_{QL} for 10^5 particles and 500 iterations. Dashed lines : asymptotic estimates (3.2) to order $K^{-1/2}$ (- - - - -) and to order K^{-1} (·····).

---

[1] The action average being performed over increasing intervals, as phase space need not be periodic with respect to I.

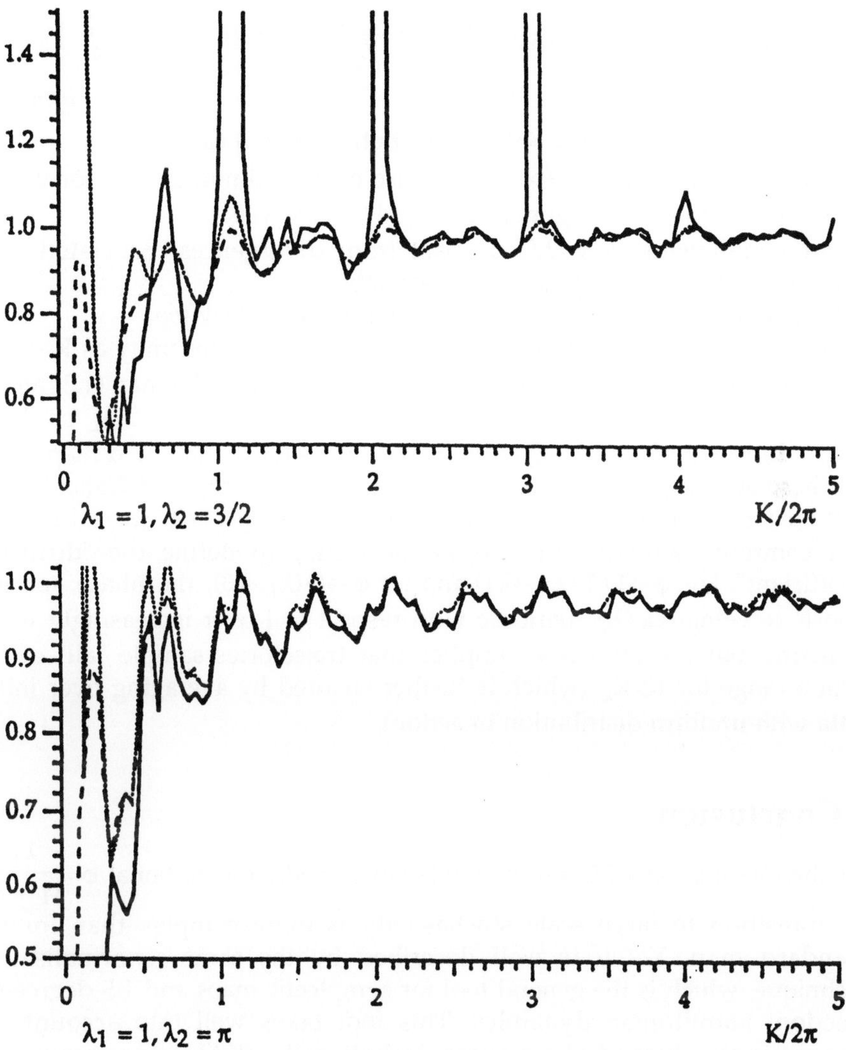

$$\lambda_1 = 1, \lambda_2 = 3/2 \qquad\qquad K/2\pi$$

$$\lambda_1 = 1, \lambda_2 = \pi \qquad\qquad K/2\pi$$

Fig. 3. Solid line : D_n/D_{QL} for 10^5 particles and 100 iterations. Dashed lines : asymptotic estimates (3.3) to order K^{-1} (- - - - -) and to order $K^{-3/2}$ ($\cdots\cdots$).

This procedure extends to Cohen's map [10] but it involves more complex Fourier space diagrams ; the dominant corrections to (3.1) yield :

$$\frac{D_{as}}{D_{QL}} = 1 - \frac{2}{\pi K \lambda_1} \cos^2\left(K\lambda_1 + \frac{\pi}{4}\right) - \frac{2}{\pi K \lambda_2} \cos^2\left(K\lambda_2 + \frac{\pi}{4}\right)$$

$$+ \frac{16}{(2\pi K)^{3/2}\sqrt{\lambda_1 \lambda_2}} \cos\left(K\lambda_1 - \frac{\pi}{4}\right) \cos\left(K\lambda_2 - \frac{\pi}{4}\right) \times$$

$$\left\{ \frac{1}{\sqrt{|\lambda_2 - \lambda_1|}} \cos\left(K|\lambda_2 - \lambda_1| - \frac{\pi}{4}\right) + \frac{1}{\sqrt{\lambda_2 + \lambda_1}} \cos\left(K(\lambda_2 + \lambda_1) - \frac{\pi}{4}\right) \right\}$$
$$+ o(K^{-3/2}) \tag{3.3}$$

for $\lambda_1 > 0$, $\lambda_2 > 0$, $\lambda_2 \neq \lambda_1$. The $K^{-1/2}$ contribution to (3.3) vanishes.

The ratio $D_n = \langle (I_n - I_0)^2 \rangle / 2n$ was also evaluated numerically for various values of λ_1, λ_2 and K. Figure 3 displays D_n / D_{QL} vs K :
• for commensurate $\varphi = 2/5$: apart from divergences associated with accelerator modes, the data agree reasonably ;
• for irrational $\varphi = 1/(\pi + 1)$: the agreement is satisfactory too.
Note that the ratio D_n / D_{QL} (like D_{as} / D_{QL}) approaches unity as $K \to \infty$, as one expects since K controls the twist and stretching of the map (1.2) and in analogy with continuous time models [1] ; note also that D_n / D_{QL} oscillates below 1 instead of oscillating around 1 as it does for the standard map.

These striking differences between the standard map and Cohen's map call for a careful discussion of the limits $\varphi \to 0$ and $\varphi \to 1/2$: these limits do not commute with the limit $n \to \infty$ necessary to define the "diffusion coefficient". For $\varphi \to 1/2$ ($\lambda_1 \to \lambda_2$) and for $\varphi \to 0$ ($\lambda_1 \to 0$), the phase space is "close" to being $(2\pi/\lambda_2)$-periodic with respect to I over increasingly wider domains, but the limit $n \to \infty$ implies that trajectories sample "all" of the action range for $K > K_c$ (which is further ensured by averaging over initial data with uniform distribution in action).

4 Conclusion

As the phase space of Cohen's map is not periodic for irrational $\varphi = \dfrac{\lambda_1}{\lambda_1 + \lambda_2}$, its transition to large scale stochasticity is more complex than for the standard map. Yet it is well described by the same renormalisation technique, which is the general tool for symplectic maps and 1.5 degree-of-freedom hamiltonian dynamics. This tool takes well into account the geometric structure of phase space, including the (KAM) gaps separating large chaotic regions in the strongly asymmetric case $\varphi \ll 1$.

The study of the "diffusion coefficient" for action in Cohen's map confirms the interest of Fourier space analysis in the limit $K \to \infty$, and even for reasonable $K > K_c$ [8,10,11,12]. It also shows that large scale behaviour of the orbits is unexpectedly sensitive to some aspects of the phase space geometry : why should D_n / D_{QL} oscillate below 1 rather than around 1 or even approach 1 from above as in some continuous time models [1,2] ?

Finally, the absence of ballistic modes in Cohen's map makes it a good test ground for anomalous transport, in particular in terms of Lévy flights.

References

1.   Didier BÉNISTI and Dominique F. ESCANDE, Scaling properties and diffusion behaviour of chaotic dynamics of a hamiltonian system with random phases (submitted for publication).
2. John R. CARY, Dominique F. ESCANDE and Alberto D. VERGA, Nonquasilinear diffusion far from the chaotic threshold, **Phys. Rev. Lett. 65** (1990) 3132-3135.
3.   Boris V. CHIRIKOV, A universal instability of many-dimensional oscillator systems, **Phys. Rep. 52** (1979) 263-379.
4.   Boris V. CHIRIKOV, Particle dynamics in magnetic traps, in *Reviews of plasma physics* **13**, B.B. Kadomtsev ed. (Consultants bureau, New York, 1987) 1-91.
5.   Robert H. COHEN, Stochastic motion of particles in mirror machines, in *Intrinsic stochasticity in plasmas* (Cargèse, 1979), G. Laval and D. Grésillon eds (Ed. de physique, Orsay, 1979), 182-191.
6.   Dominique F. ESCANDE, Mohamed Sadruddin MOHAMED-BENKADDA and Fabrice DOVEIL, Threshold of global stochasticity and universality in hamiltonian systems, **Phys. Letters 101A** (1984) 309-313.
7.   Dominique F. ESCANDE and Fabrice DOVEIL, Renormalisation method for computing the threshold of the large scale stochasticity instability in two degree of freedom hamiltonian systems, **J. Stat. Phys. 26** (1981) 257-284.
8.   Alan J. LICHTENBERG and Michael A. LIEBERMAN, *Regular and chaotic dynamics* (Springer, New York, 1992).
9.   Mohamed Sadruddin MOHAMED-BENKADDA, *Calcul de seuil de stochasticité et universalité pour des systèmes hamiltoniens : applications à la physique des plasmas*, thèse de 3ème cycle (univ. Paris sud, Orsay, 1983).
10. Brigitte R. RAGOT, *Transport dans les systèmes hamiltoniens - quelques aspects*, thèse de magistère (univ. C. Bernard–Lyon 1 et Ecole normale supérieure, Lyon, 1993).
11. A.B. RECHESTER, Marshall N. ROSENBLUTH and Roscoe B. WHITE, Fourier-space paths applied to the calculation of diffusion for the Chirikov-Taylor model, **Phys. Rev. A 23** (1981) 2664-2672.
12. A.B. RECHESTER and Roscoe B. WHITE, Calculation of the turbulent diffusion for the Chirikov-Taylor model, **Phys. Rev. Lett. 44** (1980) 1586-1589.

Blowout Bifurcations: Symmetry Breaking of Spatially Symmetric Chaotic States

Edward Ott[1], John C. Sommerer[2], Thomas M. Antonsen Jr.[1], and Shankar Venkataramani[1]

[1]University of Maryland, College Park, MD 20742, USA
[2]M. S. Eisenhower Research Center, Johns Hopkins Applied Physics Laboratory, Laurel, MD 20723 USA

Abstract. We consider a dynamical system that possesses a smooth manifold $\mathcal{M}$ on which the dynamics is chaotic. This general situation occurs in spatially symmetric extended systems that exhibit chaotic behavior of a spatially symmetric pattern. (Motion on $\mathcal{M}$ corresponds to spatial symmetry of the pattern.) Loss of stability transverse to $\mathcal{M}$ has been called a blowout bifurcation and corresponds to spatial symmetry breaking. It is shown that a blowout bifurcation is typically associated with either one of two novel types of dynamical behavior: (a) "on-off intermittency" (in which short bursts of motion far from $\mathcal{M}$ occur between epochs where the orbit is exceedingly close to $\mathcal{M}$), or (b) "riddling" of the basin of attraction of an attractor on $\mathcal{M}$ (in which all points in the basin of the attractor on $\mathcal{M}$ have arbitrarily nearby points in the basin of another attractor not on $\mathcal{M}$).

1 Introduction

We consider the basic problem of what happens when a D-dimensional dynamical system possesses a d-dimensional invariant manifold $\mathcal{M}$ $(d < D)$ on which the dynamics is chaotic. By invariant we mean that, if an initial condition is in the manifold, then the subsequent orbit remains in the manifold. The situation is shown schematically in Fig. 1. Because we take the dynamics to be chaotic in the invariant manifold $\mathcal{M}$, the dimension of $\mathcal{M}$ must be at least three if the dynamical system is a flow and at least two if the dynamical system is an invertible map. A typical initial condition on $\mathcal{M}$ is assumed to be attracted to a chaotic set $A \subset \mathcal{M}$ on which the orbit wanders ergodically; A is a chaotic attractor *for initial conditions on* $\mathcal{M}$. Whether A is an attractor for the dynamical system

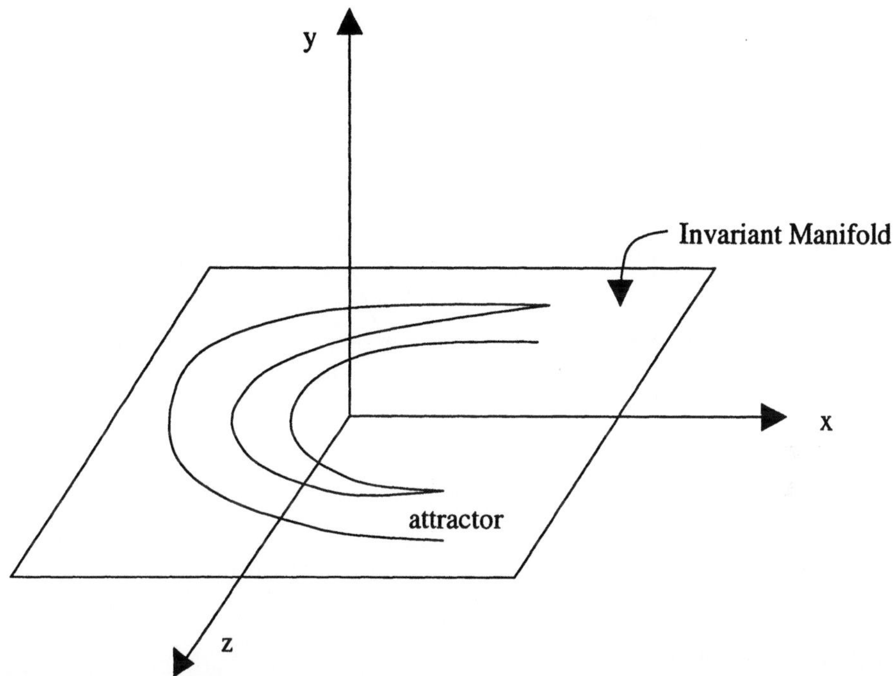

Fig. 1. Invariant manifold $\mathcal{M}$.

depends on whether initial conditions occupying a volume in the *full phase space* are attracted to $\mathcal{A}$. This, in turn, depends on whether perturbations of orbits on $\mathcal{A}$ that are transverse to $\mathcal{M}$ grow exponentially with time. Let $h_\perp$ be the largest transverse Lyapunov exponent for typical orbits on $\mathcal{A}$. Then $\mathcal{A}$ is (is not) an attractor of the dynamical system if $h_\perp < 0$ ($h_\perp > 0$). We shall be particularly interested in the situation where $h_\perp$ passes from negative to positive values as we vary some parameter p through a critical value denoted p_c. As shown in Fig. 2, we adopt the convection that $h_\perp < 0$ in $p < p_c$ and $h_\perp > 0$ in $p > p_c$. We shall find that, for p in the vicinity of p_c, the existence of an invariant manifold has very striking physical consequences. In particular, depending on the behavior of the particular system off the invariant manifold $\mathcal{M}$, either one of the following two phenomena[1-16] occurs,

(a) on-off intermittency[2,5-16]; or

(b) riddled basins of attraction[1-4].

If (a) applies, we find that, as p increases through p_c, although the attractor on $\mathcal{M}$ loses stability, a typical orbit still remains close to $\mathcal{M}$ for long stretches of time which are punctuated by short intermittent bursts in which the orbit moves far from $\mathcal{M}$. If (b) applies, there is an attractor off $\mathcal{M}$ for both $p > p_c$ and $p < p_c$, and as p decreased through p_c the

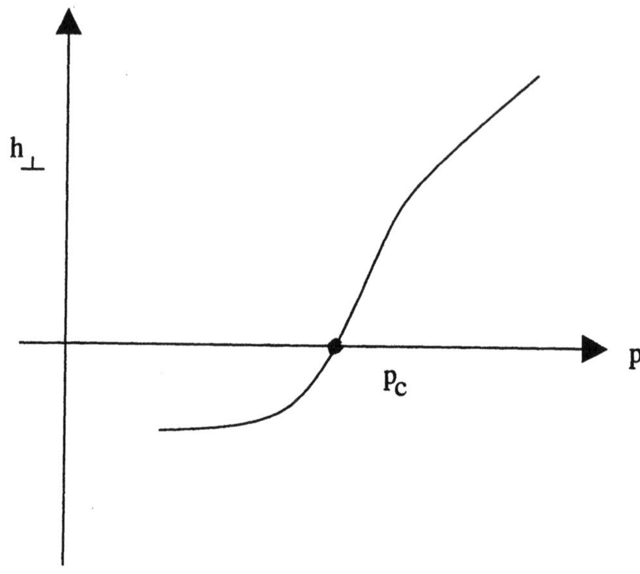

Fig. 2. $h_\perp$ versus p.

attractor in $\mathcal{M}$ is created along with its basin which is riddled. In either case, we call the transition that occurs as p passes through p_c a blowout bifurcation.[2]

We describe the unusual properties of on-off intermittency and riddled basins in Secs. 4 and 5. Section 2 is a brief discussion of physical cases in which the situation depicted in Fig. 1 may be expected to apply. Section 3 introduces a model system, used in Secs. 4 and 5, to illustrate the phenomena.

2 Physical Situations with an Invariant Manifold $\mathcal{M}$

Systems that possess chaotic dynamics in a smooth invariant manifold of lower dimension than that of the full phase space are very common. In particular, experiments with a spatial symmetry are prime candidates.

As an example, consider Rayleigh-Benard convection in a small cell, as shown in Fig. 3. The cell is assumed to be spatially symmetric about the midplane. We anticipate that for some conditions two *symmetrically* disposed rolls will form whose time dependence is chaotic. The rolls are instantaneously symmetric in that at each instant of time the velocity field is symmetric about the midplane of the apparatus.

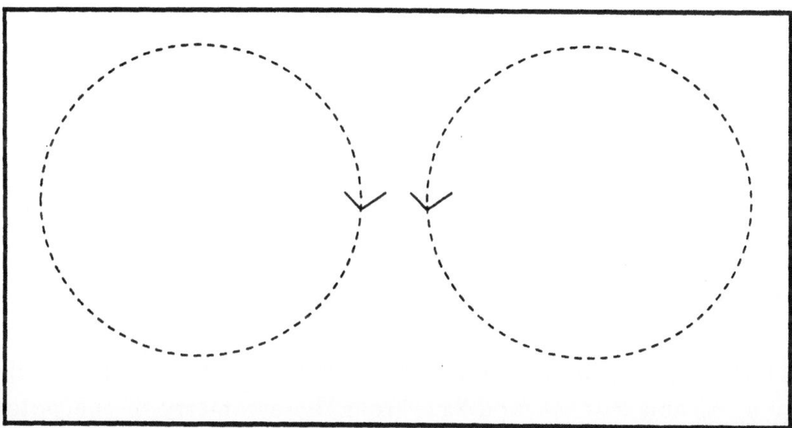

Fig. 3. Symmetric flow in a Rayleigh-Benard cell.

This symmetric chaotic motion represents motion on an invariant manifold in the full phase space, where the invariant manifold might be specified by setting the coefficients of a suitable Fourier expansion to zero if they represent spatially asymmetric motion. If, for example, the situation in Fig. 3 undergoes a blowout bifurcation to on-off intermittency with variation of a system parameter, then we should expect to see nearly spatially symmetric temporally chaotic behavior most of the time interrupted by short bursts of spatially asymmetric behavior. Although the above discussion highlights the role of spatial symmetry, we emphasize that there are also many situations where an invariant manifold $\mathcal{M}$ exists in systems without spatial symmetry.

3 Illustrative Example

In order to illustrate the occurrence of blowout bifurcations, we consider a specific example.[2] In particular, we consider the motion of a point particle of unit mass undergoing two-dimensional motion (x, y) subject to a potential

$$V(x, y) = (1 - x^2)^2 + y^2(x - p) + ky^4, \tag{1}$$

sinusoidal forcing in x, and a friction force linear in the velocity (with friction coefficient ν). The equations of motion are

$$\frac{dx}{dt} = v_x, \tag{2}$$

$$\frac{dv_x}{dt} = \nu v_x + 4x(1 - x^2) + y^2 + f_0 \sin(\omega t), \tag{3}$$

$$\frac{dy}{dt} = v_y, \tag{4}$$

$$\frac{dv_y}{dt} = -\nu v_y - 2y(x - p) - 4ky^3, \tag{5}$$

where f_0 and ω are the amplitude and frequency of the applied sinusoidal force.

The phase space for this problem is five-dimensional with coordinates x, v_x, y, v_y and $\theta = \omega t \bmod 2\pi$. From the symmetry of the potential, $V(x, y) = V(x, -y)$, we have that $y = v_y = 0$ is an invariant three-dimensional manifold in the five-dimensional phase space. (If $y = v_y = 0$ at $t = 0$, Eqs. (4) and (5) show that this remains so for all $t > 0$.) In the invariant manifold, points are specified by the coordinates x, v_x and θ, and the dynamics is governed by the equation

$$\frac{d^2x}{dt^2} + \nu\frac{dx}{dt} - 4x(1 - x^2) = f_0 \sin(\omega t). \tag{6}$$

In what follows we consider the parameters $\nu = 0.05$, $f_0 = 2.3$, and $\omega = 3.5$. For these parameters Eq. (6) has a chaotic attractor which is shown in the $\theta = 0$ surface of section in Fig. 4.

We now consider the dynamics of infinitesimal perturbations $(\delta y, \delta v_y)$ that are transverse to the invariant manifold. Taking a variation of Eqs. (4) and (5) and setting $y = 0$, we have

$$\frac{d\delta y}{dt} = \delta v_y, \tag{7}$$

$$\frac{d\delta v_y}{dt} = -\nu\delta v_y - 2\delta y(x - p), \tag{8}$$

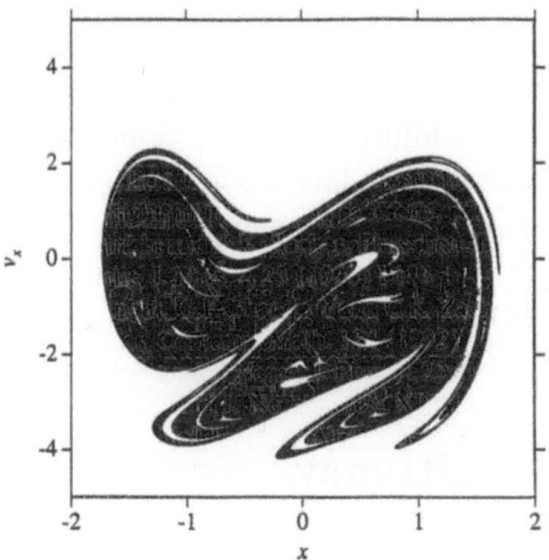

Fig. 4. Chaotic attractor for Eq. (6) in the $\theta = 0$ surface of section.

where $x(t)$ is a typical chaotic solution obtained by integration of Eq. (6). Integrating Eqs. (7) and (8), we calculate the largest Lyapunov exponent for perturbations out of the invariant manifold,

$$h_\perp = \lim_{t \to \infty} \frac{1}{t} \ell n[\delta(t)/\delta(0)], \tag{9}$$

where $\delta(t) = \{[\delta y(t)]^2 + [\delta v_y(t)]^2\}^{1/2}$. As illustrated schematically in Fig. 2, we find that $h_\perp$ increases with increasing p, passing from negative to positive values at the critical p value $-1.7887\ldots$. Thus we have a blowout bifurcation at $p = p_c = -1.7887\ldots$. Note that so far the term ky^4 in the potential (1) has been irrelevant to the discussion, and Fig. 4 applies independent of the value of k.

While irrelevant in determining the condition for the blowout bifurcation to occur, the strength k of the y^4 potential term is, nonetheless, of great importance. Because of its relatively rapid increase with increasing $|y|$, this term controls the dynamics away from the invariant plane. In particular, for $k > 0$ the ky^4 term confines the orbit, limiting its excursions in y. As k is increased the maximum size of these excursions is reduced.

What can we anticipate as the strength k of the y^4 term is increased from zero? For very large positive k one might expect that the strong confining effect of the ky^4 potential term eliminates the possibility of an attractor off the $y = v_y = 0$ manifold for $p < p_c$. Thus the bifurcation becomes nonhysteretic (because there is only one attractor), and we expect a transition from a hysteretic bifurcation (i.e., associated with a riddled basin) to a nonhysteretic bifurcation (i.e., associated with on-off intermittency) with increasing k. In fact we find numerically that the above picture is correct. Below we report numerical results for two values of k, namely $k = 0.0025$ (in which case basin riddling occurs) and $k = 0.0075$ (in which case there is on-off intermittency).

4 On-off Intermittency

Figure 5 shows a time series of $y(t)$ obtained from numerical solution of Eqs. (1)–(5) for $k = 0.0075$ and p slightly larger than p_c ($p_c = -1.175$). As p is reduced towards p_c, the frequency of bursts approaches zero. We are interested in the generic properties of this kind of bursting for $p - p_c$ small. We discuss two aspects of this.

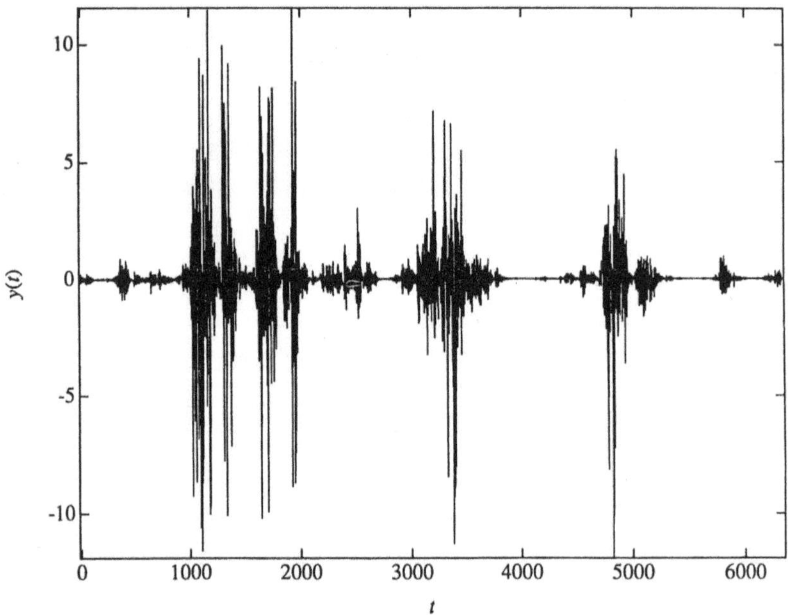

Fig. 5. y versus t.

4.1 Fractality

We have shown (Ref. 16) that the level set $y = const.$ approach a Cantor set of dimension 1/2 as $p \to p_c$ from above. In particular,
say we draw a horizontal line in Fig. 5 at $y = 1$ and determine the scaled t coordinate $\tau = h_\perp^2 t$ at the intersections of $y(t)$ with the horizontal line. We then examine these values in some interval of duration of order one (e.g., $0 \leq \tau \leq 1$). Now divide this interval into equal subintervals of length ϵ, and count the number $N(\epsilon)$ of these ϵ-intervals that contain at least one point of the level set. We have shown theoretically[16] that

$$N(\epsilon) \sim \epsilon^{-1/2}, \tag{10}$$

for ϵ in a scaling range $1 \overset{\sim}{>} \epsilon \overset{\sim}{>} h_\perp^2$. Thus in this range the level set behaves like a fractal of dimension 1/2. Figure 6 shows a numerically obtained plot verifying (10) for $y(t)$ resulting from Eqs. (1)–(5) at $k = 0.0075$ and $p = -1.788$.

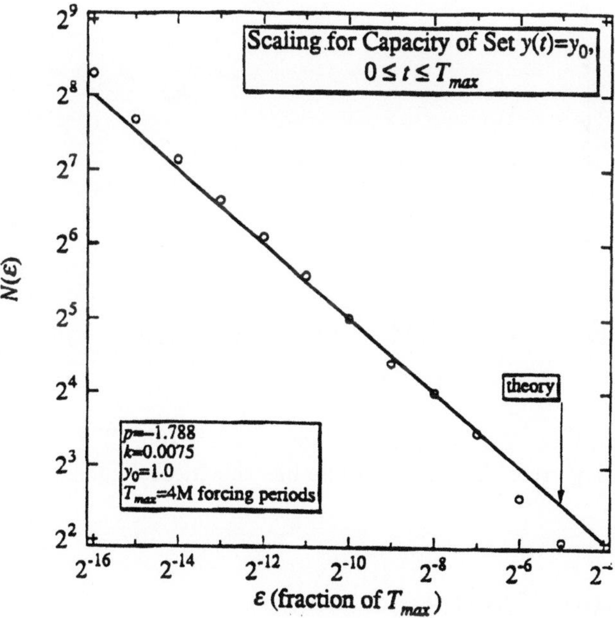

Fig. 6. Data (points) for $N(\epsilon)$ versus ϵ on log-log scale. The slope of the plotted solid line is $-1/2$.

4.2 Power Spectrum

Theory[16,10] predicts that $|y(t)|$ should have a frequency power spectrum $P(f)$ with a dependence

$$P(f) \simeq 1/f^{1/2}, \tag{11}$$

in the range

$$0(h_\perp^2) \mathrel{\tilde{>}} f \mathrel{\tilde{>}} 0(h_\perp^0).$$

Figure 7 shows a numerically obtained plot verifying (11) for the system Eqs. (1)–(5).

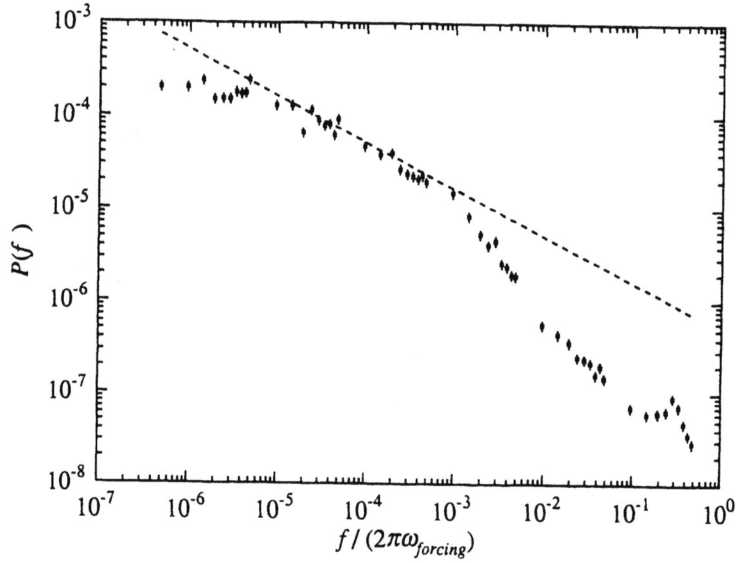

Fig. 7. Log-log plot of $P(f)$ versus f for $p = -1.780$. The data are plotted showing sampling error. The dashed line has the theoretical slope of $-1/2$.

5 Riddled Basins

If p is slightly smaller than p_c, then $h_\perp < 0$. In this case, the invariant manifold attracts a nonzero fraction of the points close to it. Let $\mathcal{W}$ be the set of all initial conditions that eventually end up on $\mathcal{M}$. Then, all the initial conditions in $\mathcal{W}$ (except possibly for those in a set of Lebesgue measure zero) will eventually end up in $\mathcal{A}$, which is the attractor embedded in $\mathcal{M}$. If we define an attractor to be a set that attracts all initial conditions in a set of positive Lebesgue measure, $\mathcal{A}$ is an attractor and $\mathcal{W}$ is its basin of attraction. This definition of an attractor differs from another, often used, definition which requires that all points in an open neighbourhood of the attractor end up on it. As we shall see, $\mathcal{A}$ does not satisfy this definition of an attractor.

The system Eqs. (1) - (5) with $p = -1.9$ and $k = 0.0025$, has a pair of periodic orbits off the invariant manifold. These orbits are symmetrically disposed with respect to the invariant manifold and are attracting. All initial conditions (except possibly for those in a set of Lebesgue measure zero) that are not in the basin of $\mathcal{A}$ are attracted to one of these periodic orbits. Let the set of all these initial conditions be $\mathcal{B}$. Then, $\mathcal{B}$ is a basin of attraction for the periodic orbits. There exist open neighbourhoods of these periodic orbits such that all initial conditions in these neighbourhoods are attracted to the periodic orbits.

5.1 Pictures of the Basin

Figure 8a shows numerical representations of the two dimensional $v_y = v_x = \theta = 0$ sections of $\mathcal{W}$, the basin of attraction of $\mathcal{A}$, and of $\mathcal{B}$, the basin of attraction of the periodic orbits. The figure is generated by numerically following the orbits of all initial conditions on a 1024 × 1024 grid. If an initial condition leads to an orbit that is attracted to one of the periodic orbits, a black dot is plotted at the location of that initial condition. Hence, to within the 1024 × 1024 resolution employed, the black region is a section of the basin $\mathcal{B}$. Similarly, the white region represents points on the grid that are in $\mathcal{W}$ *i.e*, those that end up in $\mathcal{A}$. The line $y_0 = 0$ corresponds to the invariant manifold.

We claim that the basin $\mathcal{W}$ is *riddled* in the following sense:

Let q be any point in $\mathcal{W}$ and let $S(q, \epsilon)$ be a ball of radius ϵ about q. Then, for any ϵ, however small, $S(q, \epsilon) \cap \mathcal{B}$ is nonempty; *i.e*, for every point in the basin of $\mathcal{A}$ (the white region $\mathcal{W}$), there exist points in the basin of the periodic orbits (the black region) arbitrarily close it. This implies that the interior of $\mathcal{W}$ is empty. However, $\mathcal{W}$ has a positive Lebesgue measure and so it is a *fat fractal*.[2-4]

Figures 8(b)-8(d) show numerically obtained magnifications with resolutions of 1024 × 1024, of small areas in Fig. 8(a) (see figure caption).

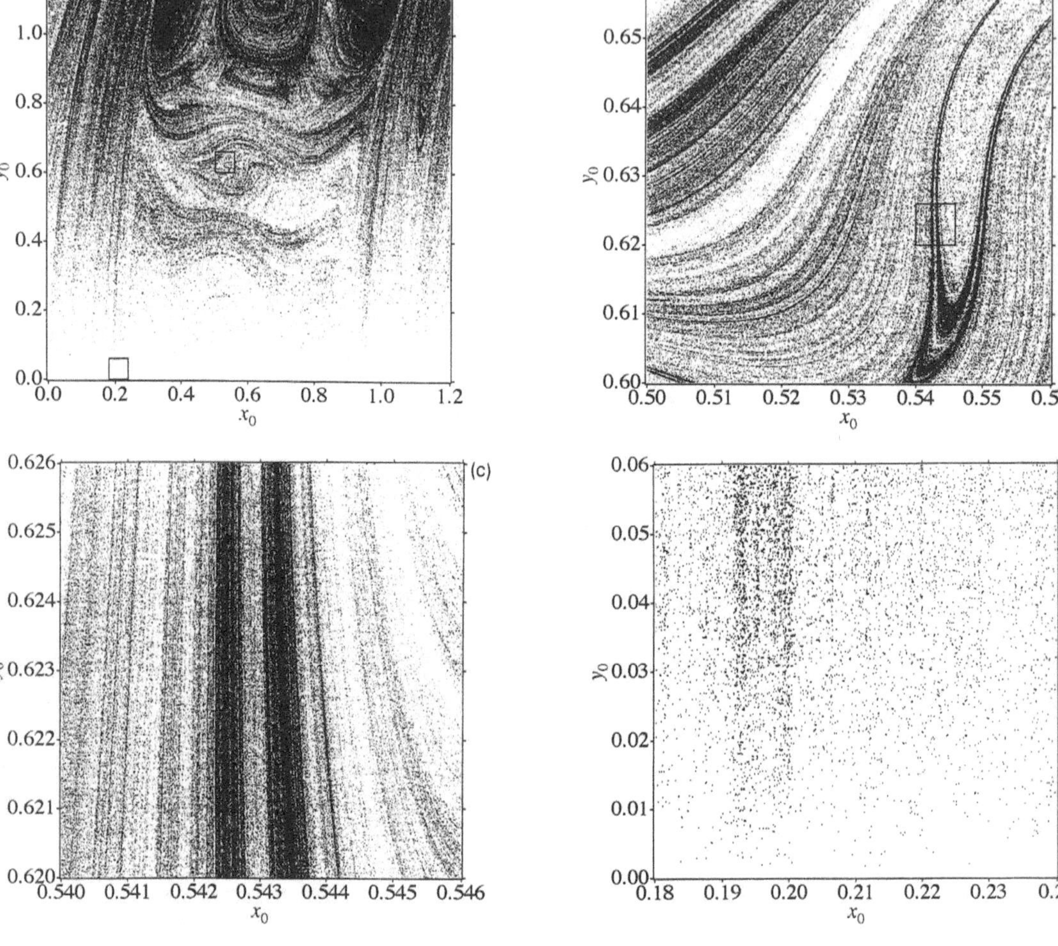

Fig. 8: $v_x = v_y = \theta = 0$ sections of the basins of attraction. Black dots correspond to the basin $\mathcal{B}$ and white regions to basin $\mathcal{W}$. (a) 1024×1024 grid. (b) Expanded view of top rectangle outlined in (a). (c) Expanded view of rectangle in (b). (d) Expanded view of bottom rectangle in (a).

We emphasize that, although Fig. 8a appears to show areas that are solid white, the implication of the above statements is that if any small apparently white region were examined under a higher resolution, black dots would always appear in that region (Figures 8b and 8c). Furthermore, for any $\epsilon > 0$ and any $q \in \mathcal{W}$, the points in $S(q, \epsilon)$ that are attracted to the periodic orbits occupy a positive fraction of the volume of $S(q, \epsilon)$. Therefore, for every initial condition in $\mathcal{W}$ there is a positive probability that an arbitrarily small random change in the initial condition will change the attractor to which the orbit goes. This has very important consequences for the reproducibility of experiments. Since the Lebesgue measure of $\mathcal{W}$ is nonzero, we have a positive probability of ending up in $\mathcal{A}$ by starting with an arbitrary inital condition *i.e*, orbits that end up in $\mathcal{A}$ are not atypical. However, unless we reproduce the initial condition exactly, we will not neccessarily end up in $\mathcal{A}$, since even the smallest perturbation of the initial conditon can take it into $\mathcal{B}$ with a positive probability.

$S(q,\epsilon)\cap\mathcal{B}$ is nonempty for any $\epsilon > 0$ even for q in $\mathcal{A}$. Thus, although $\mathcal{A}$ attracts all points from a set of positive Lebesgue measure, it does not satisfy the attractor definition which requires that all points in a neighbourhood of $\mathcal{A}$ be attracted to $\mathcal{A}$. Figure 8d shows that regions near the invariant manifold, which are apparently white in figure 8a, reveal points in $\mathcal{B}$ (black dots) when viewed under a higher resolution.

5.2 Scaling of the Measure of the points in $\mathcal{B}$

For a given y_0, let $P_*(y_0)$ be the Lebesgue measure in x of the set of initial conditions of the form $(x, y_0), 0 \le x \le 1.2$, that are in $\mathcal{B}$. This measure is evaluated numerically by drawing a horizontal line $y = y_0$ through the $v_y = v_x = \theta = 0$ section of the basins (Fig. 8),placing many evenly spaced initial x values on this line, and determining the fraction of those initial conditions in $\mathcal{B}$. Theory[4] predicts that

$$P_*(y_0) \sim y_0^{\eta} \tag{12}$$

and also yields a value for the exponent η. Figure 9 shows a numerically

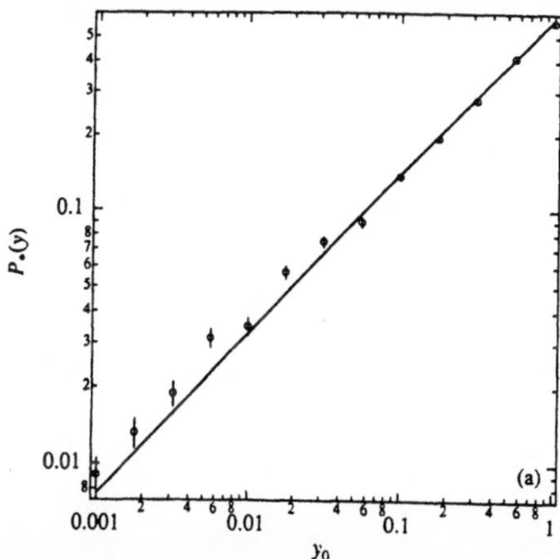

Fig 9. Log-log plot of $P_*(y_0)$ versus y_0. $p = -1.85$ and $k = 0.0025$.

obtained data for $P_*(y_0)$. The theoretical prediction is plotted as a solid straight line. The data and the theory agree reasonably well.

Acknowledgment

The work of E.O., T.M.A. and S.V. was supported by the Office of Naval Research. The work of J.C.S. was supported by the Navy Space and Naval Warfare Systems command.

References

[1] J. C. Alexander, J. A. Yorke, Z. You and I. Kan, Int. J. Bif. Chaos 2 (1992) 795.

[2] E. Ott and J. C. Sommerer, Phys. Lett. A 188 (1994) 39.

[3] J. C. Sommerer and E. Ott, Nature 365 (1993) 136.

[4] E. Ott, J. C. Sommerer, J. C. Alexander, I. Kan and J. A. Yorke, Phys. Rev. Lett. 71 (1993) 4134; Physica D (1994), to be published.

[5] A. S. Pikovsky, Z. Phys. B 55 (1984) 149.

[6] A. S. Pikovsky and P. Grassberger, J. Phys. A 24 (1991) 4587.

[7] A. S. Pikovsky, Phys. Lett. A 165 (1992) 33.

[8] H. Fujisaka and T. Yamada, Prog. Theor. Phys. 74 (1985) 919.

[9] H. Fujisaka and T. Yamada, Prog. Theor. Phys. 75 (1986) 1087.

[10] H. Fujisaka, H. Ishii, M. Inoue and T. Yamada, Prog. Theor. Phys. 76 (1986) 1198.

[11] L. Yu, E. Ott and Q. Chen, Phys. Rev. Lett. 65 (1990) 2935.

[12] L. Yu, E. Ott and Q. Chen, Physica D 53 (1992) 102.

[13] J. F. Heagy, N. Platt and S. M. Hammel, Phys. Rev. E 49 (1994) 1140.

[14] N. Platt, S. M. Hammel and J. F. Heagy, Phys. Rev. Lett. 72 (1994) 3498.

[15] N. Platt, E. A. Spiegel and C. Tresser, Phys. Rev. Lett. 70 (1993) 279.

[16] S. Venkataramani, et al. (to be published).

Lévy Description of Anomalous Diffusion in Dynamical Systems

J. Klafter[1], G. Zumofen[2], and M.F. Shlesinger[3]

[1] School of Chemistry, Tel-Aviv University, Tel-Aviv, 69978 Israel
[2] Physical Chemistry Laboratory, ETH-Zentrum, CH-8092 Zürich, Switzerland
[3] Physics Division, Office of Naval Research, Arlington, Virginia 22217-5660, USA

Abstract Anomalous diffusion properties, both enhanced and dispersive, are common to a broad spectrum of systems including dynamical systems. We review an approach based on Lévy scale-invariant distributions to describe transport in such systems. We introduce the basic ingredients that make the approach useful in describing anomalous behavior and demonstrate the applicability in the cases of the standard map, "egg-crate" potential and a novel one-dimensional iterated map which leads to a combined laminar-dispersive motion.

1. Introduction

One of the fascinating problems in the theory of chaotic systems has been the understanding of stochastic processes generated by deterministic systems and getting insight into the intricacies of the motion of a particle in such systems. In particular it has been observed that particles' orbits do not necessarily obey simple Gaussian behavior and may even exhibit "strange kinetics" [1-14], namely dispersive or enhanced diffusional motion. The study of chaotic systems has opened therefore new challenges for theories of stochastic processes, especially in the direction of generalizing Brownian motion.

Examples of anomalous diffusion in dynamical systems cover both dissipative and Hamiltonian systems. One finds diffusion anomalies in numerical studies of one-dimensional maps [5, 6, 11], of the Chirikov-Taylor standard map [1-4,14], of stochastic webs [7, 8] and in experiments on tracer diffusion in flow systems [15-17], where enhanced diffusion has been recently directly observed [17].

A new statistical description of anomalous diffusion has been recently proposed based on Lévy stable distributions. The latter generalize the central limit theorem and offer therefore the possibility to go beyond the Brownian description [10, 11, 18, 19]. This statistical description which we call the Lévy-walk approach introduces space-time coupling which, when combined with the Lévy distributions, leads to a useful description of anomalous transport and enables one to derive the corresponding propagator $P(r,t)$, the probability density to be at location r at time t. The Lévy-walk framework for enhanced diffusion has been implemented in a number of cases such as stochastic webs [7, 8], one-dimensional iterated maps [5, 11], hexagonal flows [20] and has also been introduced into calculations of turbulent flows [21, 22]. More recently Lévy walks have been used in analyzing diffusion at solid-liquid interfaces [23], predicting a regime dominated by enhanced diffusion. The well known Cauchy distribution

$$\psi(x) = \frac{1}{\pi} \frac{1}{1+x^2} \ ,$$ (1)

is an example of a probability distribution whose first moment is zero, but whose second and higher even moments are infinite. Furthermore, if N identically distributed random variables, each with a Cauchy distribution, are added, the sum $x_1 + x_2 + ... + x_N$ also has a Cauchy distribution, $P_N(x)$ with

$$P_N(x) = \frac{1}{\pi N} \frac{1}{1+(x/N)^2} \ ,$$ (2)

where $P_N(x)$ and $\psi(x)$ are related by a change of scale from x to x/N. This is the hallmark of fractals, where the whole $P_N(x)$ (the distribution of the sum) looks like $\psi(x)$ (the distribution of a single variable). In viewing the sum of random variables as adding random walk steps this scaling is equivalent to finding the distribution for the position of a random walker after one step and after N steps, i.e.

$$\psi(x) = N^\alpha P_N(x N^\beta) \ ,$$ (3)

with $\alpha = 1$ and $\beta = 1$. P. Lévy in the 1920s attacked more generally the problem of when a random walk distribution is independent (up to a scale factor) of the number of steps taken. For an unbiased one-dimensional random walk he found that

$$P_N(k) = \int\limits_{-\infty}^{\infty} P_N(x)e^{ikx}dx = \exp(-N|k|^\gamma) \ , \ 0 < \gamma \le 2 \ .$$ (4)

Here and in what follows the argument indicates in which space the function is meant to hold, e.g. $P_N(k)$ is the Fourier $(x \rightarrow k)$ transform of $P_N(x)$. Eq. (4) implies the scaling equation

$$\psi(x) = N^{1/\gamma} P_N(x N^{1/\gamma}) \ .$$ (5)

When $\gamma < 2$ the second moment of the probability diverges, so no proper characteristic scale exists and both $\psi(x)$ and $P_N(x)$ behave asymptotically as a power law $\sim |x|^{-\gamma-1}$. Again this is the signature of fractal behavior. The random-walk process which follows Eqs. (4) and (5) is known as the Lévy flight [24].

2. Lévy Flights and Lévy Walks

In this Section we look in more detail into the nature of Lévy flights and their relationship to anomalous diffusion. As mentioned, Lévy [24, 25] considered a set $\{x_i\}$ of identically distributed random variables each governed by the probability density $\psi(x_i)$. He asked the question of when the new random variable x_0, given by

$$c_0 x_0 = c_1 x_1 + ... + c_n x_N \quad , \tag{6}$$

with c_i values being constants related by an auxiliary condition

$$c_0^\gamma = c_1^\gamma + ... + c_N^\gamma \quad , \tag{7}$$

has the same probability density as the x_i. For instance, for $i > 0$ choose $\langle x_i \rangle = 0$, $c_i = 1$, and $\gamma = 2$. Then $x_0 = N^{-1/2} \sum_{i=1}^{N} x_i$. This sum of N random variables properly normalized by the coefficient $N^{-1/2}$, has the Gaussian density, or $\psi(k) = \exp(-Dk^2)$, with D a constant. For $\gamma < 2$, Lévy showed that the solution is

$$\psi(k) = \exp(-D|k|^\gamma) \quad . \tag{8}$$

This corresponds to

$$\psi(x) \sim |x|^{-\gamma-1} \quad , \quad x \to \infty \quad , \tag{9}$$

that is the moments $\langle |x|^\mu \rangle$ are finite for $\mu < \gamma$ and are infinite for $\mu \geq \gamma$; in particular the variance diverges. These processes represent a random walker (with step i given by x_i) that visits a disconnected self-similar set of points for which the mean-squared displacement $\langle x^2 \rangle$ diverges.

We now introduce a discrete space Lévy flight. Let us consider a random walk in a one-dimensional space, for simplicity; the results can be generalized to higher dimensions. Let $p(x)$ be the probability for a random-walk jump of displacement x. We choose [26, 27]

$$p(x) = \frac{n-1}{2n} \sum_{j=0}^{\infty} n^{-j} (\delta_{x,b^j} + \delta_{-x,b^j}) \quad , \tag{10}$$

with $n, b > 1$. Jumps of all orders of magnitude can occur in base b, but each successive order of magnitude in displacement occurs with an order of magnitude less probability in base n. The walker makes about n jumps of unit length before a jump of length b occurs, and a new cluster of sites visited starts to be generated. Eventually a fractal set of points is visited. To see this better, consider

$$p(k) = \frac{n-1}{n} \sum_{j=0}^{\infty} n^{-j} \cos(b^j k) \quad , \tag{11}$$

which is the Weierstrass example of a continuous but everywhere non-differentiable function for $b > n$. For small k,

$$p(k) = 1 - \frac{1}{2} k^2 \langle x^2 \rangle \quad , \tag{12}$$

if $\langle x^2 \rangle = \sum x^2 p(x)$ is finite. If $b^2 > n$, then

$$\langle x^2 \rangle = \sum_{j=0}^{\infty} (b/n)^j \quad , \tag{13}$$

diverges and Eq. (12) cannot be used. Instead note that

$$p(bk) = np(k) - (n-1)\cos k \ , \tag{14}$$

which has the following small k behavior

$$p(k) = 1 - |k|^\gamma \simeq \exp(-|k|^\gamma) \ , \tag{15}$$

with

$$\gamma = \frac{\ln n}{\ln b} \ . \tag{16}$$

The exponent γ now appears in the form of a fractal dimension. For $\gamma = 2$, as in Eq. (12), the Gaussian behavior results while for $\gamma < 2$, $\langle x^2 \rangle = \infty$ and the Lévy flights are obtained.

A particle performing a Lévy flight jumps between sites, however distant, which leads to a divergence of the mean-squared displacement. One is interested in modifying the Lévy-flight law so that the motion still follows Eqs. (8) or (9) but with a finite mean-squared displacement. We shall therefore explore a stochastic process which visits the same sites as in the Lévy flight, but with a time 'cost' which depends on the distance. For the flight we need only to specify $\psi(r)$, the probability that a move over distance r occurs according to Eq. (9). In

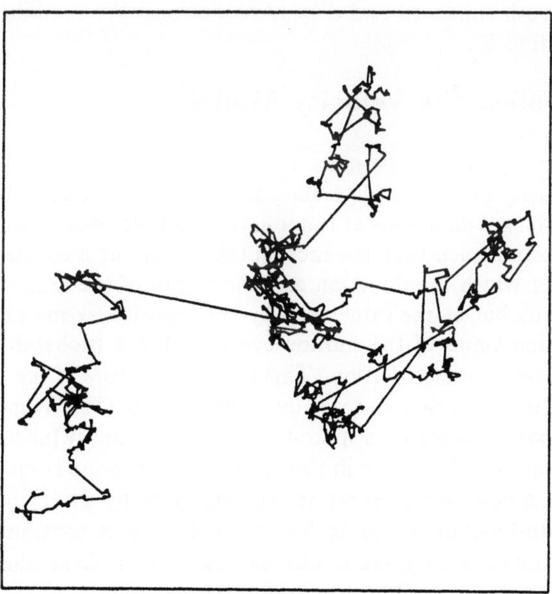

Fig. 1. The "Lévy flight": a typical trajectory of 10^4 jumps is shown with jump distances r chosen randomly from the power-law distribution $\psi(r)$, Eq. (9), with $\gamma = 1.5$. The turning points of the walk form a fractal structure of dimension γ.

the framework of the continuous time random walks (CTRW) [18, 19] we shall introduce a probability distribution $\psi(r,t)$ such that

$$\psi(r,t) = p(r|t)\psi(r) \tag{17}$$
$$p(r|t) = \delta(|r| - t) , \tag{18}$$

where $p(r|t)$ is the conditional probability to move a distance r in time t. Eqs. (17) and (18) within the CTRW approach define Lévy walks which result in finite mean-squared displacements at time t and account for the possibility of enhancement [19, 26, 28]. Alternatively, one can write Eq. (17) as

$$\psi(r,t) = \frac{1}{2}\delta(|r| - t)\psi(t) , \tag{19}$$

where now $\psi(t)$ is a time-dependent power-law. Other forms of conditional probability densities $p(r|t)$ were discussed in Ref. [18]. Lévy walks, as we see later, are characterized by a rich dynamical behavior depending on the power-law exponents of $\psi(r)$ or $\psi(t)$. In Fig. 1 a typical trajectory of a Lévy walker in two dimensions is plotted.

In the next Section we introduce the Lévy walk formulation within the velocity model [11]. It should be emphasized that the CTRW formulation allows also for the description of dispersive transport and provides therefore a unified approach for both enhanced and dispersive motions. We combine these two behaviors in Section 5.

3. Lévy Walks: the Velocity Model

We now briefly outline the main ingredients in the Lévy walk process. We choose the velocity picture in which the particle moves continuously at a constant velocity and changes directions at random [11]. Let $v(t)$ denote a stochastic time dependent function such that the motion takes place at a constant velocity for some time after which the direction and the length of the next motion event is chosen randomly but at the same velocity. The motion events are uncorrelated and the duration times of the motion events follow a probability distribution $\psi(t)$, the waiting time distribution. This idealized picture of the motion defines the probabilistic description and was introduced in the CTRW approach [11, 18, 19]. The approach is based on a space-time coupled memory function $\psi(r,t)$, the probability to move a distance r in time t in a single motion event and to stop at r for initiating a new motion event at random, given by $\psi(r,t) = \delta(|r| - t)\psi(t)$, where the delta-function accounts for the motion at a constant velocity and where length and time are given in dimensionless units. Generalizations to non-constant velocities are possible. In our analysis we use

$$\psi(t) \sim t^{-\gamma-1} , \quad \gamma > 0 . \tag{20}$$

We also introduce $\Psi(r,t)$, the probability density to move a distance r in time t

in a single motion event and not necessarily to stop at r

$$\Psi(r,t) = \frac{1}{2}\delta(|r|-t)\int_t^\infty \psi(\tau)d\tau \ .$$ (21)

$\psi(r,t)$ and $\Psi(r,t)$ are the relevant quantities for characterizing the motion. The motion consists of a sequence of these events and thus the propagator $P(r,t)$ can be cast in the following way

$$P(r,t) = \Psi(r,t) + \int_{-\infty}^{\infty}\int_0^t \psi(r',t')\Psi(r-r',t-t')dr'dt' + \ldots \ .$$ (22)

The first term denotes the probability to reach location r in time t in a single motion event. The second term is the probability to reach r at time t with one stop and so on to include all combinations of motion events. In the Fourier-Laplace space $(r \to k, t \to u)$ the convolution integrals simplify and the series in Eq. (22) can be given in a closed form as

$$P(k,u) = \frac{\Psi(k,u)}{1-\psi(k,u)} \ .$$ (23)

A more detailed derivation of $P(r,t)$ including dependence on initial conditions was reported in Refs. [11, 29].

From the propagator we calculate the time evolution of the mean-squared displacements

$$\langle r^2(t)\rangle = -\frac{\partial^2}{\partial k^2}P(k,u)\Big|_{k=0} \ .$$ (24)

Depending on γ which governs the decay of the tails of $\psi(t)$ we distinguish among three characteristic motion regimes: the ballistic-type, the intermediate-enhanced and the regular Brownian-type diffusion

$$\langle r^2(t)\rangle \sim \begin{cases} t^2 & , \quad 0 < \gamma < 1 \\ t^{3-\gamma} & , \quad 1 < \gamma < 2 \\ t & , \quad\quad 2 < \gamma \end{cases} \ .$$ (25)

Thus the Brownian motion is a special case in this Lévy-walk description. Here we concentrate on the intermediate enhanced diffusion regime $1 < \gamma < 2$ for which the asymptotic long-time behavior of the propagator has been shown to follow the scaling representation [11]

$$P(r,t) \sim f(\xi)/t^{1/\gamma} \ ,$$ (26)

where ξ is the scaling variable, $\xi = |r|/t^{1/\gamma}$, and where $f(\xi)$ is the scaling function. This function follows a Lévy distribution which to lowest order in small and large ξ values is given by

$$f(\xi) \simeq \begin{cases} \exp(-c\xi^2) & , \quad\quad \xi < 1 \\ \xi^{-\gamma-1} & , \quad 1 < \xi \, , \ |r| \lesssim t \\ 0 & , \quad\quad |r| > t \end{cases} \ .$$ (27)

where the cutoff at $|r| = t$ is a consequence of the constant velocity. The fact that $f(\xi)$ is characterized by a Lévy distribution for $|r| < t$ is a manifestation of the generalization of the central limit theorem. For the propagator corresponding to $0 < \gamma < 1$ the reader is referred to Ref. [11].

4. Lévy Walks in the Standard Map and in the "Egg-Crate" Potential

We now show that diffusion generated in the standard map provides an instructive example for the Lévy-walk statistics in conservative systems. The standard map which can be regarded as a discretized Hamiltonian for the kicked-rotor is a two-dimensional map of the form [2, 30]

$$x_{n+1} = x_n + K sin(2\pi\theta_n) \ , \quad \theta_{n+1} = \theta_n + x_{n+1} \ . \tag{28}$$

This map has been shown to have ranges of the stochasticity parameter K for which the motion along the x-direction is enhanced [31, 32]. This is due to the accelerating modes of various periods which give rise to islands of stability. These islands are encircled by cantori which separate areas of quasi- stability. Islands and cantori show some hierarchical self-similar structure [1-4,31-33]. Orbits approaching these islands tend to stick to the critical tori and become quasi stable for some time such that laminar phases result. Hence due to repeated visits to the self-similar areas of quasi-stability an intermittent behavior results which is characterized by broad distributions of sticking times [1-4] which correspond to the previously introduced $\psi(t)$, Eq. (20). These broad distributions cause the motion to be anomalous and to be a promising candidate to be described in terms of Lévy walks.

Enhanced diffusion in the standard map is observed to be dominated by accelerating modes of period p [31, 32]. The acceleration is due to performing ℓ step units along x in p iterations. For the K values studied here enhanced diffusion results from the fundamental modes for which $|\ell| = p$. This gives rise to a constant velocity $|v| \simeq |\ell|/p = 1$. Therefore sticking to the islands means being locked in a mode of constant velocity. Thus the motion is characterized by laminar phases of constant velocity interrupted by short periods of chaotic motion in the chaotic sea, which makes the space-time coupled waiting time distribution apply. The self-similar properties of the trajectory are demonstrated in Fig. 2 .

In our numerical investigations we considered the $p = 3$ and $p = 5$ (period three and period five fundamental) accelerating modes with K values 1.1 and 1.03, respectively [14, 32]. Our analysis of the waiting time distribution $\psi(t)$ calculated for orbits starting from the chaotic sea revealed that the sticking time distributions follow approximately power laws and that the asymptotic behavior starts to hold for times $t \geq 10^2$, as displayed in Fig. 3, for both $p = 3$ and $p = 5$. In the figure plotted are $\psi(t)$ and $\psi(r)$ which were calculated independently from the segments of laminarity in orbits of typical length of 10^{11} number of iterations. Here r denotes the displacement between two locations, $r = x_n - x_m$.

The statistics obtained from a few long orbits or from many shorter orbits did not significantly differ. From the figure it is evident that $\psi(t)$ and $\psi(r)$ fall on top of each other for times $t \geq 10^2$. This collapse coincides with the inset of the power-law decays of the distributions. For times $t \geq 10^2$ we observe exponents which correspond to $\gamma \simeq 1.8$ for $p = 3$ and $\gamma \simeq 1.2$ for $p = 5$. If one chooses different time-windows different exponents of values around 1.7 are obtained. Exponents in the same range were obtained in Refs. [2-4] where the Poincare recurrences were studied.

We point out that the collapse of $\psi(t)$ and $\psi(r)$ for $t \geq 10^2$ confirms the assumption that the motion is governed by fundamental accelerating modes at longer times and that the motion during the laminar phases takes place at a constant velocity. This in turn justifies the δ-function relationship in Eq. (19). In order to further investigate the sticking properties of the orbits near the islands of stability we carried out detailed computer calculations of the distribution $\phi(t)$ of exit times. $\phi(t)$ is the distribution of times it takes an orbit with initial coordinates placed close to the islands of stability to reach the chaotic sea. It turns out that $\phi(t)$ also follows a power-law, however, with an exponent approximately equal to γ being smaller than that of the corresponding $\psi(t)$ distribution by one [32].

An intuitive argument that leads to power-law behaviors of $\psi(t)$ and $\phi(t)$ can

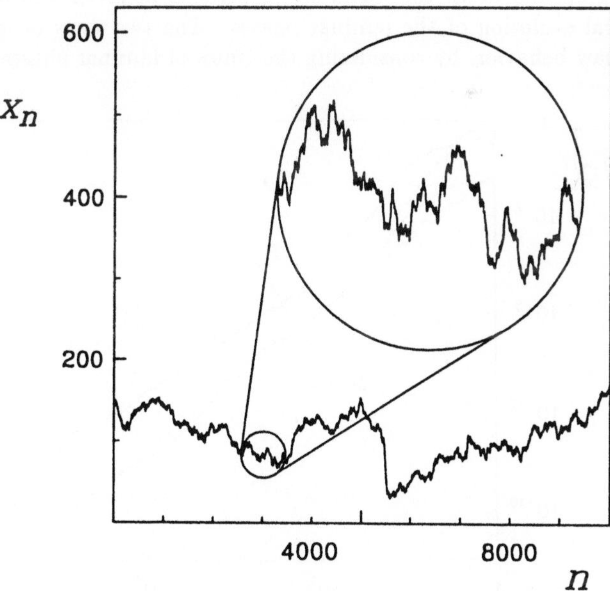

Fig. 2. Typical trajectory obtained from the iterated standard map, Eqs. (28). Plotted are x_n for 1000 iteration steps. The stochasticity parameter was set to $K = 1.04$ so that the motion is governed by the fundamental $p = |\ell| = 5$ mode. The self-similar property of the trajectory is obvious from the insert.

be given by following Hanson et al. [33] who make use of a self-similar system of
rate equations for the Markovian motion in the hierarchical structure of belts,
i.e. of zones of quasi-stability. Assuming self similarity in the probabilities to
pass across cantori and in the area sizes between them, they attained a power-
law behavior for the sojourn time of an orbit which enters the system of cantori
from the chaotic sea. They proposed a hierarchical sum of exponentials

$$\psi(t) \sim \sum_{j=1}^{\infty} (aw^2)^{-j} \exp(-w^j t) \; , \tag{29}$$

where w is the scale factor for the rate of passing and a denotes the scale factor of
the areas. The overall behavior of this sum is known to follow a power-law $\psi(t) \sim
t^{-1-\gamma}$ such that the exponent γ is related to a and w by $\gamma = 1 + \ln(a)/\ln(w)$
[32, 33]. We point out that expression (29) does not hold in general and a more
involved approach is required for a detailed analysis [6, 20, 33].

In contrast to the waiting times, for the exit times the initial coordinates
are placed at random which means that the weights $(aw^2)^j$ in Eq. (29) have
to be replaced by $(aw)^j$ which leads to an exit-time distribution $\phi(t) \sim t^{-\gamma}$.
The exponent of this power-law is in agreement with the numerical findings
mentioned above. This result supports the assumption of a hierarchy of cantori
and of the orbits initiated in the chaotic sea entering this system from the bottom
and moving up and down the hierarchy of belts, a behavior which dominates
the temporal evolution of the laminar phases. The two ways of investigating
the power-law behavior, by considering the times of laminar phases or the exit

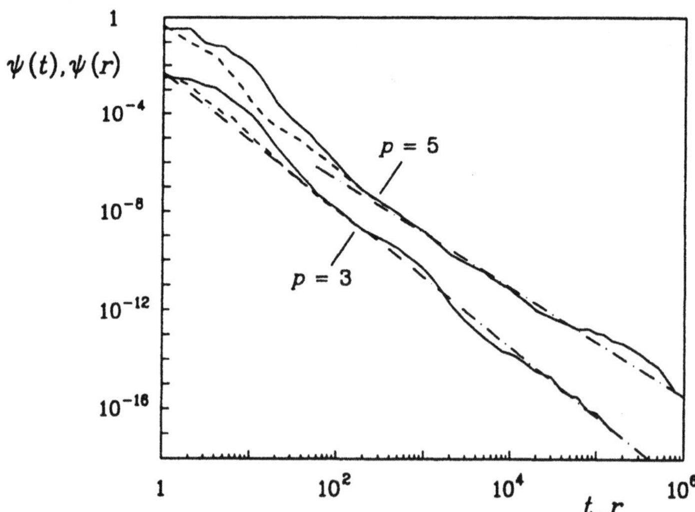

Fig. 3. Probability distributions for the standard map, Eqs. (28). Solid lines
give $\psi(t)$, dashed lines are $\psi(r)$ and dash- dotted lines indicate the power-law
behavior with exponents $\gamma = 1.8$ for $p = 3$ and $\gamma = 1.2$ for $p = 5$.

times from randomly chosen initial locations, are complementary and represent independent methods for calculating the exponent γ. We emphasize that γ values in the range $1 < \gamma < 2$, giving rise to an enhanced diffusion behavior, cause also the mean exit time to diverge which has some implications for the in interpretation of enhanced diffusion in conservative systems.

In Fig. 4 we show the propagator $P(r,t)$ for the $p = 5$ period mode in the scaling representation for the time regime $10^2 < t < 2 \times 10^3$. The numerical data are compared with the theoretical $P(r,t)$ of Eq. (26); for $\gamma = 5/3$ a reasonable fit is achieved. We notice that the curves for different times fall on top of each other; this data collapse indicates that scaling holds at least in this intermediate time regime. Furthermore, we observe that the decay in the wings follows a power-law, thus demonstrating the Lévy behavior. We thus conclude that the chaotic dynamics in the standard map is intimately related to the Lévy processes.

Another example of anomalous diffusion which follows the behavior of Lévy walks is given by the friction free motion subject to an "egg-crate" potential in two-dimensions

$$V(x,y) = A + B(\cos x + \cos y) + C \cos x \cos y \ . \tag{30}$$

In this potential the third term is responsible for the non-integrability of the

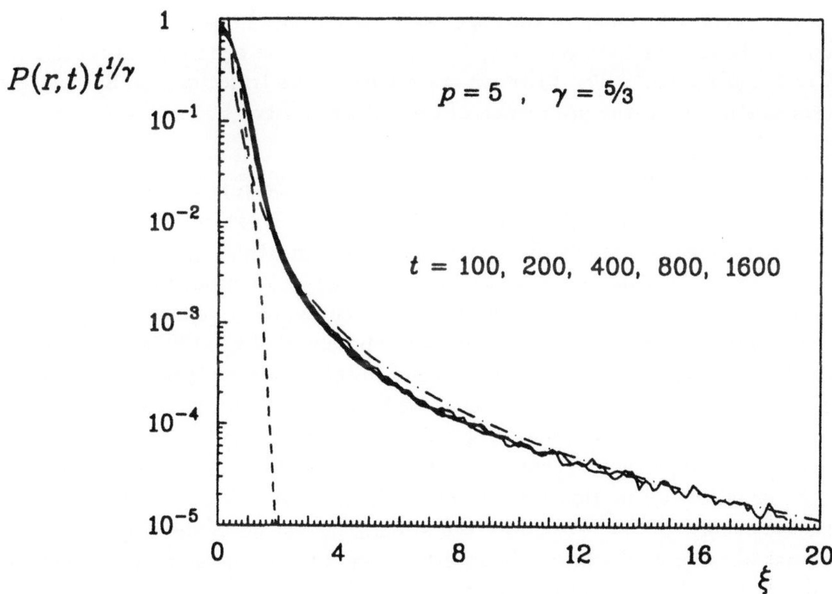

Fig. 4. The propagator for the standard map. Plotted are as solid lines the simulation results in the scaling representation for times as indicated. The stochasticity parameter was set to K=1.03 so that the $p = |\ell| = 5$ mode governs the motion. The dashed line and the dash-dotted lines give the Gaussian and the power-law approximation of the stable law with $\gamma = 5/3$.

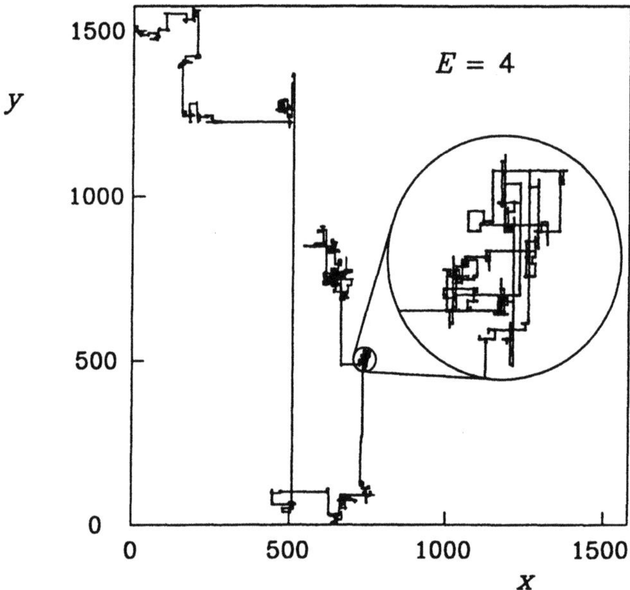

Fig. 5. The trajectory in the "egg-crate" potential. Plotted is an example of the two-dimensional trajectory $\mathbf{r}(t)$ obtained from the equations of motion for $E = 4$ and $t = 10^5$. The insert shows a part of the trajectory on an enlarged scale to strengthen the impression of the self-similarity.

corresponding Hamiltonian. This problem has been studied by Geisel et al. [6] and by Chernikov et al. [7] and has been shown to exhibit both regular and enhanced diffusion, namely mean-squared displacements which grow as $\langle r^2(t) \rangle \sim t^\alpha$, $1 \leq \alpha < 2$. This potential can bee considered for modeling the observed persistent diffusion behavior of atoms or molecules on crystal surfaces [34].

It has been shown that the motion in the potential of Eq. (30) also belongs to the class of Lévy walks and can be studied in terms of the CTRW formalism [6, 35]. The numerical investigation was based on solving numerically the equation of motion for different energies E; for details see Ref. [35].

In Fig. 5 we present a typical trajectory obtained from the numerical solution of the equations of motion for the constants being $A = 2.5$, $B = 1.5$, $C = 0.5$ and the energy $E = 4$. We notice the self-similar nature of the trajectory with laminar phases on all scales. The insert supports the impression and also shows that on the scale of a unit cell the trajectory follows a wiggly curve.

Another interesting property of the trajectories resulting from the particular shape of the "egg-crate" potential is the coexistence of phases of laminar motion with phases of localization. This aspect is better visualized in Fig. 6 where we present the two-dimensional trajectory $\mathbf{r}(t)$ as a one-dimensional trajectory $x(t) + y(t)$. This presentation is reasonable because the motion is predominantly

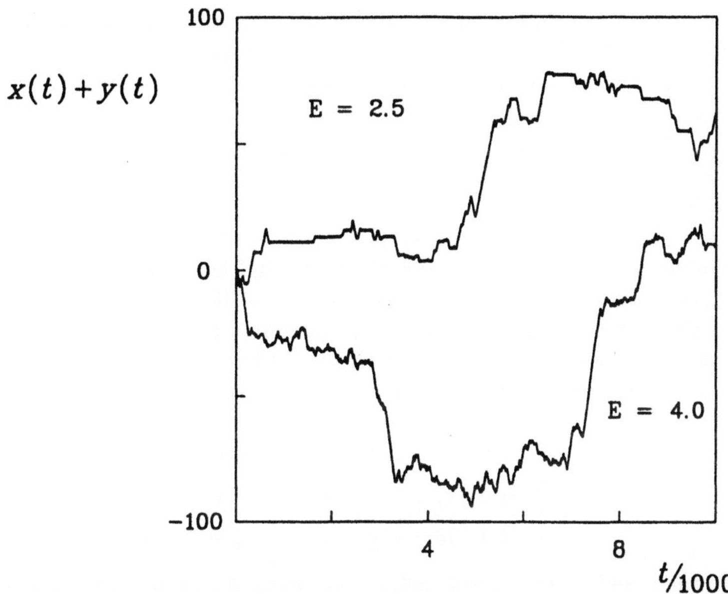

Fig. 6. Two typical one-dimensional trajectories $x(t) + y(t)$ in the "egg-crate" potential for energies as indicated.

coaxial, that is, the motion occurs either parallel to the x or to y axis. The trajectory $x(t) + y(t)$ is displayed for energies $E = 2.5$ and $E = 4$. For $E = 2.5$ we notice that phases of laminar motion are interrupted by periods of localization. For $E = 4$ these periods are hardly visible and a pattern is observed which is very similar to the one found for 1-d iterated maps [11, 36]. The case of the coexistence will be discussed in some detail in the next Section.

5. Coexistence of Laminar and Dispersive Motions

We have concentrated hitherto on one aspect of diffusion in dynamical systems, that of enhanced motion. It has been recognized however, that dispersive transport for which the mean-squared displacement grows sub-linearly with time $\langle r^2(t) \rangle \sim t^\alpha$, $0 < \alpha < 1$, is also generated by some dynamical systems [11, 37]. Dispersive transport can be described within the CTRW framework [11, 38, 39] with a broad distribution of waiting times which correspond to time spent at spatial locations in contrast to the broad distributions of time being locked in laminar phases, Eqs. (15) and (16).

An interesting situation occurs when there is a coexistence of laminar and dispersive (localized) phases. The trajectories in such cases demonstrate the interplay between the two competing motion modes and can be analyzed in terms of Lévy statistics. In a recent experiment by Solomon et al. [17] on tracer particles in a two-dimensional rotating flow it has been observed that dispersive and enhanced modes of motion can coexist in a way that a particle may perform long flights and be also intermittently trapped in space. This behavior is reminiscent of the observations for two-dimensional Hamiltonian systems discussed in the previous Sections namely: motion in the Chirikov-Taylor map and in the "egg-crate" potential. The coexistence of laminar and localized modes has been addressed by Chaikovosky and Zaslavsky [8] and Zaslavsky [40] using fractional Fokker-Planck equations.

Here we introduce a one-dimensional map, which is characterized by intermittent chaotic motion with coexisting dispersive and laminar motion events [41]. We demonstrate the applicability of the random-walk scheme with Lévy stable-law distributions in analyzing the motion generated by the map and show how the competing trends of laminar and localized phases lead to diffusional behavior, different than previously obtained, and cover the whole range of dispersive, regular and enhanced behaviors.

Fig. 7 shows an example of the one-dimensional map which is defined as

$$f(x) = \begin{cases} (1+\epsilon)x + ax^z - 1 & , \quad 0 \le x \le \frac{1}{4} \\ (1+\tilde{\epsilon})x - \tilde{a}(\frac{1}{2} - x)^{\tilde{z}} & , \quad \frac{1}{4} \le x \le \frac{1}{2} \end{cases} \quad . \tag{31}$$

Translational and inversion symmetries are inferred so that diffusional motion from one to the other unit cell is enabled. This map represents a combination of two maps introduced previously [5, 11, 37]. The map is discontinuous at the

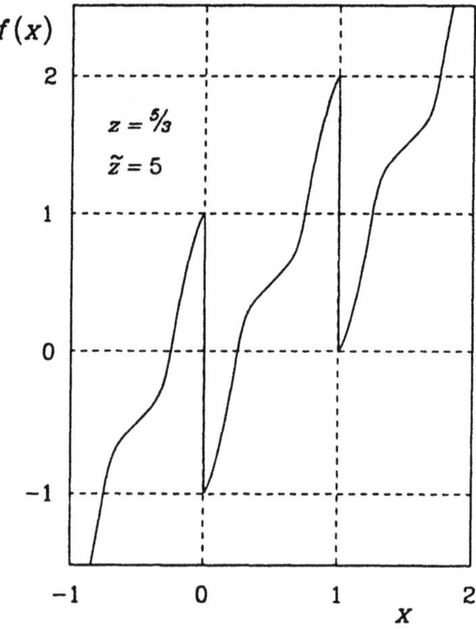

Fig. 7. The map function f(x) for the one-dimensional combined map, Eq. (31), for $z = 5/3$ and $\tilde{z} = 5$, as indicated corresponding to the exponents $\gamma = 3/2$ and $\tilde{\gamma} = 1/4$, respectively.

boundaries of each box but is continuous otherwise. For $\epsilon = \tilde{\epsilon} = 0$ the map shows marginal stable fixed points at $x = 0$ and $x = \frac{1}{2}$. The neighborhood of the fixed point at $x = 0$ is responsible for the laminarity, while the fixed point at $x = \frac{1}{2}$ gives rise to localization. The exponents z and $\tilde{z}$ determine the characteristic behaviors of the two types of events, free flights and localization, respectively. The prefactors a and $\tilde{a}$ are weights and are chosen so that the map $f(x)$ is continuous at $x = \frac{1}{4}$ including the first derivative. For $\epsilon = \tilde{\epsilon} = 0$ we used $a = 4^z \tilde{z}/(z + \tilde{z})$ and $\tilde{a} = 4^{\tilde{z}} z/(z + \tilde{z})$. ϵ and $\tilde{\epsilon}$ were considered to be small quantities and to differ from zero in the cases $z > 2$ and $\tilde{z} > 2$, respectively, in order to avoid problems in the numerical realization of the statistical analyses. This relatively simple map displays a rich spectrum of behaviors and a unique interplay of the two modes of motion.

In Fig. 8 we present two typical trajectories generated by the map function Eq. (31). Again r denotes the displacement between two locations: $r = x_n - x_m$. The interchange of laminar and dispersive (localization, no motion) behavior is evident. The corresponding distributions of flight (laminar) and dispersive (localization) times, as shown in Fig. 9, follow approximately power laws. For

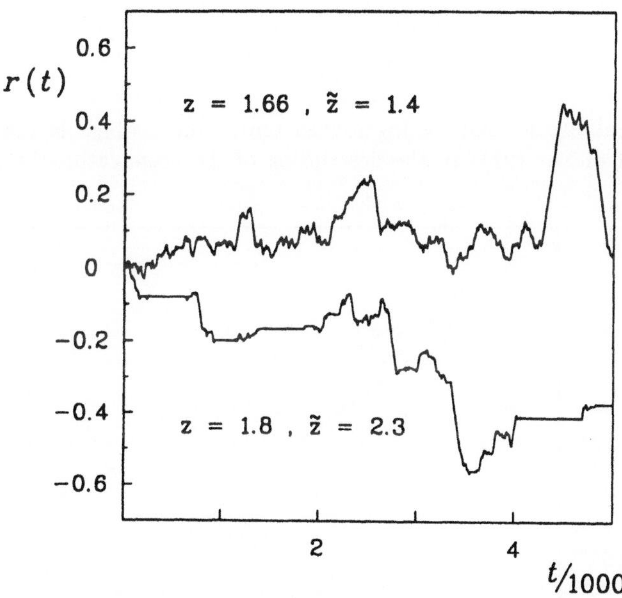

Fig. 8. Two typical trajectories from the iterated combined map, Eq. (31). The upper trajectory is obtained for values $z = 1.66$ ($\gamma = 1.5$) and $\tilde{z} = 1.4$ ($\tilde{\gamma} = 2.5$) which give rise to enhanced diffusion with $\alpha = 1.5$. The lower trajectory is obtained for values $z = 1.8$ ($\gamma = 1.25$) and $\tilde{z} = 2.33$) ($\tilde{\gamma} = 0.75$) giving rise to enhanced diffusion again with $\alpha = 1.5$.

the laminarity and for the localization times we observe

$$\psi(t) \sim t^{-\gamma-1} \quad \text{and} \quad \tilde{\psi}(t) \sim t^{-\tilde{\gamma}-1} \ , \tag{32}$$

respectively, with the exponents $\gamma = (z-1)^{-1}$ and $\tilde{\gamma} = (\tilde{z}-1)^{-1}$. The observed trajectories and the coexistence of the two motional modes resemble the behavior, although in a completely different type system, that has been reported by Solomon et al. [17]. In this respect our map generates statistical properties that are amenable to experimental observation. What relate the different cases are the underlying Lévy stochastic processes which we now outline.

We choose the velocity picture in which the particle moves continuously at a constant velocity, changes directions at random and in which the motion is occasionally interrupted by phases of spatial localization. This means that the particle does not move at a constant velocity at all times but that the phases of laminar motion are intermittently interrupted by periods of no motion on the scale of typically one box. The probability distribution $\psi(r,t)$ to move a distance r in time t in a single motion event during the laminar phase, and to stop at r for initiating a new motion event at random, is given by Eq. (19). For the dispersive case we note that

$$\tilde{\Psi}(t) = \int\limits_{t}^{\infty} \tilde{\psi}(\tau)d\tau \tag{33}$$

is the probability for not having moved until time t which is the dispersive counterpart of Eq. (21). In the description of the propagator, the probability

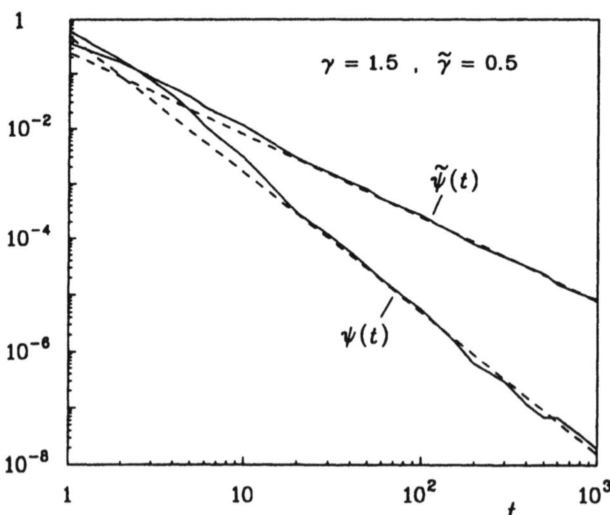

Fig. 9. The probability distributions $\psi(t)$ and $\tilde{\psi}(t)$ for $z = 5/3$ and $\tilde{z} = 3$ corresponding to $\gamma = 3/2$ and $\tilde{\gamma} = 1/2$, as indicated. Simulations results are given by solid lines, the dashed lines indicate the power-laws of Eq. (32).

density to be at location r at time t, we assume that the observation starts with an event of motion at constant velocity and we can thus write

$$P(r,t) = \Psi(r,t) + \int\limits_0^t \psi(r,t')\tilde{\Psi}(t-t')dt'$$

$$+ \int\limits_{-\infty}^{\infty} dr' \int\limits_0^t dt' \int\limits_0^{t'} dt'' \psi(r',t'')\tilde{\psi}(t'-t'')\Psi(r-r',t-t') + \dots \; , \qquad (34)$$

where the first term denotes the probability to reach location r in time t in a single motion event. The second term is the probability to reach r at an earlier time and to stay localized until time t. The third term is the probability to reach r in time t in two motion events interrupted by one period of localization. The sum has to be extended over all possible combinations of motion events interrupted by periods of localization. Taking the Fourier-Laplace transform and summing over even and odd terms independently we obtain

$$P(k,u) = \frac{\Psi(k,u) + \tilde{\Psi}(u)\psi(k,u)}{1 - \tilde{\psi}(u)\psi(k,u)} \; . \qquad (35)$$

A similar expression is obtained when the walks are initiated by a localization event followed by motion at constant velocity. For the analysis of iterated maps in terms of CTRWs we have demonstrated that stationary conditions are an important issue [29], here we do not consider this situation.

For the description of the mean-squared displacement we make use of Eq. (24) and obtain for the leading term in the asymptotic behavior a power law $\langle r^2(t)\rangle \sim t^\alpha$, with the exponent α depending on γ and $\tilde{\gamma}$ as

$$\alpha = \left\{ \begin{array}{ll} 2 + \min\{\tilde{\gamma}, 1\} - \min\{2, \gamma\} \; , & \gamma > 1 \\ 2 + \min\{\tilde{\gamma}, \gamma\} - \gamma & , \; 0 < \gamma < 1 \end{array} \right. \qquad (36)$$

For clarity, we concentrate on the range $1 < \gamma < 2$ which corresponds to the intermediate enhanced diffusion regime if localization is disregarded [11]. When localization is included the combined effects lead to

$$\langle r^2(t)\rangle \sim \left\{ \begin{array}{ll} t^{2+\tilde{\gamma}-\gamma} & , \; 1 < \gamma < 2 \; , \; \tilde{\gamma} < 1 \\ t^{3-\gamma} & , \; 1 < \gamma < 2 \; , \; \tilde{\gamma} > 1 \end{array} \right. \qquad (37)$$

Equation (36) indicates that, depending on the two exponents γ and $\tilde{\gamma}$, the motion shows enhanced, regular, or dispersive behavior. We recall that the exponents γ and $\tilde{\gamma}$ are related to those of the combined map by $\gamma = (z-1)^{-1}$ and $\tilde{\gamma} = (\tilde{z}-1)^{-1}$. It should be noted that for $\tilde{\gamma} > 1$, for which $\tilde{\psi}(t)$ in Eq. (32) has a finite first moment, the enhancement reduces to the result obtained by the original Lévy-walk scheme [11, 28]. In Fig. 10 we show some numerical results for the time evolution of the mean-squared displacements. Plotted is the ratio $\langle r^2(t)\rangle/t$ for various γ and $\tilde{\gamma}$ values. The denominator t has been chosen to strengthen the impression of the deviation from simple Brownian motion. The results follow reasonably the predicted power-law behaviors.

For the tracer motion in a rotating annulus, mentioned above [17], the exponents $\mu = 2.3 \pm 0.2$ for the flights, $\nu = 1.6 \pm 0.3$ for the sticking and $\alpha = 1.65 \pm 0.15$ for the mean-squared displacement are reported which were determined from independent measurements. According to the present convention we have $\gamma = \mu - 1 = 1.3 \pm 0.2$ and $\tilde{\gamma} = \nu - 1 = 0.6 \pm 0.3$ so that Eq. (36) applies and therefore $\alpha = 1.3 \pm 0.5$ which is consistent with the experimental value.

We also discuss the propagator for the intermediate-enhanced diffusion regime, $1 < \gamma < 2$. We distinguish between the two cases $\tilde{\gamma} > 1$ and $\tilde{\gamma} < 1$. In the former case the small (k, u) expansion of the propagator gives to lowest order

$$P(k, u) \sim \frac{1}{u + c|k|^{\gamma}} \, , \tag{38}$$

which yields the known Lévy stable distribution [11]

$$P(r, t) \sim \begin{cases} t^{-1/\gamma} L_{\gamma}(\xi) & , \quad r < t \\ 0 & , \quad r > t \end{cases} \, , \tag{39}$$

where ξ is the scaling variable $\xi = cr/t^{1/\gamma}$ and the cutoff is due to the δ-function correlation in Eq. (19). Here the laminar phase dominates the transport properties.

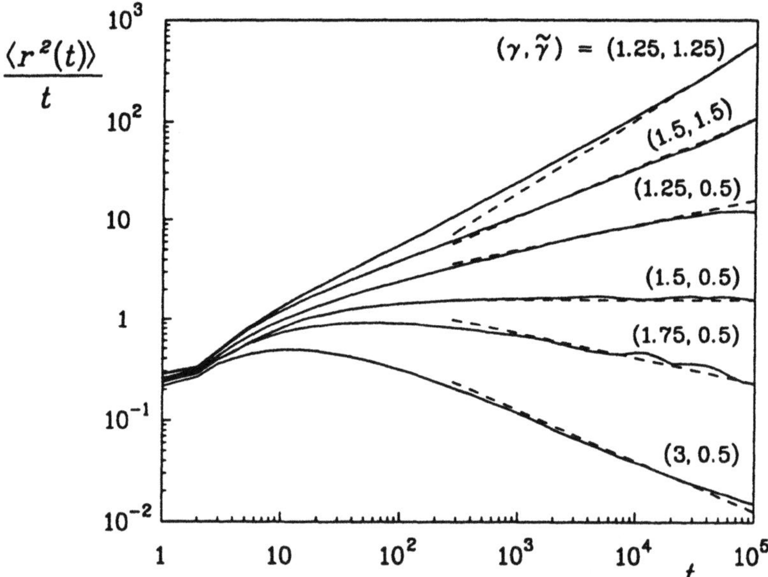

Fig. 10. The time evolution of the mean-squared displacement obtained from the iterated combined map, Eq. (32) for various enhanced and dispersive cases. Plotted are the ratios $\langle r^2(t) \rangle / t$ as solid lines for γ and $\tilde{\gamma}$ as indicated. The dashed lines give the predicted slopes according to Eqs. (36) and (37).

A different behavior is derived for $1 < \gamma < 2$ and $\tilde{\gamma} < 1$, where the small (k, u) expansion of the propagator gives

$$P(k, u) \sim \frac{1}{u^{1-\tilde{\gamma}}} \frac{1}{u^{\tilde{\gamma}} + c|k|^{\gamma}} \ . \tag{40}$$

This leads to the approximate scaling form

$$P(r, t) \sim \begin{cases} t^{-\tilde{\gamma}/\gamma} Q(\xi) & , \ \xi \to 0 \\ t^{-\tilde{\gamma}/\gamma} \xi^{-\gamma - 1} & , \ r \lesssim t \\ 0 & , \ r > t \end{cases} , \tag{41}$$

where Q is a scaling function which, depending on γ and $\tilde{\gamma}$, shows a cusp at the origin [29, 11]. The scaling variable is $\xi = r/t^{\tilde{\gamma}/\gamma}$. In the case of $\gamma > 2$ and $\tilde{\gamma} < 1$ we recover the dispersive behavior treated in Ref. [38]. Here the role of localization is pronounced and may even dominate, a situation which we did not observe in Hamiltonian systems [38]. From the scaling in Eqs. (39) and (41) we notice that for the average of the absolute value of the displacement we have $\langle |r(t)| \rangle \sim t^{\min\{1, \tilde{\gamma}\}/\gamma}$ for $1 < \gamma < 2$ in agreement with the results in Ref. [20, 40].

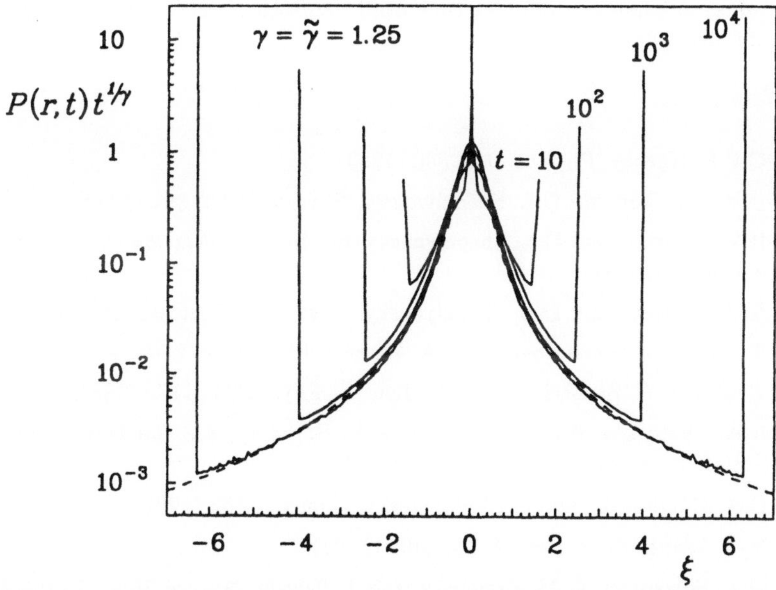

Fig. 11. The propagator $P(r, t)$ for the iterated combined map of Eq. (32). The numerical results are given as solid lines for times t and γ and $\tilde{\gamma}$ values, as indicated, in the scaling representation with the scaling variable $\xi = r/^{1/\gamma}$. The dashed line is the stable law $cL_{\gamma}(cr)$ with c introduced as an adjustable parameter. The sharp peaks are due to stationary conditions considered in the averaging procedure.

In Fig. (11) we present results obtained for the propagator $P(r,t)$. The numerical data are plotted for $\gamma = \tilde{\gamma} = 1.25$ in the scaling representation. Because of the finite mean trapping time the propagator, as expected, follows the stable law, Eq. (39), for $|r| < t$. The collapse of the lines for various times onto a single master curve indicates that scaling holds. The peaks at $r \simeq 0$ and $|r| \simeq t$ are attributed to stationary condition effects which result from the method of averaging [11, 29].

In summary chaotic dynamics in various dynamical systems have been shown to be intimately related to the Lévy processes. We have analyzed the transport in the Lévy walk framework which provides a useful random-walk description of enhanced as well as other diffusional behaviors. Lévy processes have been reported in other situations different from those mentioned above which leads one to believe that this generalization of Brownian motion may become as ubiquitous as its predecessor.

Acknowledgment

We thank Dr. G. Poupart for helpful discussions and F. Weber for technical assistance. A grant of computer time from the Rechenzentrum der ETH-Zürich is gratefully acknowledged.

References

[1] C.F.F. Karney, Physica D **8**, 360 (1983).

[2] B.V. Chirikov and D.L. Shepelyansky, Physica D **13**, 394 (1984).

[3] B.V. Chirikov and D.L. Shepelyansky, in *"Renormalization Group"*, World Scientific (1988).

[4] B.V. Chirikov and D.L. Shepelyansky. Phys. Rev. Lett. **61**, 1039 (1988).

[5] T. Geisel, J. Nierwetberg, and A. Zacherl, Phys. Rev. Lett. **54**, 616 (1985).

[6] T. Geisel, A. Zacherl, and G. Radons, Z. Phys. **B71**, 117 (1988).

[7] A.A. Chernikov, B.A. Petrovichev, A.V. Rogalsky, R.Z. Sagdeev, and G.M. Zaslavsky, Phys. Lett. **A144**, 127 (1990).

[8] D.K. Chaikovsky and G.M. Zaslavsky, Chaos **1**, 463 (1991).

[9] X.-J. Wang, Phys. Rev. A **45**, 8407 (1992).

[10] M.F. Shlesinger, G.M. Zaslavsky, and J. Klafter, Nature **263**, 31 (1993).

[11] G. Zumofen and J. Klafter, Phys. Rev. **E47**, 851 (1993).

[12] A.N. Jannacopoulos and G. Rowlands, J. Phys. **A26**, 6231 (1993).

[13] J.-P. Bouchaud and A. Georges,Phys. Rep. **195**, 127 (1990).

[14] R. Ishizaki, H. Hata, T. Horita, and H. Mori, Prog. Theor. Phys. **84**, 179 (1990); R. Ishizaki, T. Horita, T. Kobayashi, and H. Mori, Prog. Theor. Phys. **85**, 1013 (1991).

[15] O. Cardoso and P. Tabeling, Europhys. Lett. **7**, 225 (1988).

[16] R. Ramshankhar, D. Berlin, and J.P. Gollub, Phys. Fluids **A2**, 1955 (1990).

[17] T.H. Solomon, E.R. Weeks, and H.L. Swinney, Phys. Rev. Lett. **71**, 3975 (1993).

[18] M.F. Shlesinger, J. Klafter, and Y.M. Wong, J. Stat. Phys. **27**, 499 (1982).

[19] M.F. Shlesinger and J. Klafter, Phys. Rev. Lett. **54**, 2551 (1985).

[20] G.M. Zaslavsky, D. Stevens, and H. Weitzner, Phys. Rev. **E48**,1683 (1993).

[21] F. Hayot, Phys. Rev. A **43**, 806 (1991).

[22] J.A. Viecelli, Phys. Fluids **A2**, 2036 (1990).

[23] O.V. Bychuck and B. O'Shaugnessy, J. Physique (Paris) **4**, 1135 (1994).

[24] B. Mandelbrot, *"The Fractal Geometry in Nature"* (Freeman, San Francisco, 1982).

[25] P. Lévy, "Théorie de l'addition des variables aléatoires", Gauthier Villars, Paris 1937).

[26] M.F. Shlesinger and J. Klafter, in *"On Growth and Form"*, (ed. H.E. Stanley and N. Ostrowsky, Kluwer, Boston, 1986).

[27] E.W. Montroll and M.F. Shlesinger, in *"Studies in Statistical Mechanics"*, XI, (ed. J.L. Lebowitz and E.W. Montroll, North Holland, Amsterdam, 1984), p1.

[28] M.F. Shlesinger, B.J. West, and J. Klafter, Phys. Rev. Lett. **58**, 1100 (1987).

[29] J. Klafter and G. Zumofen, Physica **A196**, 102 (1993).

[30] B.V. Chirikov, Phys. Rep. **52**, 264 (1979).

[31] Y.H. Ichikawa, T. Kamimura, and T. Hatori, Physica **D29**, 247 (1987).

[32] G. Zumofen and J. Klafter, Europhys. Lett. **25**, 565 (1994).

[33] J.D. Hanson, J.R. Cary, and J.D. Meiss, J. Stat. Phys. **39**, 27 (1985).

[34] D.S. Sholl and R.T. Skodje, Physica **D71**, 168 (1994).

[35] J. Klafter and G. Zumofen, Phys. Rev. **E49**, 4873 (1994).

[36] G. Zumofen, J. Klafter, and A. Blumen, Phys. Rev. E **47**, 2183 (1993).

[37] T. Geisel, S. Thomae, Phys. Rev. Lett. **52**, 1936 (1984).

[38] J. Klafter and G. Zumofen, J. Phys. Chem. **98**, 7366 (1994).

[39] A. Blumen, J. Klafter, and G. Zumofen, in *"Optical Spectroscopy of Glasses"*, ed. I. Zschokke (Reidel, Dordrecht, 1986), p.199.

[40] G.M. Zaslavsky, Chaos, **4**, 1 (1994).

[41] G. Zumofen and J. Klafter, Phys. Rev. E, xxx, (1995).

From Lévy Flights to the Fractional Kinetic Equation for Dynamical Chaos

G.M. Zaslavsky[1,2]
[1]Courant Institute of Mathematical Sciences
New York University, New York, NY 10012
[2]and Physics Department, New York University
New York, NY 10003

Abstract: Chaotic dynamics of Hamiltonian systems can be described by the random process which resembles the Lévy-type flights and trappings in the phase space of a system. The probability distribution function satisfies the fractional in space and time generalization of the Fokker-Planck-Kolmogorov equation. Orders of the fractional derivatives in space and time can be connected to the Pesin's dimensions of the trajectories. A new look on the problem of Maxwell's Demon is discussed in the context of the anomalous ("strange") kinetics.

1 Introduction

Lévy flights (or Lévy process) [1] are known as a random walk process with infinite second moment $\langle x^2 \rangle$ of the displacement x. There are other characteristic features of the Lévy flights such as a power-like tail of the probability distribution function

$$P(x) \sim 1/x^{1+\alpha} , \qquad 0 < \alpha < 2 , \qquad (x \to \infty) \qquad (1.1)$$

absence of a characteristic scale of random walks, connection of the parameter α to the fractal dimension of the space of random walk, etc. [2]. Initially it appeared that the Lévy process could be applied mainly to the understanding of turbulence [3]. However as time went by, more and more phenomena were found to have properties similar to the Lévy process (see reviews [2,4-7]). In this article we consider kinetic description of the Hamiltonian dynamical chaos which displays the characteristic properties of the Lévy process, but in a more complicated form.

We would like to touch on three main topics:

(1) The scaling properties of the continuous time random walk (CTRW) [8,9] will be discussed only briefly, mainly in the context of the problem

of anomalous transport for which

$$\langle x^2 \rangle \sim t^{2\mu}, \qquad (t \to \infty,\ 0 < \mu < 1) \tag{1.2}$$

The transport exponent μ may differ from the value $\mu = 1/2$ for Gaussian process. Non-obvious feature of (1.2) is that $\langle x^2 \rangle < \infty$ for finite t despite the original Lévy process has $\langle x^2 \rangle = \infty$. In this part of the article the Weierstrass random walk [10] will be mentioned as it gives an insight in how the scaling organizing of a random orbit can reveal the Lévy process. The "strange kinetics" motion has been introduced in [7] to describe a broad class of processes that lead to the anomalous transport of the (1.2) type with $\mu \neq 1/2$.

(2) The main, second, part of the article considers the fractional kinetics in the form of the Fractional Fokker-Planck-Kolmogorov equation (FFPK). It has been introduced in a set of our publications [11,12] a new type of kinetics which occurs in Hamiltonian systems with chaotic dynamics, as opposed to the regular kinetics described by the Fokker-Planck-Kolmogorov (FPK) diffusional type equation. The evidence of the existence of an unusual (non-Gaussian) kinetics in Hamiltonian chaotic dynamics has appeared fairly long ago, starting from the results for the standard map system [13]. Other physical models of the strange kinetics can be found in [14-16], as well as in this volume. The phenomena of the anomalous transport have been observed in a number of experiments [17-19] and now the experimental technique even permits not only to get quantitative results for the transport exponent μ but also to get the values of local fractional dimensions of chaotic orbits [19] (see also this volume).

(3) In the third part of the article we discusss how the existence of the strange (fractal) kinetics influences the statistical behavior of a system and how it is related to the Maxwell's Demon problem.

2 Trappings, Flights, and Transport Exponent

Let $P(x,t)$ be a probability density to find a particle in the position x at time t. The idea of the Montroll-Weiss kinetic equation for the CTRW is to express $P(x,t)$ through the two basic probabilities: $\psi(t)$-probability density to have the time interval t between two consequent steps (time distributions), and $w(x)$-probability density to have the distance x between two consequent steps (step distributions). Laplace transform in time and Fourier transform in space can be used

$$\psi(u) = \int_0^\infty \psi(t)e^{-ut}dt$$

$$w(k) = \int_{-\infty}^\infty w(x)e^{ikx}dx \tag{2.1}$$

$$P(k,u) = \int_0^\infty dt \int_{-\infty}^\infty dx\; P(x,t)e^{-ut+ikx}$$

The typical problem of particle kinetics is to consider asymptotics of the function $P(x,t)$ for $x \to \infty$, $t \to \infty$ which correspond to the $k \to 0$, $u \to 0$ asymptotics for $P(k,u)$. For this case one can use expansions

$$\psi(u) = 1 + t_0 u - Bu^\beta$$

$$w(k) = 1 + ix_0 k - Ak^\alpha \tag{2.2}$$

where $\alpha, \beta \leq 2$ and A, B, t_0, x_0 are some constants. Applying the expansions (2.2) to the general solution of the Markov equation, obtained in [8]

$$P(k,u) = \frac{1 - \psi(u)}{u(1 - \psi(u)w(k))} \tag{2.3}$$

one can get different kinds of processes depending on the character of the local properties of the transition probabilities in the (u,k)-space. For example if $x_0 = 0$, $B = 0$, $A \neq 0$ and $\alpha = 2$ then $\psi(u) = 1 + t_0 u$, $w(k) = 1 - Ak^2$, and

$$P(k,u) = t_0(1 - Ak^2)$$

which corresponds to the kinetic equation for the Gaussian process. Many different situations have been considered in [2]. Here we would like to discuss only the case of $x_0, t_0 = 0$, $\beta \neq 1$, $\alpha < 2$ [20].

From the equations (2.2)-(2.3) one has

$$P(k,u) = \frac{Bu^{\beta-1}}{Bu^\beta + Ak^\alpha} \tag{2.4}$$

The values of the exponents (α, β) can be established in the following way. The step length distribution function is [2]

$$w(x) \sim 1/x^{1+\alpha} \tag{2.5}$$

and therefore the mean length of the step $\langle x \rangle$ is finite for $1 < \alpha < 2$ and infinite for $\alpha < 1$. At the same time $\langle x^2 \rangle$ is infinite for $\alpha < 2$.

The distribution of time between steps can be characterized in the same way

$$\psi(t) \sim 1/t^{1+\beta} \qquad (2.6)$$

and the mean time between steps (hopping time)

$$\langle t \rangle = \int_0^\infty t\psi(t)dt \qquad (2.7)$$

is finite for $\beta > 1$ and is infinite for $\beta < 1$. The $\langle t \rangle$ is finite also if there is exponential (for example Poissonian) distribution instead of (2.6).

For $\beta < 1$ we have a trap and $1 < \alpha < 2$ we have a flight. General description of a process is a combination of traps and flights. There is no characteristic time scale or length scale in the case of a trap or flight.

One can operate directly with the moments of full probability density $P(x,t)$. For example using (2.1),(2.4) we can get directly [20]

$$\langle x^2 \rangle = \int_{-\infty}^\infty x^2 P(x,t)dx$$

$$= \frac{1}{2\pi i}\,\frac{1}{2\pi}\int_{c-i\infty}^{c+i\infty} du \int_{-\infty}^\infty dk \int_{-\infty}^\infty dx\, e^{ut-ikx} x^2 \frac{Bu^{\beta-1}}{Bu^\beta + Ak^\alpha} \sim t^{2\beta/\alpha} \quad (2.8)$$

which gives for the transport exponent in (1.2)

$$\mu = \beta/\alpha \qquad (2.9)$$

This important formula expresses the transport properties through the local exponents (α, β) of the probabilities of step length and time. It will be shown below how the local exponents (α, β) can be obtained from the dynamics in a straightforward way.

3 Weierstrass Random Walk

This section outlines the results of [10] (see also in [2]). Let us consider a random walk process in which a walker's step can be $\pm b^m$ with the probability a_m where m is an integer. Then the probability that the walker makes a step of the length x is

$$p(x) = \text{const} \sum_{m=0}^\infty a_m(\delta_{x,b^m} + \delta_{x,-b^m}) \qquad (3.1)$$

where $\delta_{x,y}$ is the Kronecker symbol. The jumps in this example could be $0, \pm b, \pm b^2$ and so on. Weierstrass random walk can be defined as a process with probability density (3.1) and with $a_m = a^{-m}$ where a is a constant:

$$p(x) = \frac{a-1}{2a} \sum_{m=0}^{\infty} a^{-m} (\delta_{x,b^m} + \delta_{x,-b^m}) \tag{3.2}$$

In the process (3.2) the steps are powers of a constant, as are the probabilities of the steps. Such a process has been used to describe the so-called St. Petersburg paradox of Bernoulli [2].

The characteristic function of (3.2) is

$$p(k) = \sum_x e^{ikx} p(x) = \frac{a-1}{a} \sum_{m=0}^{\infty} a^{-m} \cos(b^m k) \tag{3.3}$$

and coincides with definition of the Weierstrass function. There exists a renormalization transform for (3.3)

$$p(k) = \frac{1}{a} p(bk) + \frac{a-1}{a} \cos k \tag{3.4}$$

which permits to present $p(k)$ in the form

$$p(k) = p_n(k) + p_s(k) \tag{3.5}$$

with analytic part $p_n(k)$ in the neighborhood of $k = 0$, and singular part $p_s(k)$ at $k = 0$. The expressions for the p_s, p_n are known [10] and we present here only $p_s(k)$:

$$p_s(k) = |k|^{1/\mu_W} Q(k) \tag{3.6}$$

where $Q(k)$ is a periodic function of $\ln k$ with period $\ln b$, and

$$\mu_W = \ln b / \ln a \tag{3.7}$$

There is now a direct connection between the process $p(k)$ and the time-dependent random walk. Nevertheless some speculations can be made [10,21] to construct a master equation type of kinetics which leads to the transport exponent $\mu = \mu_W/2$ [21]. We are not discussing this here because more general and adequate construction will be used in the following sections. Nevertheless it will be seen that the property of having power-type steps of the wandering process, and the corresponding power-type tails for the probabilities is the characteristic feature of the anomalous kinetics in dynamical chaos.

4 Islands in the Stochastic Sea

In real Hamiltonian systems exhibiting chaotic dynamics, the phase space can be described in a simplified way as an area of chaotic motion (stochastic sea) with islands imbedded into the sea. A chaotic trajectory in the stochastic sea cannot penetrate into the islands, and avoids them in an extremely complicated way which we describe below. Motion inside the islands is mainly regular (quasiperiodic) although there is an infinite number of thin filaments isolated from the stochastic sea (stochastic layers or webs) with the chaotic dynamics inside the filaments. For many cases particle motion in the stochastic sea can be described by the FPK kinetic equation with acceptable accuracy. Even the influence of the excluded volume due to presence of the islands can be satisfactorily considered by modifying the diffusional coefficient, but without changing the main structure of the kinetic equation [22,23].

New understanding of the kinetic description of chaotic dynamics comes from more accurate analysis of particle motion in the vicinity of the islands' boundary. Let us say that, roughly speaking, each island can be surrounded by a thin annulus-like strip called boundary layer. It has been long known that when a particle hits the boundary layer, it can spend a long time doing many revolutions around the island, before it leaves the layer. There are numerous publications of simulations which confirm that chaotic trajectories can be divided into two parts: the normal (n) part which represents wandering in the stochastic sea sufficiently far from the islands boundary, and the anomalous (an) part representing trappings inside the islands' boundary strips.

The main conjecture of the new kinetic theory of chaos is that the long time asymptotics is defined by the anomalous part of trajectories. The reason for this was formulated in [20] in a fairly simple form. Let us present the mean square displacement in a form

$$\langle x^2 \rangle = c \langle x^2 \rangle_n + (1 - c)\langle x^2 \rangle_{an} \tag{4.1}$$

where the averaging over both the normal parts of orbits and the anomalous ones is performed in any approximate way, and c or $(1-c)$ are corresponding measures of these parts. One can even assume that $1 - c \ll 1$. Then we have asymptotically

$$\langle x^2 \rangle_n \sim t , \qquad \langle x^2 \rangle_{an} \sim t^{2\mu} \tag{4.2}$$

with μ close to one (for example). For any small value of $(1 - c)$ one can expect that

$$(1 - c)t^{2\mu} > ct , \qquad \mu > 1/2 \tag{4.3}$$

if t is sufficiently big. This effect has not been observed for a long

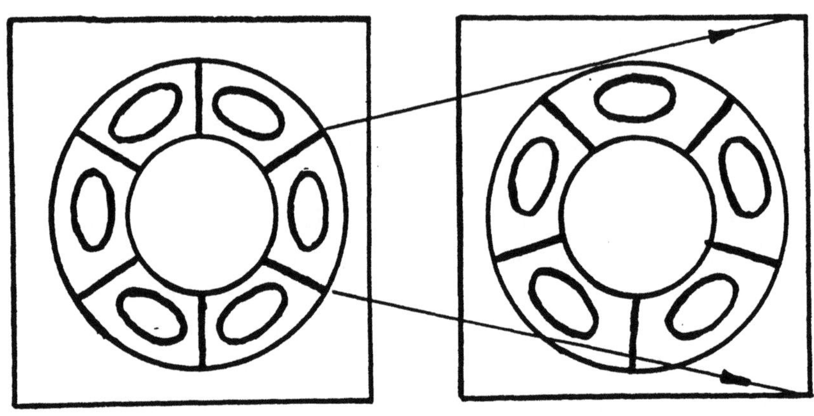

Fig. 1

time. One can suppose that it was overshadowed by the errors of simulation which produce a noise-like effect. Such a possibility was demonstrated in [21].

Another conjecture, based mainly on simulations, was formulated in [15,20]. It consists in that the anomalous transport is the most effective when the control parameter (perturbation parameter) is in the vicinity of a bifurcation value. This is similar to critical phenomena where the transport is anomalous near a critical point. We can add to these arguments that experimental observation of the anomalous transport in [18,19] shows that it is dependent on how close the observed system is to some critical regimes.

Having all this in mind we can describe a more detailed pattern of the islands' stickiness. A schematic picture is given in Fig. 1 where each island has in its vicinity a number of subislands which are surrounded by small subislands as well. This picture can be continued to infinity and it will be considered as approximately self-similar. Self-similarity of the islands-around-islands pattern was considered in [24-25] and in more detail in [21,26].

5 Space-Time Self-Similarity

In the description of the self-similar pattern of chaotic trajectories in the islands boundary layer we use the results of [12]. Let us surround an island with the annulus (Fig. 1) which contains a number of subislands inside it, and let us do a partition of the annulus as it is shown in Fig. 1 so that each element (curved rectangle) includes one subisland of the first generation. This partitioning can be continued to the infinity.

Let n is an order of generation and q_n is the number of islands of the n-th generation. Then existence of the self-similarity means

$$q_n = q_0 \lambda_q^n = q_0 \exp(n/\alpha_q)$$

$$\alpha_q = 1/\ln \lambda_q \ . \tag{5.1}$$

Here q_0 is a constant corresponding to the initial number of islands of the same self-similarity law, and λ_q is a scaling parameter, α_q is a scaling exponent that corresponds to the island's number. In real systems a slow function of $n, \bar{q}(n)$, may replace the q_0 in (5.1).

By analogy one can write down the self-similarity laws for the islands area

$$S_n = \bar{S}(n)\lambda_S^n = \bar{S}(n)\exp(n/\alpha_S) \ ,$$

$$\alpha_S = 1/\ln \lambda_S \ , \tag{5.2}$$

for the Lyapunov exponents near the island boundary

$$\sigma_n = \bar{\sigma}(n)\lambda_\sigma^n = \bar{\sigma}(n)\exp(n/\alpha_\sigma)$$

$$\alpha_\sigma = 1/\ln \lambda_\sigma \ , \tag{5.3}$$

for the period of rotation of the last invariant curve of the islands

$$T_n = \bar{T}(n)\lambda_T^n = \bar{T}(n)\exp(n/\alpha_T)$$

$$\alpha_T = 1/\ln \lambda_T \ , \tag{5.4}$$

and for the lengths of flights

$$\ell_n = \bar{\ell}(n)\lambda_\ell^n = \bar{\ell}(n)\exp(n/\alpha_\ell)$$

$$\alpha_\ell = 1/\ln \lambda_\ell \ , \tag{5.5}$$

where $\bar{S}(n), \bar{\sigma}(n), \bar{T}(n), \bar{\ell}(n)$ could be functions that are slowly (non-exponentially) varying with n.

For the motion with 1 1/2 degrees of freedom we suppose that only two scalings define the motion in the boundary layer of islands: space and time scales. This means that from the set $(\lambda_q, \lambda_S, \lambda_\ell, \lambda_T, \lambda_\sigma)$ only two are independent, and we can consider for example λ_ℓ, λ_T to characterize the space and time behavior of orbits correspondingly.

There is a simple way to find the connection between λ_ℓ and λ_S. Let us specify flight in a more explicit way. Any island can be considered as a section of an invariant torus or cylinder. An orbit trapped into the narrow boundary layer of the annulus around an island is the projection or Poincaré section of the trajectory that wind the torus or cylinder. This means that trappings and flights are complimentary notions depending on how the trajectory is considered. Let us imagine the flux of particles with initial conditions that fill a curved rectangle of the annulus in the boundary layer of the initial island (Fig. 1a), and let all these orbits switch almost simultaneously to the next generation of the boundary layer. The boundary layer of the next generation becomes thinner but the flux should be the same due to the phase volume preservation for Hamiltonian dynamics. From this conclusion we have

$$\mathcal{P}_n \ell_n = \mathcal{P}_{n\pm1} \ell_{n\pm1} \tag{5.6}$$

where $\mathcal{P}_n$ is the perimeter of the n-the generation island. Using $\mathcal{P}_n \sim S_n^{1/2}$ we obtain from (5.2),(5.5),(5.6) that

$$\lambda_\ell \lambda_S^{1/2} = 1 , \qquad \alpha_\ell = -2\alpha_S \tag{5.7}$$

which states that the smaller the island diameter the longer the flight along the corresponding torus.

6 Scaling Exponents and Fractal Dimensions

In the previous section different kinds of scaling exponents were introduced:
$\alpha_q, \alpha_S, \alpha_\ell, \alpha_T, \alpha_\sigma$. All of them can be considered as certain fractal dimensions if a generalized notion of the Pesin's dimension [27,28] is applied.

Let the set of events which are of interest to us is included into a metric space M for which the distance between any two elements $x_1, x_2 \subset M$ is defined and let us introduce arbitrary partition of M by a set of the partition elements $\{u_j\}$. We can select only such elements u_j which cover any nonzero number of events imbedded into M. Then

$$\cup_{j=1}^{N} u_j \subset M \tag{6.1}$$

when N is the number of the elements of partition. As a set of events we can take an appropriate set of the trajectory points in the phase space, for example the points of the Poincaré section. As a set of covering elements we can consider curved rectangles of the annulus that represent a boundary layer of the island of some generation.

Consider a sum

$$\mathcal{I} = \lim_{N \to \infty} \sum_{j=1}^{\infty} \xi(u_j)\phi^{\alpha}(u_j) \tag{6.2}$$

where ξ and ϕ are some functions. According to Pesin [27] the limit (6.2) exists independently of the way of partitioning and choice of functions ξ, ϕ, and for the appropriately selected ξ and ϕ we have

$$\mathcal{I}(\xi, \phi) = \begin{cases} 1 & \alpha = \alpha_0 \\ 0 & \alpha > \alpha_0 \\ \infty & \alpha < \alpha_0 \end{cases} \tag{6.3}$$

The value α_0 will be called Pesin's dimension.

The equations (6.2),(6.3) can be rewritten in the form [27]

$$\frac{1}{N} \sum_{j=1}^{N} \xi(u_j)\phi^{\alpha_0}(u_j)\langle\langle\xi(u_j)\phi^{\alpha_0}(u_j)\rangle\rangle \sim \frac{1}{N} \tag{6.4}$$

where the angle brackets denote mean value and the limit $n \to \infty$ should be performed. For example if $\xi \equiv 1$ and $\phi(u_j)$ is diameter of u_j we have

$$\langle\langle(\text{diam } u_j)^{\alpha_0}\rangle\rangle \sim \frac{1}{N} \tag{6.5}$$

and α_0 in this case is simply the Hausdorff dimension. Taking diam $u_j = \epsilon$ we obtain

$$N(\epsilon) \sim \epsilon^{-\alpha_0} \tag{6.6}$$

and in (6.6) α_0 is box dimension.

For the case $\xi \equiv 1$ and $\phi^{\alpha}(\text{ diam } u_j)$ in (6.2) instead of $\phi^{\alpha}(u_j)$ we have Caratheodory's dimension definition. The advantage of the Pesin's formula is that dimension-like exponent can be obtained for a wide range of variables with different physical meaning. In [27] the dimension was introduced for the Poincaré recurrences. From the numerical evidence the exponents $\alpha_S, \alpha_\ell, \alpha_T, \alpha_q, \alpha_\sigma$ introduced in Sec. 5 can be treated as Pesin's dimension up to a certain multiplier.

Let us take for example the periods of the last invariant curve of the set of islands. The partitioning introduced in Sec. 5 gives the number of islands of the n-th generation:

$$N(n) = q_n = q_0 \exp(n/\alpha_q) \tag{6.7}$$

in correspondence with (5.1). Consider (6.4) in the form that is adjusted to our partition:

$$\xi(u_j(n))\phi^{\alpha_{0T}}(u_j(n)) \sim 1/N(n) \tag{6.8}$$

If $\phi(u_j(n))$ is simply T_n then due to (5.4) and (6.7) the equation (6.8) can be transformed to

$$\xi(u_j(n)) \exp(n\alpha_{0T}/\alpha_T) \sim q_0 \exp(-n/\alpha_q)$$

and in the limit $n \to \infty$ we obtain

$$\alpha_{0T} = -\alpha_T/\alpha_q \tag{6.9}$$

which means that α_T is Pesin's dimension normalized to the factor $(-1/\alpha_q)$.

In full analogy to (6.9) one can obtain from (5.1),(5.2),(5.3),(5.5) the corresponding relations

$$\alpha_{0q} = 1$$
$$\alpha_{0S} = \alpha_S/\alpha_q$$
$$\alpha_{0\sigma} = -\alpha_\sigma/\alpha_q$$
$$\alpha_{0\ell} = -\alpha_\ell/\alpha_q \tag{6.10}$$

Self-similarity and the corresponding dimension α_S for the island areas were introduced in [24] by Meiss for the island-around-island structure. It is important comment that non-exponential, slow variation of the weight coefficients $\xi(u_j)$ in (6.2) probably fits the real situation of the dynamical chaos better than the case of $\xi(u_j) = $ const.

7 Fractional Kinetic Equation

In this section a fractional generalization of the FPK equation (we call it FFPK) will be discussed. Two, space (α_ℓ) and time (α_T) scaling exponents can be used to derive the FFPK. Here we give a brief description of the derivation (see [12] for details).

Starting with the standard Markov equation

$$W(x_3, t_3; x_1, t_1) = \int dx_2 \, W(x_3, t_3; x_2, t_2) W(x_2, t_2, x_1, t_1) \qquad (7.1)$$

for the transition probability density W we assume that

$$W(x, t; y, t_0) = W(x, y; t - t_0) \qquad (7.2)$$

For small $\Delta t = t - t_0$ there is an obvious relation

$$\lim_{\Delta t \to 0} W(x, y; \Delta t) = \delta(x - y) \qquad (7.3)$$

which plays the role of the initial condition.

For $t \gg t_0$ denote

$$W(x, t; y, t_0) \equiv_{t \gg t_0} P(x, t) \qquad (7.4)$$

The difference in notations $W(x, y; \Delta t)$ and $P(x, t)$ is to underline the small time scale probability distribution $(\Delta t \to 0)$ and large time scale distribution $(t \to \infty)$.

With new notations the initial equaiton (7.1) can be rewritten in the form

$$P(x, t + \Delta t) - P(x, t) = \int dy [W(y, x; \Delta t) - \delta(y - x)] P(y, t) \qquad (7.5)$$

where we extract $P(x, t)$ from the left and right sides. The main idea in deriving the kinetic equation follows the original work by Kolmogorov [29] with two differences: instead of the Taylor series expansion for P and W the fractional derivatives expansion is applied, and the delta-function expansion for W is used instead of the functional scheme introduced in [29].

For the $P(x, t + \Delta t)$ we can write down a generalized expansion

$$P(x, t + \Delta t) = P(x, t) + (\Delta t)^\beta \frac{d^\beta P(x, t)}{dt^\beta} \qquad (7.6)$$

where $0 < \beta < 1$ and the next terms of the expansion are omitted. The definition of the fractional derivative can be found in [30]. We are using the Schwartz's formula

$$\mathcal{D}^\alpha f(x) \equiv \frac{d^\alpha f}{dx^\alpha} = \frac{x_+^{-\alpha - 1}}{\Gamma(-\alpha)} \star f(x) , \qquad (\alpha > 0) \qquad (7.7)$$

where Γ is the gamma-function, sign $(\star)$ means convolution

$$g(x) \star f(x) = \int_{-\infty}^{x} dy \ g(x-y)f(y) \tag{7.8}$$

and

$$x_+^{-\alpha-1} = \begin{cases} x^{-\alpha-1} & x > 0 \\ 0 & x < 0 \end{cases} \tag{7.9}$$

In the case of Re $\alpha < 0$ equation (7.7) defines the fractional integration. The following properties of the fractional derivatives are useful

$$\mathcal{D}^\alpha \mathcal{D}^\beta f(x) = \mathcal{D}^{\alpha+\beta} f(x)$$

$$\mathcal{D}^\alpha [\delta^{(k)}(x)] = \frac{x_+^{k-\alpha-1}}{\Gamma(-k-\alpha)} \tag{7.10}$$

$$\left(\frac{d^\alpha f(x)}{dx^\alpha} \cdot g(x) \right) = \left(f(x) \cdot \frac{d^\alpha g(x)}{d(-x)^\alpha} \right)$$

where the inner product is used in the last equatoin

$$(f_1(x) \cdot f_2(x)) \equiv \int_{-\infty}^{\infty} dx \ f_1(x)f_2(x) \tag{7.11}$$

It appears natural to use the fractional calculus to describe self-similar wandering along the fractal space-time set which will be demonstrated in the next section.

Due to the condition (7.3) the expansion of $W(x,y;\Delta t)$ over δ-function and its derivatives can be used

$$W(x,y;\Delta t) =_{\Delta t \to 0} \delta(x-y) + A(y;\Delta t)\delta^{(\alpha)}(x-y) + \frac{1}{2}B(y;\Delta t)\delta^{(2\alpha)}(x-y) \tag{7.12}$$

where the coefficients A, B can be defined for $0 < \alpha < 1$ by multiplying (7.12) by $1, |x-y|^\alpha, |x-y|^{2\alpha}$ concievably, and integrating over y

$$\int dy \ W(x,y;\Delta t) = 1$$

$$\int dy |x-y|^\alpha W(x,y;\Delta t) = A(x;\Delta t) \equiv \langle\langle |x-y|^\alpha \rangle\rangle$$

$$\int dy |x-y|^\alpha W(x,y;\Delta t) = B(x;\Delta t) \equiv \langle\langle |x-y|^{2\alpha} \rangle\rangle$$

where the first line is simply the normalizatoin condition and two other ones give the definition of A, B as moments of the transitional probability W.

Using the expansions (7.6),(7.12) and auxiliary equations (7.10),(7.13), we can transform the basic equation (7.5) applying the limit $\Delta t \to 0$

$$\frac{\partial^\beta P(x,t)}{\partial t^\beta} = \frac{\partial^\alpha}{\partial(-x)^\alpha} \left(\mathcal{A}(x)P(x,t)) + \frac{1}{2} \frac{\partial^{2\alpha}}{\partial(-x)^{2\alpha}}(\mathcal{B}(x)P(x,t) \right) \quad (7.14)$$

which is FFPK equation (see for details in [12]). Here

$$\mathcal{A}(x) = \lim_{\Delta t \to 0} \frac{1}{(\Delta t)^\beta} A(x; \Delta t) = \lim_{\Delta t \to 0} \frac{\langle\langle |\Delta x|^\alpha \rangle\rangle}{(\Delta t)^\beta}$$

$$\mathcal{B}(x) = \lim_{\Delta t \to 0} \frac{1}{(\Delta t)^\beta} B(x; \Delta t) = \lim_{\Delta t \to 0} \frac{\langle\langle |\Delta x|^{2\alpha} \rangle\rangle}{(\Delta t)^\beta} \quad (7.15)$$

The FFPK equation is reduced to the regular FPK equation if $\alpha = \beta = 1$. The equation (7.15) presents the important conditions for the derivation of the FFPK equation, assuming the existence of finite non-zero limits if $\Delta t \to 0$. One can consider these conditions as generalized Kolmogorov conditions for the FPK equation with $\alpha = \beta = 1$.

8 Renormalization Group of Kinetics (RGK)

The derivation of the FFPK equation is performed in a rather formal way, and here we are going to link the equation to the chaotic dynamics. Consider a process

$$\ell_n \; \overrightarrow{} \; \ell_{n\pm1} \quad (8.1)$$

in which a particle trajectory switches from the flight of the n-th order generation to the $(n \pm 1)$-th order and backward. The n-th order flights were introduced in Sec. 5 as a long time trajectory part winding around a torus of the n-th generation torus. The corresponding characteristic time scale of the flights $t_n \sim T_n$ where T_n is the period of the last invariant curve of the n-th order island.

Let us start with the assumption that there exists a local self-similarity in space-time for the process described by the FFPK equation (7.14). The self-similarity can be described by the invariance of the Eq. (7.14) relative to the transform

$$\hat{R}_k : \; \Delta x_{n+1} = \lambda_x \Delta x_n \, , \qquad \Delta t_{n+1} = \lambda_t \Delta t_n \quad (8.2)$$

with some scaling parameters λ_x, λ_t. We call (8.2) RGK-transform. Using (8.2) and the definition (7.15) gives

$$\hat{R}_k \mathcal{A} = (\lambda_x^\alpha / \lambda_t^\beta)\mathcal{A}$$

$$\hat{R}_k \mathcal{B} = (\lambda_x^{2\alpha} / \lambda_t^\beta)\mathcal{B} \tag{8.3}$$

or applying $\hat{R}_k$ m times

$$\hat{R}_k^m \mathcal{A} = (\lambda_x^\alpha / \lambda_t^\beta)^m \mathcal{A}$$

$$\hat{R}_k^m \mathcal{B} = (\lambda_x^{2\alpha} / \lambda_t^\beta)^m \mathcal{B} \tag{8.4}$$

The limit $m \to \infty$ provides the existence of a fixed point under the condition

$$\frac{\alpha}{\beta} = \frac{\ln \lambda_t}{\ln \lambda_x} \tag{8.5}$$

or

$$\frac{2\alpha}{\beta} = \frac{\ln \lambda_t}{\ln \lambda_x} \tag{8.6}$$

and the choice of (8.5) or (8.6) depends on the actual values of $\mathcal{A}$ or $\mathcal{B}$.

Using formula (8.5) or (8.6) one can express the fractional exponents of derivatives in the FFPK equation through the scaling parameters of the local self-similarity of trajectories in the phase space. Now we can use the conjecture:

$$\lambda_t = \lambda_T , \qquad \lambda_x = \lambda_\ell \tag{8.7}$$

to describe the trapping-flight pattern of the dynamical chaos of Hamiltonian systems [21,26]. From (8.7) and (5.4),(5.5) we have

$$1/\mu_0 \equiv \alpha/\beta = \alpha_\ell/\alpha_T = \frac{\ln \lambda_T}{\ln \lambda_\ell} \tag{8.8}$$

Using notations (6.9),(6.10) and (5.7) we transform (8.8) into the form

$$\mu_0 = \beta/\alpha = \alpha_{0T}/\alpha_{0\ell} \tag{8.9}$$

where α_{0T} and $\alpha_{0\ell}$ are the corresponding Pesin's dimensions.

9 Transport Exponent

In the Sec. 1 the transport exponent μ has been introduced to describe the large time asymptotics of the second moment $< x^2 >$. For a self-similar kinetics like the one described by the FFPK equation (7.14), one can expect a power-like self-similarity for different moments $< |x|^\gamma >$ where γ does not have to be an integer. The corresponding asymptotics can be obtained directly from the equation (7.14) where only one term in the right hand side will be taken into account ($\mathcal{A}$ or $\mathcal{B}$) for simplicity.

Multiplying equation (7.14) by x^α and its integrating we obtain for the case when the term with $\mathcal{B}$ can be neglected

$$< x^\alpha > \sim t^\beta , \qquad (x > 0) \tag{9.1}$$

or

$$< |x| > \sim t^{\beta/\alpha} \tag{9.2}$$

which gives

$$\mu = \beta/\alpha = \mu_0 = \frac{\ln \lambda_\ell}{\ln \lambda_T} \tag{9.3}$$

where we use (8.8). In full analogy to (9.2),(9.3) we obtain

$$\mu = \beta/2\alpha = \mu_0 = \frac{\ln \lambda_\ell}{\ln \lambda_T} \tag{9.4}$$

when $\mathcal{B}$ is significant and the term with $\mathcal{A}$ can be neglected.

Formulas (9.3),(9.4) give the transport exponent as a function of the local scaling properties of trajectories in the phase space, i.e. μ is expressed through the λ_ℓ and λ_T. For $\alpha = \beta = 1$ we have the standard Gaussian result from (9.4).

10 Maxwell's Demon and Fractional Kinetics

In his 1871 book "Theory of Heat", James Clerk Maxwell described an intelligent device which later was named *Maxwell's Demon* by William Thomson (1974). The initial description of Maxwell's design was very simple. There are two equal cameras with gas and there is a small hole in the division of the cameras (Fig. 2). The Demon is operating near a hole in such a way that low energy particles are allowed to pass from B to A, and swifter particles are allowed to pass from A to B. After such an operation the gas, initially being in the thermal equilibrium ($T_A = T_B$), raises the temperature of B without expenditure of work in contradiction to the second law of thermodynamics. There is no precise description of Maxwell's Demon and its operational system and the discussion of

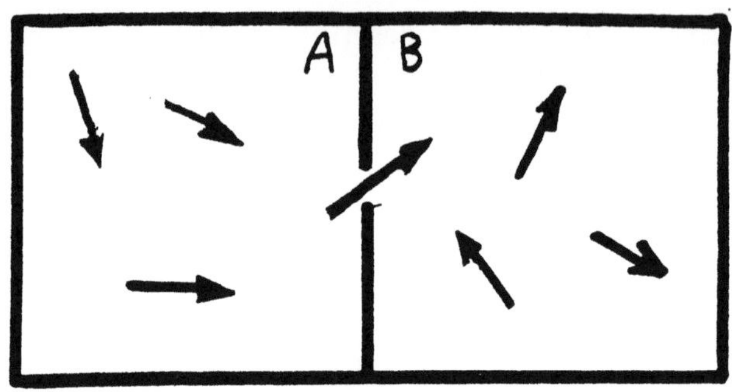

Fig. 2

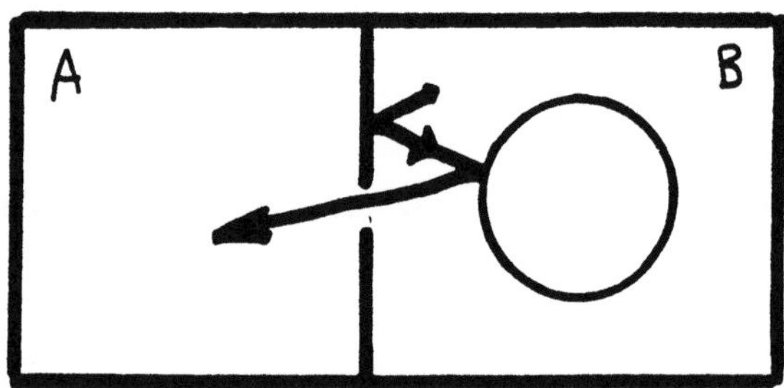

a

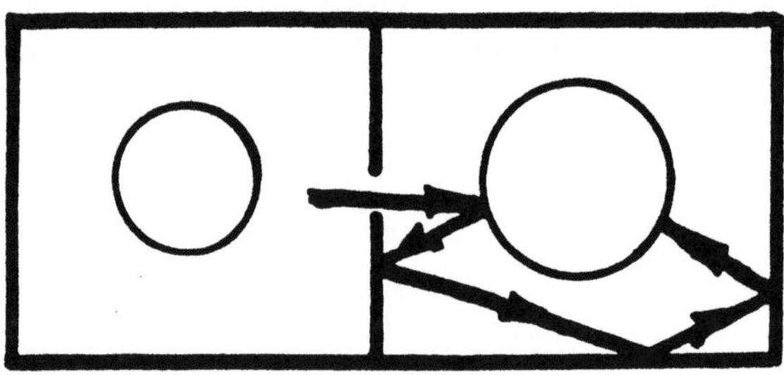

b

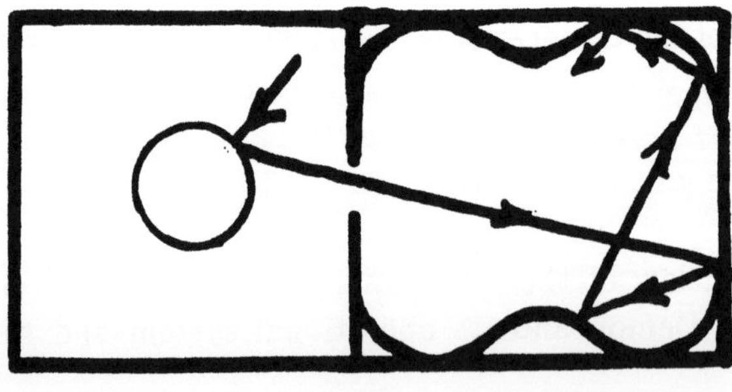

c

Fig. 3

the possibility of the Maxwell's Demon existence still continues involving more and more sophisticated analysis using classical and quantum physics [31]. More information on the problem of Maxwell's Demon can be obtained from [32].

Understanding of the dynamical chaos made possible new insights into the origin of the statistical laws [33], and in particular, a new look at the famous Zermelo's and Locshmidt's paradoxes. In this section we would like to present a new way to design Maxwell's Demon using fundamental properties of the dynamical chaos.

Maxwell's design of the Demon is fairly general and one can imagine any kind of nonequilibrium state created by the Demon's operation, for example the difference in the gas density or pressure between the cameras. But the important issue abandoned by Maxwell deliberately was the dynamical way of the consideration of the entire structure cameras + demon: "In dealing with masses of matter, while we do not perceive the individual molecules, we are compelled to adopt what I have described as the statistical method of calculation, and to abandon the strict dynamical method, in which we follow every motion by the calculus" [34]. Now we have the possibility to look at the problem of the Maxwell's Demon existence from the dynamical point of view.

Let us consider the initial two cameras with a hole in the cameras' division (Fig. 2), but without a Demon. No temperature is assumed. For a general initial condition a particle dynamics in the cameras is ergodic but not mixing. Imagine that one of the cameras (say B) is the Sinai's billiard (Fig. 3a) and let volumes $V_A = V_B'$ where prime means that the ball volume is excluded. Then the motion is ergodic and mixing in both cameras (!). This mixing motion is equivalent to the thermolized motion with some distribution functions $P_A(\tau), P_B(\tau)$ of time τ that the particle spends in a camera between two consequent passings the hole. The flux through the hole should be zero because of the Liouville theorem on the phase volume preservation. One can expect the existence of finite mean times

$$\tau_A = \lim_{t \to \infty} \int_0^t d\tau \, \tau \, P_A(\tau)$$

$$\tau_B = \lim_{t \to \infty} \int_0^t d\tau \, \tau \, P_B(\tau) \tag{10.1}$$

of the particle participation in a camera A or B between two consequent entrances (exits) of the camera. For a system with "good" mixing properties the limit in (10.1) exists and it is equal to the phase space average over the particles ensemble. We can also consider the situation when two different Sinai's billiards correspond to camers A and B (Fig. 3b). The actual times that a particle spend in each of the camer A or B are equal $(\tau_A = \tau_B)$ if the volumes are equal: $V_A' = V_B'$.

The actual Hamiltonian systems of general case are different from the Sinai's billiard. They possess strange kinetics as it was described in the previous sections. In particular, by changing the camera shape (Fig. 3c) or introducing a potential field in the camera we receive dynamical traps and flights which drastically change the time-averaged variables and destroy good properties of the convergency mean time values to their phase averaged values. As it was seen in Sec. 2 the mean time τ_B can be infinite. The more we are observing the system, the longer living fluctuations occur which traps a particle into camera B producing a nonequilibrium state. In this way Maxwell's Demon can be considered as a field which generates chaotic dynamics with "wrinkles of memory" in the phase space. A particle can be trapped into an almost regular motion of a very long duration. Speaking about the time during which there was a deviation of the system's state from equilibrium, we can deduce that this time tend to infinity, on average, if the system possesses the anomalous kinetics property.

This result will be discussed in more detail in a separate publication.

Acknowledgments

I am very thankful to V. Afraimovich for the discussions and for the possibility to get his results before their publication. This work was supported by the U.S. Department of the Navy, Grant No. N00014-93-1-0218.

REFERENCES

1. P. Lévy, Theorie de l'Addition des Variables Aleatoires (Gauthier-Villiers, Paris, 1937).

2. E.W. Montroll and M.F. Shlesinger, in Studies in Statistical Mechanics, Edited by J. Lebowitz and E. Montroll (North-Holland, Amsterdam, 1984), Vol. 11, p. 1.

3. B. Mandelbrot, The Fractal Geometry of Nature (W.H. Freeman, San Francisco, 1982); B.B. Mandelbrot and J.W. Van Ness, SIAM Rev. **10**, 422 (1968).

4. J.P. Bouchaud and A. Georges, Phys. Rep. **195**, 127 (1990).

5. M.F. Shlesinger, Physica D **38**, 304 (1989).

6. M.F. Shlesinger, Ann. Rev. Phys. Chem. **39**, 269 (1988).

7. M.F. Shlesinger, G.M. Zaslavsky, and J. Klafter, Nature **363**, 31 (1993).

8. E.W. Montroll and G.H. Weiss, J. Math. Phys. **6**, 167 (1965).

9. J. Klafter, A. Blumen, and M.F. Shlesinger, Phys. Rev. A **35**, 3081 (1987).

10. B.D. Hughes, M.F. Shlesinger, and E.W. Montroll, Proc. Natl. Acad. Sci. USA **78**, 3287 (1981); B.D. Hughes, E.W. Montroll, and M.F. Shlesinger, J. Stat. Phys. **28**, 111 (1982).

11. G.M. Zaslavsky, in Topological Aspects of the Dynamics of Fluids and Plasmas, Edited by H.K. Moffatt, et al. (Kluwer, Dordrecht, 1992), p. 481.

12. G.M. Zaslavsky, Chaos 4, 25 (1994); Physica D **76**, 110 (1994).

13. C.F.F. Karney, Physica D **8**, 360 (1983); V.V. Beloshapkin and G.M. Zaslavsky, Phys. Lett. A **97**, 121 (1983); B.V. Chirikov and D.L. Shepelyanski, Physica D **13**, 394 (1984).

14. T. Geisel and S. Thomae, Phys. REv. Lett. **52**, 1936 (1984); T. Geisel and J. Nierwetberg, Phys. Rev. Lett. **48**, 7 (1982); Z. Phys. B **56**, 59 (1984); T. Geisel, J. Nierwetberg, and A. Zacherl, Phys. Rev. Lett. **54**, 616 (1985); T. Geisel, A. Zacherl, and G. Radons, Phys. Rev. Lett. **59**, 2503 (1987); Z. Phys. B **71**, 117 (1988).

15. A.A. Chernikov, B.A. Petrovichev, A.V. Rogalsky, R.Z. Sagdeev, and G.M. Zaslavsky, Phys. Lett. A **144**, 127 (1990); D.K. Chaikovsky and G.M. Zaslavsky, Chaos **1**, 463 (1991); G.M. Zaslavsky, R.Z. Sagdeev, D.K. Chaikovsky, and A.A. Chernikov, Sov. Phys. JETP **68**, 995 (1989).

16. R. Ishizaki, T. Morita, T. Kobayashi, and H. Mori, Progr. Theor. Phys. **85**, 1013 (1991); S. Benkadda, Y. Elskens, and B. Ragot, Phys. Rev. Lett. **72**, 2859 (1994); J. Klafter, G. Zumofen, and M. Shlesinger (see this volume).

17. O. Cardoso and P. Tabeling, Europhys. Lett. **7**, 225 (1988).

18. R. Ramshankar, D. Berlin, and J.P. Gollub, Phys. Fluids A **2**, 1955 (1990); R. Ramshankar and J.P. Gollub, Phys. Fluids A **3**, 1344 (1991).

19. T.H. Solomon, E.R. Weeks, and H.L. Swinney, Phys. Rev. Lett. **71**, 3975 (1993) (see also this volume).

20. V.V. Afanas'ev, R.Z. Sagdeev, and G.M. Zaslavsky, Chaos **1**, 143 (1991).

21. G.M. Zaslavsky, D. Stevens, and H. Weitzner, Phys. Rev. E **48**, 1683 (1993).

22. A.B. Rechester, M.N. Rosenbluth, and R.B. White, Phys. Rev. Lett. **42**, 1249 (1979).

23. I. Dana, N.W. Murray, and IC. Percival, Phys. Rev. Lett. **62**, 233 (1989); J.D. Meiss and E. Ott, Phys. Rev. Lett. **55**, 2741 (1985).

24. J.D. Meiss, Phys. Rev. A **34**, 2375 (1986); Rev. Mod. Phys. **64**, 795 (1992); J.D. Hanson, J.R. Cary, and J.D. Meiss, J. Stat. Phys. **39**, 327 (1985).

25. B.V. Chirikov, Chaos, Solitons, Fractals **1**, 79 (1991).

26. G.M. Zaslavsky and S.S. Abdullaev (to be published).

27. Ya. Pesin, Russian Math. Surveys **43**, 111 (1988).

28. V. Afraimovich and Ya. Pesin (to be published).

29. A.N. Kolmogorov, Uspekhi Mat. Nauk **5**, 5 (1938).

30. I.M. Gelfand and G.E. Shilov, Generalized Functions (Academic Press, New York, 1964), Vol. 1; K.B. Oldham and J. Spanies, The Fractional Calculus (Academic Press, New York, 1974).

31. Maxwell's Demon, Editors H.S. Leff and A.F. Rex (Princeton Univ. Press, Princeton 1990).

32. H.S. Leff and A.F. Rex, see [31], pp. 1-32.

33. G.M. Zaslavsky, Chaos in Dynamic Systems (Harwood Acad. Publ., NY, 1985).

34. In quotation we follow to [31], p. 4.

PART 4:

LÉVY FLIGHTS AND STATISTICAL MECHANICS

MORE LEVY DISTRIBUTIONS IN PHYSICS

J.P. Bouchaud

Service de Physique de l'Etat Condensé, CEA, Orme des Merisiers, 91
191 Gif s/ Yvette CEDEX

Abstract: We give a brief survey of some important results concerning power-law distributions and extreme statistics. We then review several recent physical realizations of Lévy statistics, in quite different contexts (Diffusion in micelles, laser cooling, aging in glassy systems, etc.), as well as new theoretical progress, such as the theory of random matrices with broadly distributed elements, developed by Pierre Cizeau. Finally, we describe interesting new paths (in particular in finance) and open problems.

0 INTRODUCTION

Despite quite a number of early insightful studies [1-5], the fact that many natural phenomena must be described by power law statistics has only been fully accepted in the past decade (see e.g. [7,8]). Correspondingly, an intense activity has developed in order to understand both the physical content of the mathematical tools devised by P. Lévy and others and the origin of these ubiquitous power law tails. This has lead to several interesting ideas, in particular the seminal concept of 'self-organized criticality' [9].

As a matter of fact, it is in economy and finance that these power law distribution were first noted by Pareto and, in the early sixties, by Mandelbrot and Fama [5,6]. Their proliferation in physics started in the early eighties and the trend is still gaining momentum – as witnessed by the present proceedings. Interestingly, this activity is now bouncing back towards finance, with a rapidly growing impact [10-12].

The research connected to these Lévy statistics roughly follows two main paths: a *descriptive* point of view, which consist in recognizing physical situations where broad distributions come into play and extreme events are of crucial importance, and a more *prospective* one, where one analyses the consequence of broad distributions on models well studied within a gaussian framework and extends the available mathematical toolbox associated to these Lévy distributions. The aim of this paper is to review briefly various pieces of work related to Lévy statistics in which I have been fortunate enough to play a part in recent years. In the first part, I shall recall a few important results on power-law distributions. Then three different physical situations will be scrutinized using Lévy's goggles : laser cooling using 'dark resonances', aging in glassy materials and superdiffusion in giant micelles. In section 5, the theory of 'Lévy matrices' (i.e. random matrices with power law distributed elements) will be summarized. A few open problems and perspectives will then be evoked in the conclusion.

1 POWER LAWS TAILS : WHERE TO SEE THEM AND HOW TO TAME THEM

We shall focus in the sequel on random variables x, which we shall call shall $[\mu]$-variables, distributed according to a certain distribution $\rho(x)$ decaying as $\frac{x_0^\mu}{x^{1+\mu}}$ when x becomes large. We take x's to be all positive for simplicity; x_0 can be interpreted as a typical scale for x. We also stress that the following discussion is only intended to be qualitative, more rigourous statements can be found in [2,3,4,7].

When can one expect to observe physical quantities distributed according to such laws ? There is now a score of 'non trivial' examples of systems at a critical point where a length scale, or time scale diverges leaving the system in self-similar state. The canonical example is the cluster size distribution at the percolation point. This is a fined-tuned criticality, to be contrasted with 'self-organized' criticality where complex systems spontaneously evolve towards scale invariant states. This seems to be the case, out of an ever expanding list, of earthquakes, avalanches, traffic jams, vortex lines pinned by impurities, etc. In these examples, the index μ appearing in the tail of the distribution is a critical exponent which in general is hard to compute and captures the 'universality class' of the phenomenon. An interesting example is the distribution of the annual turnover of the 100 largest European companies in 1992 ($\mu \simeq 2$, see Fig. 1 below), for which no theory seems to exist at present.

There is another class of systems where these power law tails have a simpler origin (although well concealed in certain cases), which is a mere *change of variables*. Suppose that a certain variable y is uniformly distributed, say, between 0 and 1. It may occur, as we shall discuss below, that the variable of physical interest is $x = y^{-\frac{1}{\mu}}$. It is a immediate to see that x is then a $[\mu]$-variable ('Inversion scenario'). Another scenario, which we shall call the 'exponential conspiration', is when a certain quantity E has an exponentially small probability to be large ($P(E) \sim \exp{-\frac{E}{E_0}}$), but that the relevant information is in a variable $\tau = \tau_0 \exp{\frac{E}{kT}}$ (obviously random notations). Then one can easily show that τ is a $[\mu]$-variable with $\mu = \frac{kT}{E_0}$.

Finally, random walks (which are in fact the simplest 'critical objects' [13], offer many power-laws. A well known example is the first return time of a one dimensional random walk, which is a $[\frac{1}{2}]$-variable.

The main property of a $[\mu]$-variable is that all its moments $m_q = <x^q>$ with $q \geq \mu$ are infinite. Of course, these power-law tails are usually truncated, for physical reasons, beyond a certain cut-off value $x_{\max}$. The divergence of $< x^q >$ for $q \geq \mu$ then means that m_q depends sensitively on the precise value of the extreme $x_{\max}$, while for $q < \mu$, m_q is only determined by the 'body' of the distribution.

Suppose that one orders a set of N $[\mu]$-variables x_i in decreasing order. An important result is that the nth one is of order

$$x_n \propto x_0 \left(\frac{N}{n}\right)^{\frac{1}{\mu}} \tag{1}$$

which means that:

i) The largest out of N $[\mu]$-variables is of order $x_0 N^{\frac{1}{\mu}}$.

ii) The difference between the largest and the second largest is also of order $x_0 N^{\frac{1}{\mu}}$: as $N \longrightarrow \infty$ (or $\mu \longrightarrow 0$) the hierachisation of the sequence of x_i becomes more and more pronounced.

iii) The smallest variable is of order x_0 (this gives a clear meaning to the scale x_0.)

iv) Finally, Eq. (1) gives an operational way to recognize $[\mu]$-variables from an ordered set of data : as shown in Fig. 1, plotting the value of the nth variable as a function of the 'rank' n in log-log coordinates should give a straight line with slope $-\frac{1}{\mu}$.

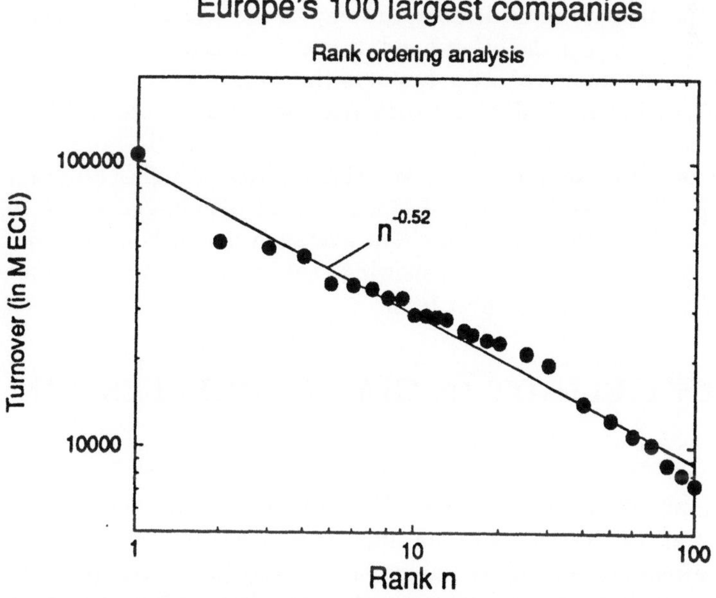

Figure 1 : Rank ordering of Europe 100 largest companies' turnover

Let us now focus on *sums* of N (independent) [μ]-variables, which gives rise to Lévy's famous theorems and stable laws. Define $S = \sum_{i=1}^{N} x_i$: S could be the total mass released by N avalanches, or the total time spent in N traps, or else the total loss of a portfolio containing N assets, etc. If $\mu > 2$, the mean square of x, m_2, is finite and the 'usual' central limit theorem applies : the rescaled variable $\frac{(S-m_1)}{\sqrt{m_2 N}}$ converges, for N large, towards a Gaussian variable. This convergence can however be very slow if μ is close to 2 : for finite N, the *tail* of the distribution of S is still a power-law with an exponent μ, and the the 'anomalous' (non gaussian) weight contained in this tail only decays as $\frac{N^{1-\frac{\mu}{2}}}{(\log N)^{\frac{\mu}{2}}}$ [12]. For $1 < \mu < 2$, the rescaled variable $\frac{(S-m_1)}{N^{\frac{1}{\mu}}}$, or for $\mu < 1$ the rescaled variable $\frac{S}{N^{\frac{1}{\mu}}}$ is distributed, for N large, according to a *Lévy stable distribution*, the Fourier transform of which can be simply expressed [2,3,7]. (These Lévy distributions can also be written in terms of H-functions, see e.g.[14]). The main feature of these Lévy distributions is that they decay themselves, for large arguments, as power-laws with the same index μ as one started with.

A striking fact is that, for $\mu < 1$, the sum S grows as $N^{\frac{1}{\mu}}$, i.e., faster than N but precisely as the largest element of the series of N x_i (see Eq. 1): this is a key feature of Lévy sums, *that the whole sum is, in a first approximation, given by its largest term*. This idea can be made quantitative by introducing a 'participation' (or 'intermittency') ratio. Let $\pi_i \equiv \frac{x_i}{S}$ be the natural weight of x_i in the sum S. Then, obviously, $\sum_{i=1}^{N} \pi_i = 1$. The quantity $Y = \sum_{i=1}^{N} \pi_i^2$ is the 'average' of the weights and behaves very differently when $\mu > 1$ or $\mu < 1$: in the former case, all x_i participate in a 'democratic' way to the whole sum – hence $\pi_i \sim \frac{1}{N}$ and $Y \sim \frac{1}{N} \longrightarrow 0$. On the other hand, if $\mu < 1$, then only a few x_i share the whole meal and Y remains of order 1 even in the limit $N \longrightarrow \infty$. Interestingly, Y is *non self-averaging*, i.e. it does not converge towards a well defined limit for N large, but rather probes a rather non-trivial distribution, diverging near $Y = 1$, with a narrow spike around $Y = \frac{1}{2}$ and singularities of higher order for all values of $Y = \frac{1}{n}$. This was first realized in the context of spin-glasses [15,16], for reasons which will become clear below. The average value of Y is easy to compute : one finds (for $\mu < 1$) $\overline{Y} = 1 - \mu$, showing that in the limit $\mu = 0$, the whole sum S is really equal to the largest term.

2 LEVY FLIGHT IN GIANT MICELLES [17]

A certain class of surfactant molecules aggregate to form, in adequate conditions, very long, worm like, micelles [18]. These objects are very much like polymers (albeit at a different scale); an important difference with the latter is however that micelles break and recombine very rapidly (a typical time is $\tau^* \sim 10^{-3}$ sec.) and in a random manner. This has the immediate consequence that a solution of those micelles cannot be monodisperse - long and short chains coexist. A simple mean-field theory predicts an exponential distribution of (curvilinear) lengths : $P(\ell) \propto \exp -[\frac{\ell}{\ell_0}]$, where ℓ_0 depends both on temperature and concentration.

In order to probe the dynamical properties of such assemblies, a common technique is to incorporate fluorescent molecules to the micelles. The fluorescence is monitored through laser fringes with a certain modulation length $\frac{2\pi}{q}$. The measured intensity I_q directly yields the Fourier transform of the 'diffusion front', $P(R,t)$, which is the probability to find a fluorescent molecule at site R after time t, knowing that it was at $R = 0$ at time $t = 0$. In the case of a simple Brownian diffusion, one thus observes an exponentially decaying signal $I_q \propto \exp[-Dq^2 t]$, which allows one to measure the diffusion constant D. In the case of giant micelles, the surprising result is that I_q indeed decays exponentially with time, but with an anomalous q dependence : $I_q \propto \exp[-Dq^\mu t]$ with $\mu \simeq 1.5$ [17]. This means that the fluorescent probes have an superdiffusive motion: $R \propto t^{\frac{1}{\mu}} \gg \sqrt{t}$.

The interpretation goes as follows. During time τ^* the fluorescent molecule sits on a micelle of a given length ℓ_i and thus undergoes a normal (Brownian) diffusive motion with a diffusion constant $D(\ell_i)$ borrowed from polymer theory : $D(\ell_i) = D_0 \ell_i^{-2\beta}$, with $\beta \simeq 1$ (reptation). After τ^*, the molecule 'changes horse' and rides either a slower one ($\ell_{i+1} > \ell_i$) or a much faster, short one. The motion of the molecule can thus be thought as a succession of hops every τ^*, with hopping length $r_i \simeq \sqrt{D(\ell_i)\tau^*}$. The probability of r_i is simply obtained from that of ℓ_i. One finds:

$$P(r) = \ell P(\ell)\frac{d\ell}{dr} \propto \exp[-\frac{r_0^{\frac{\mu}{2}}}{r^{\frac{\mu}{2}}\ell_0}]\frac{1}{r^{1+\mu}} \underset{r \text{ large}}{\simeq} r^{-1-\mu} \tag{2}$$

with $r_0 \equiv \sqrt{D_0\tau^*}$ and $\mu \equiv \frac{2}{\beta}$ ('Inversion'). [The extra factor ℓ comes from the fact that it is more probable to end up on a long chain than on a short one in proportion of their length]. The total displacement of the molecule is obtained as the sum over $N = \frac{t}{\tau^*}$ individual hops : $R = \sum_{i=1}^{N} r_i$. For $\beta \geq 1$, it is thus a *Lévy flight* (the presence of the exponential in Eq. (2) gives rise to an *effective* value of μ which can be smaller than $\frac{2}{\beta}$ and close to the one observed experimentally). R is almost entirely travelled on the fastest (shortest) micelle encountered before time t. Furthermore, $P(R,t)$ should a Lévy distribution of index μ centered around $R = 0$ (since the jumps are isotropic). But by definition, the Fourier transform of a symetrical Lévy distribution is $\exp[-Dq^\mu t]$ – exactly what is measured experimentally ! This was, to my knowledge, the first time that a Lévy flight was fully characterized experimentally : not only a superdiffusive behaviour was observed, but also the full diffusion front – which is indeed a Lévy distribution.

3 LASER COOLING AND LEVY 'FIGHTS' [19]

Atomic cooling using lasers is a rapidly developing field – extraordinary low temperatures (below 1μK) are now routinely obtained in gaseous phases. A very sucessful method is the 'Velocity Selective Coherent Population Trapping' proposed in [20]. Atoms absord and reemit photons continuously at a certain rate Γ, and exchange momentum with the laser field since each photon carries a momentum $\hbar k$. Atoms thus perform a random walk in momentum space. Using ingeniously the internal degrees of freedom of the atoms, one can devise a situation where this rate strongly depends on the momentum of the atoms. In particular, if the momentum p is small, this rate goes to zero as $\Gamma(p) \propto p^2$. Said differently, *if an atom happens to have a very small velocity, it will remain so for a very long time* $\tau = \Gamma(p)^{-1}$. The process by which an atom reaches a very small velocity state is spontaneous emission, which is a random process. It is thus reasonable to suppose that the sprinkling probability P_0 is uniform around $p = 0$. Thus the probability that an atom is 'trapped' around $p = 0$ for a (long) time τ is obtained through:

$$P(\tau) = P(p)\frac{dp}{d\tau} \propto \frac{P_0}{\tau^{3/2}} \tag{3}$$

(we have assumed a one dimensional situation, which is indeed a common experimental case.) Hence, the trapping time distribution is a [1/2]-variable.

In order to predict what is going to happen [19], one also needs to know how long the atoms will spend out of the $p = 0$ 'trap'. This very much depends on the experimental situation. In the simplest case, the atoms freely diffuse outside the trapping region. The probability to come back to the trap after a time $\hat{\tau}$ is then the probability of first return to the origin of a one dimensional random walk. As mentioned above, $\hat{\tau}$ is also a [1/2]-variable ! In more favorable cases, it is possible to *confine* the motion – and hence to truncate the distribution of $\hat{\tau}$ beyond a certain $\hat{\tau}_{\max}$ Still an other (unfavourable) situation is when the Doppler detuning becomes important. Atoms than are out of resonance and the exchange of momentum slows down. One can then show [19] that $\hat{\tau}$ is a $[\frac{1}{4}]$ variable.

The total experimental time t_w is related to the number N of trapping/detrapping events through $t_w = \sum_{i=1}^{N} \tau_i + \sum_{i=1}^{N} \hat{\tau}_i$. The first sum grows as N^2, while the second grows as N for a confined motion, and as N^2 for free diffusion. The battle between these two sums ('Lévy fight') determines the fraction f of atoms asymptotically trapped in around $p = 0$. One finds that $f = 1$ in the confined case, but tends to a constant less than 1 for the unconfined case (see [19] for a more detailed discussion). One can also predict how these limits are reached when $t_w \longrightarrow \infty$.

Another quantity of interest is the momentum distribution after time t_w. Since $\sum_{i=1}^{N} \tau_i \sim \tau_{\max} \sim t_w$, the most likely momentum is such that $\Gamma(p^*) \sim t_w^{-1}$ or $p^* \sim t_w^{-1/2}$: the momentum distribution $\mathcal{P}(p)$ narrows as the inverse square root of time – this is the cooling mechanism. A more refined analysis reveals that this momentum distribution is however very far from a Maxwellian, and is given by :

$$\mathcal{P}(p) \propto \begin{cases} \frac{1}{\sqrt{t_w p^2}} & \text{for } p \gg p^* \\ \sqrt{t_w} & \text{for } p \ll p^* \end{cases} \tag{4}$$

Note that $\mathcal{P}(p)$ never reaches a stationary state for large t_w : the very same feature will also be present in the model of aging that we shall present now, and has its root in the fact that the trapping times have an infinite mean.

4 AGING IN GLASSY SYSTEMS [21,23]

One of the most striking features of glassy dynamics is the *aging* phenomenon, that is, the fact that most physical properties strongly depend on the time t_w elapsed since the quench from high temperature into the glass phase. For example, the a.c. susceptibility of a spin-glass at a frequency ω decays as $(\omega t_w)^{x-1}$, where x is an exponent less than 1. In fact, the response function $R(t, t')$ does not depend on the *difference $t - t'$*, as in usual equilibrium dynamics, but rather on the *ratio $\frac{t}{t'}$*. Similar effects were observed in other contexts, such as the mechanical or dielectric properties of glasses.

The theoretical investigation of this peculiar type of dynamics is only quite recent, but extremely active [21,22,23]. A simple picture, based on mean-field solutions of the spin-glass problem, was proposed in [21]. A remarkable result found within these mean-field theories [15] is that the (free-)energy distribution of the low-lying states is *exponential*:

$$P(\epsilon) = \frac{x}{T} \exp[-\frac{x|\epsilon|}{T}], \quad \epsilon < 0 \tag{5}$$

with a certain number x which happens to be ≤ 1 in the whole spin-glass phase. Now it is reasonable to assume that the energy barrier Δ that the system will have to cross in order to leave a particular metastable state is related to ϵ; the simplest possibility is that $\Delta = -\epsilon$. The time spent in a certain metastable state is thus given by $\tau = \tau_0 \exp \frac{|\epsilon|}{T}$. The 'exponential conspiracy' then comes into play and generates a power-law distribution $P(\tau) \propto \frac{\tau_0^x}{\tau^{1+x}}$ for these 'trapping' times, where the traps must now be thought of as deep valleys in phase space.

The key point to notice is that the probability to *observe* the system in a given trap is not simply $P(\tau)$ since it is *a priori* more probable to find the system in a state in which it remains for a long time. [Compare with the remark after Eq. (2)]. The naive guess is that the probability to observe the system in a given trap is $\mathcal{P}(\tau, t_w) \propto \tau P(\tau)$. However, when the index x is less than 1, the integral $\int_{\tau_0}^{\infty} d\tau \, \tau P(\tau)$ diverges. The regularization comes from the experimental time scale t_w itself : as usual by now, the sum of the N successive trapping times is of the order of the largest one encountered, itself being of order t_w. More precisely, one can show that:

$$\mathcal{P}(\tau, t_w) = \frac{\tau_0^{x-1}}{\tau^x} g(\frac{\tau}{t_w}) \quad \text{for } x > 1 \tag{6-a}$$

$$\mathcal{P}(\tau, t_w) = \frac{t_w^{x-1}}{\tau^x} g(\frac{\tau}{t_w}) \quad \text{for} \quad x < 1 \tag{6-b}$$

where $g(u)$ is a certain cut-off function decaying as $g(u) \sim u^{-1}$ for large u. Note that for $x < 1$, the microscopic time τ_0 totally disappears from $\mathcal{P}(\tau, t_w)$: t_w is the only relevant time scale for the dynamics, as observed experimentally. Furthermore, the relaxation function is expected (from Eq. (6-b)) to decay as $1 - C(\frac{t}{t_w})^{1-x}$ for $t \ll t_w$, corresponding to the relaxation of 'fast' traps $\tau < t$, and as $(\frac{t_w}{t})^x$ for $t \gg t_w$, corresponding to the fraction of very 'deep' traps still surviving after time t. Such a behaviour is indeed observed in many glassy systems : the initial decay $1 - C(\frac{t}{t_w})^{1-x}$ is better known as a *stretched exponential* $\exp -C(\frac{t}{t_w})^{1-x}$. A somewhat more refined version of this simple picture is needed to account for the data in a quantitative manner [23]. In particular, the $(\omega t_w)^{x-1}$ behaviour of the a.c. susceptibility mentioned above can be accounted for very naturally within this framework. Note that the persistence of a power-law distribution of relaxation times *above the glass temperature* and thus with $x > 1$ has also been reported (see e.g. [24,25])

In summary, we have thus proposed a model in which aging occurs because of the very broad nature of the relaxation time distribution. Correspondingly, the 'equilibrium' state is non-normalizable and thus equilibrium is never fully achieved : only a time dependent, partial equilibrium is reached. This is what we have called 'weak ergodicity breaking' [21,22-b,23]. Let us mention that power-laws seem to be a generic in random systems : for example, the susceptibility of pinned lines (vortex, polymers, etc.) is also a $[\mu]$-variable in the low temperature phase [26,27].

5 THEORY OF LEVY MATRICES

Central limit theorems are at the heart of statistical physics. The theorems on *sums* of random variables, extended by Lévy to $[\mu]$-variables, has been amply discussed above, with many illustrative examples. Such theorems also appear when discussing *non-linear* combinations of random variables, as for example in the polymer in random medium problem [28], or in the theory of random matrices. It is now well known that under rather mild hypothesis, the spectrum of a real, symetrical, random matrix is given by Wigner's semi-circle law; furthermore, the correlation between the positions of the eigenvalues are to a large degree universal [29]. This has various important physical consequences, in particular for electronic transport in mesoscopic devices [30,31].

One of the hypothesis used to derive these results is that the elements of the random matrix have a finite variance. With Pierre Cizeau [32], we have seeked to extend the theory of Gaussian random matrices to 'Lévy matrices', much as the usual central limit theorem had to be extended to treat variables with infinite variance. Apart from this rather mathematical motivation, our interest in the problem again came from spin-glasses [33]. Suppose that one throws at random spins in a three dimensional space. The interaction between these spins is often of the RKKY type, or of dipolar origin. In both cases, this interaction decays as r^{-3}, where r is the distance between spins. If J_{ij} denotes the interaction between spin i and spin j, the probability distribution of J_{ij} induced by the randomness of the spins positions and of the oscillatory nature of the RKKY interaction is obtained from:

$$P(J) = P(r)r^2 \frac{dr}{dJ} \underset{J \text{ large}}{\propto} J^{-1-\mu} \tag{7}$$

with $\mu = 1$ (see [34,35]). Thus, the couplings J_{ij} are again $[\mu]$-variables; the nature of the eigenstates of J then appears quite naturally when one studies the magnetic properties of such systems. Our results are as follows [32]:

• We have calculated exactly the density of eigenvalues for Lévy matrices. In contrast with Gaussian matrices, this spectrum extends over the whole real axis.

• We have found *two* separate localisation transitions in the spectrum, related to two different localisation criteria. The first one focuses on the participation ratio Y, which we already encountered above: if π_i is the weight of an eigenfunction on site i, then Y is defined as $\sum_i \pi_i^2$. For extended states $Y \longrightarrow 0$ for large systems, while Y remains finite for localized states. Another criterion is based on the behaviour of the imaginary part of the self energy near the real energy axis, and is directly related to the transport properties of the system. In general, both criterion give the same localisation transition. In the case of Lévy matrices, however, there exist an intermediate 'mixed' phase where states are both 'extended' and 'localised', depending on one's standpoint.

• Finally, the level spacing distribution was studied. For $1 < \mu < 2$ and in the 'localised' phase, this level spacing distribution does not appear to belong to one of the canonical form (Wigner or Poisson), but is in a new universality class, still to be characterized.

6 CONCLUSION - OPEN PROBLEMS

We have thus discussed several recent examples of Lévy statistics in physics, which add to an already flourishing list – one should cite, among other examples, the conductivity of amorphous samples [36], or of one dimensional conductors [37,38], diffusion in dynamical systems [39,8], in convection rolls [40,41] , etc. The present book is hopefully more comprehensive than this small inventory. The example of laser cooling is particularly interesting because detailed and helpful predictions on possible experimental set-ups could be made using Lévy distributions, which were much harder to reach using the traditional methods of atomic physics.

There are several other subjects and themes which I would like to touch upon before closing. One is theoretical finance, and in particular the problem of optimal portfolios in a Lévy world [6,12]. It is usually thought that diversification reduces risk. This is not the case is the assets are $[\mu]$-variables with $\mu < 1$ - on the contrary, because the largest loss is dominant, diversification increases the total risk ! Diversification for $1 < \mu < 2$ is quite different from diversification in a Gaussian world, and is discussed in detail in [12].

Another interesting proposal is the 'Lévy' stimulated annealing procedure [42] : the idea is to make a 'Lévy flight' in phase space in order to avoid getting trapped in deep local minima, thereby improving the efficiency of the search for a global minimum. A related work is that of Fogedby, who discusses Lévy flights in random potentials [43]. His main conclusion is that, in situations where Brownian motions are significantly slowed down, disorder is inefficient to trap Lévy flights.

A suggestion which I find particularly challenging is that of Hilfer in the context of second order phase transitions [14]. He claims that the distribution of the total magnetisation of a large volume *at the critical point* is of the Lévy type (it is obviously a Gaussian in the paramagnetic phase). His prediction is in remarkable agreement with numerical simulation, and raises important problems, such as the universality of the renormalized coupling constant at the critical point – which has been used in many numerical studies of critical phenomena.

Finally, I would like to draw the attention to some members of the Lévy family, which are, to my eyes, unduly banned. These are the laws which Fourier transform is $\exp -C|q|^\mu$ *with μ greater than 2*. These laws are stable upon convolution, are normalisable, but have the awkward property of not being of a constant sign. It is nevertheless an safe bet, in the spirit of Gnedenko and Kolmogorov's prophecy [2] for 'standard' Lévy distributions, to anticipate that these distributions will play, in due course, an important role.

Acknowledgments I take the opportunity to thank many collaborators with whom I have had great pleasure to work over the past few years on these matters, in particular J.P. Aguilar, A. Aspect, F. Bardou, P. Cizeau, C. Cohen-Tannoudji, D. S. Dean, J. Hammann, F. Ladieu, D. Langevin, A. Ott, D. Sornette, W. Urbach, E. Vincent and Ch. Walter.

References

1. V. Pareto, Cours d'économie politique. Reprinted as a volume of *Oeuvres Complètes*. (Droz, Geneva, 1896-1965).
2. B. V. Gnedenko, A.N. Kolmogorov, *Limit distributions for sum of independent random variables*, Addison Wesley, Reading, MA, (1954).
3. P. Lévy, Théorie de l'addition des variables aléatoires (Gauthier Villars, Paris, 1937-1954).
4. E. Montroll, M. Shlesinger, 'On the Wonderful world of Random Walks', in Nonequilibrium phenomena II, From stochastic to hydrodynamics, Studies in statistical mechanics XI (J.L. Lebowitz and E.W. Montroll, eds) (North Holland, Amsterdam, 1984).
5. B.B. Mandelbrot, Journal of Business, **36** 394 (1963), B.B. Mandelbrot, 'The Fractal Geometry of Nature' (Freeman, San Francisco, 1983)

6.   E. Fama, Management Science **11** 404-419 (1965).

7.   for a review see e.g J.P. Bouchaud, A. Georges, Phys. Rep. **195** 127 (1990).

8.   M. Shlesinger, G. Zaslavsky, J. Klafter, Nature (London) **363** 31 (1993)

9.   P. Bak, C. Tang, K. Wiesenfeld, Phys. Rev. Lett. **59** 381 (1987), Phys. Rev. A 38, 364 (1988); P. Bak, K. Chen, J.A. Scheinkman and M. Woodford, Ricerche Economiche, **47** 3-30 (1993); D. Sornette, Les phénomènes critiques auto-organisés, Images de la Physique 1993, édition du CNRS.

10.   Ch. Walter, 'L'utilisation des lois Lévy-stables en finance : une solution possible au problème posé par les discontinuités des trajectoires boursières', Bulletin de l'Institut des actuaires français, **349** 3 (1990) and **350** 4 (1991); Ch. Walter, 'Lévy-stable distributions and fractal structure on the Paris market : an empirical examination', Proc. of the first AFIR colloquium (Paris, April 1990), **3** 241; Ch. Walter, Thèse (1994), unpublished.

11.   R. Mantegna, Physica A **179** 232 (1991), R. Mantegna, H.E. Stanley, submitted to Phys. Rev. Lett. (June 1994).

12.   J.P. Bouchaud, D. Sornette, C. Walter, J. P. Aguilar, 'Taming large events : Optimal portfolio theory for strongly fluctuating assets.', submitted to J. Finance, (October 1994)

13.   C. Itzikson, J. M. Drouffe, 'Statistical field theory', Cambridge University Press (1990) (Chapter I).

14.   R. Hilfer, 'Absence of Hyperscaling violations for phase transitions with positive specific heat exponent', Z. Phys. B (1994), in print.

15.   M. Mézard, G. Parisi, M.A. Virasoro, "Spin Glass Theory and Beyond", (World Scientific, Singapore 1987)

16.   B. Derrida, 'Non-Self averaging effects in sum of random variables', in 'On Three Levels', Edited by Fannes et al, page 125 Plenum Press, NY, 1994

17.   A.Ott, J.P.Bouchaud, D.Langevin and W.Urbach, Phys. Rev. Lett. **65**, 2201 (1990)

18.   M. E. Cates. S. J Candau, J. Phys. Cond. Mat. **2** 6869 (1990)

19.   F.Bardou, J.P.Bouchaud, O. Emile, A. Aspect and C.Cohen-Tannoudji, Phys. Rev. Lett. **72** 203 (1994)

20.   A. Aspect, E. Arimondo, R. Kaiser, N. Vansteenkiste and C. Cohen-Tannoudji, Phys. Rev. Lett. **61**, 826 (1988), and J.O.S.A. **B6**, 2112, (1989)

21.   a- J.P. Bouchaud, J. Physique I (France) **2**, 1705 (1992), b- J.P. Bouchaud, E. Vincent, J. Hammann, J. Physique I (France) **4**, 139 (1994) and refs. therein.

22.   a- V.S. Dotsenko, M. V. Feigelmann, L.B. Ioffe, Spin-Glasses and related problems, Soviet Scientific Reviews, vol. 15 (Harwood, 1990), b- L. Cugliandolo, J. Kurchan, Phys. Rev. Lett. **71** (1993) 173, c- S. Franz, M. Mézard, Europhys. Lett. **26** (1994) 209, Physica **A209** (1994) 1, d- L. Cugliandolo, J. Kurchan, J. Phys A **27** 5749 (1994), e- H. Rieger, J. Phys. A **26** L615 (1993).

23.   J.P. Bouchaud, D. S. Dean, submitted to J. Physique I (France), September 1994.

24.   P. Doussineau, Y. Farssi, C. Frénois, A. Levelut, J. Toulouse, S. Ziolkiewicz, J. Phys. I (France) **4** 1217 (1994)

25.   T. Odagaki, J. Matsui, Y. Hiwatari, Physica A **204** 464 (1994).

26.   M. Mézard, G. Parisi, J. Physique I **1** 809 (1991)

27.   T. Hwa, D.S. Fisher, preprint (1993)

28.   see e.g. B. Derrida, R. B. Griffiths, J. Stat. Phys. **8** 111 (1989)

29.   M. L. Mehta. *Random matrices and the Statistical Theory of Energy Levels.* Academic Press, New York,(91)

30. A. D. Stone, P. A. Mello, K. Muttalib, J. L. Pichard, *Mesoscopic phenomena in solids*, Eds. B. Altshuler, P. A. Lee, R. Webb, North Holland (90).

31. C. W. J. Beenakker, Phys. Rev. Lett. **70** 1155 (1993), Phys. Rev. B **47** 15763 (1993), Phys. Rev. B **49** 2205 (1994).

32. P. Cizeau, J. P. Bouchaud, Phys Rev E, to appear (1994)

33. P. Cizeau, J. P. Bouchaud, J. Phys. A **26** L187 (1993)

34. The first discussion of a similiar mechanism is astronomy seems to be by S. Chadraseckhar, Rev. Mod. Phys **15** 1 (1943), ch. IV.

35. J. Klafter, G. Zumofen, preprint (1993)

36. H. Pfister, H. Scher, Adv. Phys. **27** 747 (1978)

37. J. Bernasconi, H. Beyeler, S. Strassler, S. Alexander, Phys. Rev. Lett. **42** 819 (1979)

38. F. Ladieu, J.P. Bouchaud, J. Physique (France) I **3** 2311 (1993)

39. T. Geisel, J. Nierwetberg, A. Zacherl, Phys. Rev. Lett. **54** 616 (1985)

40. O. Cardoso, P. Tabeling, Europhys. Lett. **7** 225 (1988)

41. T.H. Solomon, E.R. Weeks, H.L. Swinney, Phys. Rev. Lett. **71** 3975 (1993)

42. C.A. Tsallis, D. Stariolo, preprint (1994), T.J.P Penna, preprint (1994)

43. H. Fogedby, preprint submitted to Phys. Rev. Lett. (1994)

Aspects of Lévy flights in a quenched random force field

Hans C. Fogedby
Institute of Physics and Astronomy, University of Aarhus
8000 Aarhus C, Denmark

Abstract: We consider Lévy flights characterized by the step index f in a quenched isotropic short range random force field. By means of a dynamic renormalization group analysis we find that the dynamic exponent z for $f < 2$ locks onto f, independent of dimension and *independent* of the presence of weak quenched disorder. The critical dimension for $f < 2$ is given by $d_c = 2f - 2$. For $d < d_c$ the disorder is *relevant*, corresponding to a non-trivial fixed point for the force correlation function. We also discuss the behavior of the subleading diffusive term.

Keywords: Lévy flights, long range, rare events, dynamical exponent, scaling, subdiffusion, superdiffusion, quenched environments, random force field, Gaussian distribution, short range, dynamic renormalization group, fixed points, force correlations

There is a current interest in the dynamics of fluctuating manifolds in quenched random environments [1]. This fundamental issue in modern condensed matter physics is encountered in problems as diverse as vortex motion in high temperature superconductors, moving interfaces in porous media, and random field magnets and spin glasses.

The simplest case is that of a random walker in a random environment, corresponding to a zero dimensional fluctuating manifold. This problem has been treated extensively in the literature [2, 3, 4] and many results are known.

In the case of ordinary Brownian motion, characterized by a finite mean square step, in a pure environment without disorder, the central limit theorem [5] implies that the statistics of the walk is given by a Gaussian distribution with a mean square deviation proportional to the number of steps or, equivalently, the elapsed time, i.e., the mean square displacement

$$\langle r^2(t) \rangle \propto Dt^{2/z}, \tag{1}$$

where the dynamic exponent $z = 2$ for Brownian walk and D is the diffusion coefficient.

There are, however, many interesting processes in nature which are characterized by an anomalous diffusion with $z \neq 2$ due to the statistical properties of the environments [3, 4]. Examples are found in chaotic systems [6], turbulence [7, 8], flow in fractal geometries [9], and Lévy flights [10, 11], which generally lead to enhanced diffusion or superdiffusion with $z < 2$. We note that the ballistic case corresponds to $z = 1$. The other case of subdiffusion or dispersive behaviour with $z > 2$ is encountered in various constrained systems like doped crystals, glasses or fractals [12, 13, 14, 15, 16].

Independent of the spatial dimension d, ordinary Brownian motion traces out a manifold of fractal dimension $d_F = 2$ [17]. In the presence of a quenched disordered force field in d dimensions the Brownian walk is unaffected for $d > d_F$, i.e., for d larger than the critical dimension $d_c = d_F$ the walk is transparent and the dynamic exponent z locks onto the value 2 for the pure case. Below the critical dimension $d_c = 2$ the long time characteristics of the walk is changed to subdiffusive behaviour with $z > 2$ [18, 19, 20]. In $d = 1$, $\langle r^2(t) \rangle \propto [\log t]^4$, independent of the strength of the quenched disorder [21].

Lévy flights constitute an interesting generalization of ordinary Brownian walks. Here the step size is drawn from a Lévy distribution characterized by the step index f [5, 11]. The Lévy distribution has a long range algebraic tail corresponding to large but infrequent steps, so-called *rare events*. This step distribution has the interesting property that the central limit theorem does not hold in its usual form. For $f > 2$ the second moment or mean square deviation of the step distribution is finite, the central limit theorem holds and the dynamic exponent z for the Lévy walk locks onto 2, corresponding to ordinary diffusive behaviour; however, for $f < 2$ the mean square step deviation diverges, the rare large step events prevail and determine the long time behaviour, and the dynamic exponent z depends on the microscopic step index f according to $z = f$ $(f < 2)$, indicating anomalous enhanced diffusion, that is superdiffusion [11, 22, 23]. The "built in" superdiffusive characteristics of Lévy flights have been used to model a variety of physical processes such as self diffusion in micelle systems [24], and transport in heterogeneous rocks [22].

In the present article we consider Lévy flights in the presence of a quenched random force field and examine the interplay between the "built in" superdiffusive behaviour of the Lévy flights and the pinning effect of the random environment generally leading to subdiffusive behaviour. Generalizing the discussion in refs.[18, 20, 25] we find that in the case of enhanced diffusion for $f < 2$ the dynamic exponent $z = f$, independent of the presence of weak disorder. On the other hand, we can still identify a critical dimension $d_c = 2f - 2$, depending on the step index f for $f < 2$. Below d_c the quenched disorder becomes *relevant* as indicated by the emergence of a non-trivial fixed point in the renormalization group analysis. We also discuss the behaviour of the subleading diffusive term.

It is convenient to discuss Lévy flights in terms of a Langevin equation with "power law" noise [26, 27]. In an arbitrary drift force field $\mathbf{F}(\mathbf{r})$, representing the quenched disordered environment, the equation takes the form

$$\frac{d\mathbf{r}(t)}{dt} = \mathbf{F}(\mathbf{r}(t)) + \eta(t). \tag{2}$$

Here η is the instantly correlated power law white noise with the isotropic distribution,

$$p(\eta)d^d\eta \propto \eta^{-1-f}d\eta, \tag{3}$$

characterized by the step index f [28]. In order to ensure normalizability we have introduced a lower cutoff $\eta \sim a$ of the order of a microscopic length a and chosen $f > 0$. For $f > 2$ the second moment, $\langle \eta^2 \rangle = \int p(\eta)\eta^2 d^d\eta$, is finite and a characteristic step size is given by the root mean square deviation $\sqrt{\langle \eta^2 \rangle}$. For $1 < f < 2$ the second moment diverges but the mean step, $\langle \eta \rangle$, is finite. In the interval $0 < f < 1$ the first moment diverges and even a mean step size is not defined [11].

The power law noise η, describing the consecutive Lévy steps, drives the position $\mathbf{r}$ of the walker. For the quenched force field $\mathbf{F}(\mathbf{r})$ we assume a Gaussian distribution,

$$p(\mathbf{F}) \propto \exp\left[-\frac{1}{2} \int d^d r d^d r' F^\alpha(\mathbf{r}) \Delta^{\alpha\beta}(\mathbf{r},\mathbf{r}')^{-1} F^\beta(\mathbf{r}')\right] \tag{4}$$

$$\langle F^\alpha(\mathbf{r}) F^\beta(\mathbf{r}') \rangle_F = \Delta^{\alpha\beta}(\mathbf{r} - \mathbf{r}'). \tag{5}$$

Here $\Delta^{\alpha\beta}(\mathbf{r}-\mathbf{r}')$ is the force correlation function expressing the range and vector nature of $\mathbf{F}(\mathbf{r})$ and $\langle \ldots \rangle_F$ indicates an average over the force field.

In the absence of the force field, i.e. $\mathbf{F}(\mathbf{r}) = 0$, the distribution for the position of the walker is easily inferred. From the definition $P(\mathbf{r},t) = \langle \delta(\mathbf{r} - \mathbf{r}(t)) \rangle$, the solution of Eq. (2), and averaging according to Eq. (3), we find the scaling form [11, 17],

$$P(\mathbf{r},t) = \int \frac{d^d k}{(2\pi)^d} \exp\left(i\mathbf{k}\mathbf{r} - D_1 k^\mu \,|\,t\,|\right) = |\,t\,|^{-\frac{d}{\mu}} G(r/\,|\,t\,|^{\frac{1}{\mu}}). \tag{6}$$

D_1 is a diffusion coefficient setting the time scale. This procedure is equivalent to applying the central limit theorem to a sum of random variables with the Lévy distribution in Eq. (3) [5, 11, 17].

The scaling exponent μ depends on the step index f, characterizing the Lévy distribution. For $f > 2$, i.e., the case of a finite mean square step, μ locks onto the value 2 and the scaling function $G(x)$ takes the Gaussian form for ordinary Brownian walk, $G(x) = \exp(-x^2)$. This is a consequence of the central limit theorem which here leads to universal behaviour [5]. For $f < 2$ the scaling exponent $\mu = f$ and the scaling function $G(x)$ can only be given explicitly in terms of known functions for $\mu = 1$, the ballistic case, where we find the Cauchy distribution $G(x) = (1 + x^2)^{-((d+1)/2)}$. It is, however, easy to show that $G(x) \to const.$ for $x \to 0$ and $G(x) \to 0$ for $x \to \infty$.

From the distribution in Eq. (6) we deduce the scaling form for the mean square displacement of the walker in time t. In the scaling regime [29] we have

$$\langle r^2(t) \rangle \propto \int P(\mathbf{r}, t)\, r^2 d^d r \propto t^{\frac{2}{\mu}} \tag{7}$$

and from Eq. (1) $z = \mu$. We also note that the fractal dimension of a Lévy flight is $d_F = \mu$ [17]. In Fig. 1 we have plotted μ as a function of f.

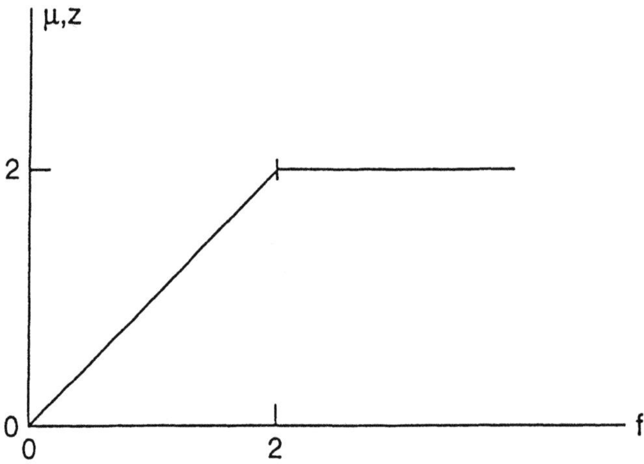

Figure 1: Plot of the scaling index μ and the dynamic exponent z as functions of the Lévy step index f. For $f > 2$ we have normal Brownian diffusion and we obtain corrections to z below $d = 2$; for $0 < f < 2$ we have anomalous Lévy superdiffusion.

In the presence of the quenched force field given by Eqs. (4) and (5) it is convenient to recast the problem given by the Langevin equation (2) in terms of the associated Fokker-Planck equation,

$$\frac{\partial P(\mathbf{r}, t)}{\partial t} = -\nabla(\mathbf{F}(\mathbf{r})P(\mathbf{r}, t)) + D_1 \nabla^\mu P(\mathbf{r}, t) + D_2 \nabla^2 P(\mathbf{r}, t), \tag{8}$$

including the ordinary diffusion term, $D_2 \nabla^2 P$, originating from the low η part of the distribution $p(\eta)$. Here the first term on the right hand side of Eq. (8) is the usual drift term due to the motion of the walker in the force field, the second term follows from Eq. (6) and the assumption of independent contributions to the probability current. The "fractional" gradient operator ∇^μ is the Fourier transform of $-k^\mu$ and is a spatially non-local integral operator reflecting the long range Lévy steps; for $\mu = 2$ it reduces to the usual Laplace operator describing ordinary diffusion, see also [27].

The Fokker-Planck equation in Eq. (8) together with the distribution in Eqs. (4) and (5) defines the problem of a random Lévy walker in a quenched random force field. The "microscopic" Lévy steps occurring on a fast time scale represented by the noise term in the Langevin equation (2) has now been absorbed and replaced by the anomalous diffusion term in the Fokker-Planck equation (8). The remaining randomness due to the quenched force field is assumed static compared to the time scale of $\mathbf{r}(t)$.

There are a variety of techniques available in order to treat the random Fokker-Planck equation (8). Applying the Martin-Siggia-Rose formalism in functional form [30, 31, 32, 33] and using either the replica method [4] or an explicit causal time dependence [18, 32, 34], one can average over the quenched force field and construct an effective field theory. A more direct method, which we shall adhere to in the present discussion, amounts to an expansion of the Fokker-Planck equation (8) in powers of the force field and an average over products of $\mathbf{F}(\mathbf{r})$ according to the distribution in Eqs. (4) and (5) [25].

Defining the Fourier transform

$$P(\mathbf{k},\omega) = \int d^d r dt \exp{(i\omega t - i\mathbf{k}\mathbf{r})} P(\mathbf{r},t)\theta(t), \qquad (9)$$

where $\theta(t)$ is the step function, we obtain, introducing the dimensionless coupling strength λ for the vertex, the Fokker-Planck equation

$$(-i\omega + D_1 k^\mu + D_2 k^2)P(\mathbf{k},\omega) = P_o(\mathbf{k}) - i\lambda \mathbf{k} \int \frac{d^d p}{(2\pi)^d} \mathbf{F}(\mathbf{k} - \mathbf{p})P(\mathbf{p},\omega). \quad (10)$$

Here the force field is averaged according to Eqs. (4) and (5), or

$$\langle F^\alpha(\mathbf{k})F^\beta(\mathbf{p})\rangle_F = \Delta^{\alpha\beta}(\mathbf{k})(2\pi)^d \delta(\mathbf{k} + \mathbf{p}) \qquad (11)$$

for all pairwise force contractions (Wick's theorem for the Gaussian distribution); $P_o(\mathbf{k}) = P(\mathbf{k}, t = 0)$ is the initial distribution. For simplicity we consider the case of isotropic zero range force correlations, i.e., $\Delta^{\alpha\beta}(\mathbf{k}) = \Delta\delta^{\alpha\beta}$; the general case has been discussed in refs.[4, 25, 37]. We introduce a microscopic UV cut off and assume $0 < k, p < 1$. Iterating Eq. (10), averaging according to Eq. (11), and identifying self energy, vertex, and force correlation corrections to first order in Δ, we find divergent contributions to the subleading term $D_2 k^2$ for $d < \mu$ and to the vertex λ for $d < 2\mu - 2$. In order to disentangle the breakdown of primitive perturbation theory and deduce the scaling properties of the force averaged distribution $\langle P(\mathbf{r},t)\rangle_F$ and the mean square displacement $\langle\langle r^2(t)\rangle\rangle_F$, we carry out a renormalization group analysis, following the momentum shell integration method [25, 35, 36]. Averaging over the force in the shell $e^{-\ell} < k,p < 1$, we thus obtain the "corrected" Fokker-Planck equation

$$(-i\omega + D_1 k^\mu + (D_2 + \delta D_2)k^2)P(\mathbf{k},\omega) = P_o(\mathbf{k}) - i(\lambda + \delta\lambda)\mathbf{k} \int \frac{d^d p}{(2\pi)^d} \mathbf{F}(\mathbf{k}-\mathbf{p})P(\mathbf{p},\omega).$$

(12)

and the force correlation function

$$\langle F^\alpha(\mathbf{k})F^\beta(\mathbf{p})\rangle_F = (\Delta + \delta\Delta)\delta^{\alpha\beta}(2\pi)^d\delta(\mathbf{k}+\mathbf{p})$$

(13)

for $0 < k,p < e^{-\ell}$. Note that there is no correction to the leading Lévy term. For small values of the scale parameter ℓ the corrections δD_2, $\delta\lambda$, and $\delta\Delta$ are proportional to ℓ. From the diagrammatic contributions to δD, $\delta\lambda$, and $\delta\Delta$ given in refs. [25], evaluated in the Lévy case, we obtain

$$\delta D_2 = A' \frac{D_1(d-\mu) + D_2(d-2)}{(D_1 + D_2)^2}\lambda^2\Delta\ell,$$

(14)

$$\delta\lambda = -C' \frac{\lambda^3\Delta}{(D_1 + D_2)^2}\ell,$$

(15)

$$\delta\Delta = -B' \frac{\lambda^2\Delta^2}{(D_1 + D_2)^2}\ell,$$

(16)

where A', B', and C' are geometric factors associated with the area of a d-dimensional unit sphere. In order to derive the "renormalized" Fokker-Planck equation we introduce scaled quantities, $\mathbf{k}' = \mathbf{k}e^\ell$, $\mathbf{p}' = \mathbf{p}e^\ell$, $\omega' = \omega e^{\alpha(\ell)}$, $P'(\mathbf{k}',\omega') = P(\mathbf{k},\omega)e^{-\alpha(\ell)}$, and $\mathbf{F}'(\mathbf{k}') = \mathbf{F}(\mathbf{k})e^{-\beta(\ell)}$ such that $0 < k',p' < 1$.

From the renormalized Fokker-Planck equation and force correlation function, adjusting $\beta(\ell)$ so that $\lambda = 1$, setting $\alpha(\ell) = \int_o^\ell z(\ell')d\ell'$, and choosing $z(\ell) = \mu$ in order to fix $D_1(\ell) = D_1$, we read off the renormalization group equations for D_2 and Δ,

$$\frac{dD_2}{d\ell} = (\mu - 2)D_2 + A\frac{D_1(d-\mu) + D_2(d-2)}{(D_1 + D_2)^2}\Delta$$

(17)

$$\frac{d\Delta}{d\ell} = (2\mu - d - 2)\Delta - B\frac{\Delta^2}{(D_1 + D_2)^2}.$$

(18)

Here $A = (1/2d)S_d/(2\pi)^d$, $B = (3/d - 1)S_d/(2\pi)^d$, and $S_d = 2\pi^{d/2}/\Gamma(d/2)$ the surface area of a d-dimensional sphere.

Proceeding with the discussion of Eqs. (17) and (18) we note that for $\mu < 1 + d/2$ Eqs. (17) and (18) have the trivial fixed points $D_2^* = 0$ and $\Delta^* = 0$, indicating that (i) the subleading term, $D_2 k^2$, scales to zero compared with the leading Lévy term and (ii) the quenched disorder, characterized by Δ, is *irrelevant*. The long range Lévy steps predominate and control the scaling behavior. In a plot of $D_2(\ell)$ versus $\Delta(\ell)$ the size of the linear scaling regime, i.e., the region where the trajectories flow to the fixed point with constant slope, depends on μ and becomes largest for $\mu = d$, precisely the case where perturbation theory begins to yield a divergent contribution to D_2. For $1 + d/2 < \mu < 2$ nontrivial fixed points emerge for D_2 and Δ,

$$D_2^* = (A/B)\frac{D_1(d - \mu)(2\mu - d - 2)}{2 - \mu + (A/B)(2\mu - d - 2)(2 - d)} \tag{19}$$

$$\Delta^* = (1/B)(2\mu - d - 2)(D_1 + D_2^*)^2 \tag{20}$$

The fixed point D_2^* indicates that the subleading diffusive term $D_2 k^2$ now yields a contribution compared to the Lévy term $D_1 k^\mu$. The fixed point Δ^* shows that for d less than the critical dimension $d_c = 2\mu - 2$ the quenched disorder becomes *relevant*. The fixed point value of the diffusion coefficient, D_2^*, is negative since the pinning environment tends to reduce the ordinary diffusion from $D_2^* = 0$ for $\mu < 1 + d/2$. We also note that unlike the case of Brownian motion the critical dimension $d_c = 2\mu - 2$ is less than the fractal dimension $d_F = \mu$. For $\mu \to 2$ it follows from Eq. (19) that $D_2^* \to -D_1$ so that the Lévy term $D_1 k^\mu$ precisely cancels with the diffusive term $D_2 k^2$ in the Fokker Planck equation (10); this is consistent with the fact that there is no correction to first loop order or more precisely to first order in $d_c - d$ in the Brownian case [28]. In Fig. 2 we have shown the position of the fixed point D_2^* as a function of μ.

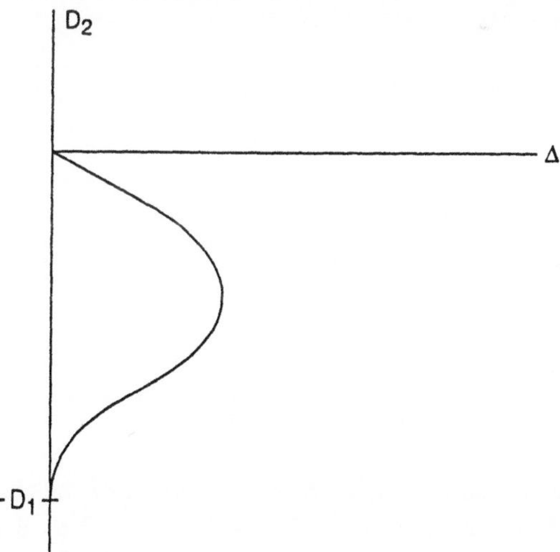

Figure 2: Plot of the diffusive fixed point D_2^* as a function of the force correlation function fixed point Δ^*.

In order to derive the scaling properties of the force averaged distribution $\langle P(\mathbf{k}, \omega)\rangle_F$ and the means square displacement $\langle\langle r^2(t)\rangle\rangle_F$ we use the methods discussed in refs. [25, 35, 36]. From the derivation of the renormalization group equations we infer the scaling relation $\langle P(\mathbf{k}, \omega, \Delta)\rangle_F = e^{\alpha(\ell)}\langle P(\mathbf{k}e^\ell, \omega e^{\alpha(\ell)}, \Delta(\ell))\rangle_F$.

In the vicinity of either fixed point we have, setting $\alpha(\ell) \propto \mu\ell$, $\langle P(\mathbf{k},\omega,\Delta)\rangle_F = e^{\mu\ell}\langle P(ke^\ell, \omega e^{\mu\ell}, \Delta^*)\rangle_F$. Choosing $ke^\ell \sim 1$ we obtain the scaling form

$$\langle P(\mathbf{k},\omega,\Delta)\rangle_F = k^{-\mu}L(k/\omega^{1/\mu}), \qquad (21)$$

where $L(x)$ is a scaling function. From Eq. (21) follows directly (see [29])

$$\langle\langle r^2(t)\rangle\rangle_F \propto t^{2/\mu} = t^{2/z}. \qquad (22)$$

For Lévy flights in random quenched environment we have shown that the dynamic exponent z locks onto the scaling index μ, depending on the Lévy step index f, *independent* of the presence of weak quenched disorder. The long range superdiffusive behaviour characteristic of Lévy flights enables the walker to escape the inhomogeneous pinning environment and the long time behaviour is the same as in the pure case. We have also identified a critical dimension $d_c = 2\mu - 2$, depending on the scaling exponent μ. For $d < d_c$ the weak disorder becomes *relevant* as shown by the emergence of a nontrivial fixed point. In Fig. 3 we have shown the critical dimension d_c as a function of μ.

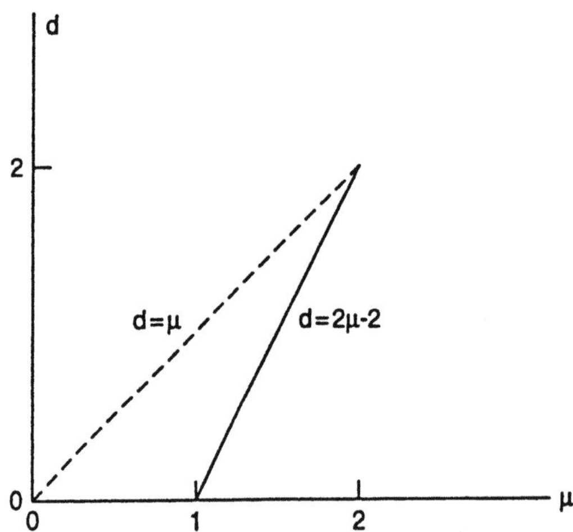

Figure 3: Plot of the critical dimension d_c as a function of the scaling index μ. For $\mu = 2$ we have the Brownian case $d_c = 2$; for $1 < \mu < 2$ d_c depends linearly on μ. Note that $d : c = 0$ in the ballistic case for $\mu = 1$. For $d < \mu$ indicated by the dashed line $d = \mu$ there are divergent contributions to the subleading term $D_2 k^2$.

Bouchaud [38] has given a heuristic argument yielding the critical dimension $d_c = \mu$ in the Lévy case. This result is at variance with the critical dimension $d_c = 2\mu - 2$ given here based on a renormalization group analysis. Note, however, that perturbation theory yields a divergent contribution to the subleading diffusive term for $d = \mu$. We have not entirely appreciated the discrepancy between Bouchaud's argument and the present analysis. It clearly would be of interest to construe a qualitative heuristic argument for the critical dimension $d_c = 2\mu - 2$ given here and the insensitivity of the dynamic exponent $z = \mu$ to the weak quenched disorder.

The present analysis opens up several avenues to pursue such as the interplay between the temporal and spatial features of Lévy flights in quenched environments and the role played by the range and vector nature of the random force field. These problems will be dealt with in a future publication.

The author wishes to thank L. Mikheev, A. Svane, and K. Bækgaard Lauritsen for helpful discussions. This work was supported by the Danish Research Council, Grant No. 11-9001.

References

[1] D.S. Fisher in "Phase Transitions and Relaxations in Systems with Competing Energy Scales", NATO ASI Series, eds. T. Riste and D. Sherrington (Kluwer Academic Publishers, London, 1993)

[2] S. Alexander, J. Bernasconi, W.R. Schneider, and R. Orbach, Rev. Mod. Phys. **53**, 175 (1981)

[3] S. Havlin and D. ben-Avraham, Adv. Phys. **36**, 695 (1987)

[4] J.-P. Bouchaud and A. Georges, Phys. Rep. **195**, 127 (1990)

[5] W. Feller, *An Introduction to Probability Theory and its Applications* (Wiley, N.Y. 1971)

[6] T. Geisel, J. Nierwetberg, and A. Zacherl, Phys. Rev. Lett. **54**, 616 (1985),

[7] S. Grossman, F. Wegner, and K.H. Hoffmann, J. Phys. (Paris) Lett. **46**, L575 (1985)

[8] T. Bohr and A. Pikovsky, Phys. Rev. Lett. **70**, 2892 (1993)

[9] M. Sahimi, J. Phys. **A20**, L1293 (1987)

[10] M.F. Shlesinger and J. Klafter, Phys. Rev. Lett. **54**, 2551 (1985)

[11] H.C. Fogedby, T. Bohr, and H.J. Jensen, J. Stat. Phys. **66**, 583 (1992)

[12] H. Scher and E.W. Montroll, Phys. Rev. **B12**,2455 (1974)

[13] M.F. Shlesinger, J. Stat. Phys. **10**, 421 (1974)

[14] S. Alexander and R. Orbach, J. Phys. (Paris) Lett. **43**, L625 (1982)

[15] A. Blumen, J. Klafter, B.S.White, and G. Zumofen, Phys. Rev. Lett. **53**, 1301 (1984)

[16] A. Blumen, J. Klafter, and G. Zumofen in *Optical Spectroscopy of Glasses*, edited by T. Zschokke (Reidel, Dordrecht, 1986), p. 199

[17] B.D. Hughes, M.F. Shlesinger, and E.W. Montroll, Proc. Natl. Acad. Sci. USA **78**, 3287 (1981)

[18] D.S. Fisher, Phys. Rev. **A30**, 960 (1984)

[19] B. Derrida and J.M. Luck, Phys. Rev. **B28**, 7183 (1983)

[20] J.M. Luck, Nucl. Phys. **B225**, 169 (1983)

[21] Ya. G. Sinai, in: Lecture Notes in Physics, Vol 153, eds. R. Schrader, R. Seiler, and D. Uhlenbrock (Springer,Berlin,1981) and Ya. G. Sinai, Theor.Prob.Appl.**27**, 247 (1982)

[22] J. Klafter, A. Blumen, G. Zumofen, and M.F. Shlesinger, Physica A **168**, 637 (1990)

[23] A. Blumen, G. Zumofen, J. Klafter, Phys. Rev. **A40**, 3964 (1989)

[24] A. Ott, J.-P. Bouchaud, D. Langevin, and W. Urbach, Phys. Rev. Lett. **65**, 2201 (1990)

[25] J. Aronovitz and D.R. Nelson, Phys. Rev. **A30**, 1948 (1984)

[26] Power law noise has also been used in order to model disorder for moving interfaces, Y.-C. Zhang, J. Phys. (Paris) **51**, 2129 (1990)

[27] H.C. Fogedby, Phys. Rev. **E50**, 1657 (1994)

[28] The Langevin equation (2) is the continuum limit of the corresponding difference equation defined for a discrete time step Δ. For $f > 1$ this limit is attained letting the lower cut-off for $p(\eta)$ approach infinity; for $f < 1$ the cut-off must go to zero. Note that these "renormalizations" are not observable; the problem at hand is defined by the Langevin equation (2) or, correspondingly, the Fokker Planck equation (8).

[29] For Lévy flights the scaling regime must be defined with some care. Since for $f < 2$, $\langle \eta^2 \rangle$ diverges, it clearly follows that $\langle r^2(t) \rangle = \langle \sum_i \eta_i^2 \rangle$ diverges as well. However, considering Lévy flights within a volume of linear extent L, it follows from Eq. (6) that for (i) $L \gg t^{1/\mu}$, $\langle r^2 \rangle_L \propto tL^{2-\mu}$, i.e., diverging for $L \to \infty$, and for (ii) $L \ll t^{1/\mu}$, $\langle r^2 \rangle_L \propto t^{2/\mu}$. The latter regime is the scaling regime in the present discussion. Further discussion of these points is deferred to a future publication.

[30] P.C. Martin, E. Siggia, and H. Rose, Phys. Rev. **A8**, 423 (1973)

[31] C. De Dominicis and L. Peliti, Phys. Rev. **B18**, 353 (1978)

[32] C. De Dominicis, Phys. Rev. **B18**, 4913 (1978)

[33] R. Bauch, H.K. Janssen, and H. Wagner, Z. Physik **B24**, 113 (1976)

[34] L. Peliti and Y.C. Zhang, J. Phys. **A18**, L709 (1985)

[35] D. Forster, D.R. Nelson, and M.J. Stephen, Phys. Rev. Lett. **36**, 867 (1976)

[36] D. Forster, D.R. Nelson, and M.J. Stephen, Phys. Rev. **A16**, 732 (1977)

[37] D.S. Fisher, D. Friedan, Z. Qui, S.J. Shenker, and S.H. Shenker, Phys. Rev. **A31**, 3841 (1985)

[38] J.-P. Bouchard, A. Comtet, A. Georges, and P. Le Doussal, J. Phys. (Paris) **48**, 1445 (1987), **49**, 369 (1988) (erratum)

Universality of Escape from a Half-space for Symmetrical Random Walks

Uriel Frisch[1] and Hélène Frisch[1]

[1] C.N.R.S., Observatoire de Nice, B.P. 229, 06304 Nice Cedex 4, France
e–mail: uriel@obs-nice.fr

Abstract. A new proof is given that a one–dimensional random walk starting from the origin, with independent steps having an *even* p.d.f. $K(x)$, has a probability p_n, of never visiting the negative half-space for the n first steps, which is *universal*, i.e. independent of $K(x)$.

Keywords. Random walks, radiative transfer

1 Introduction

Recently, Sinai (1992 a, b) proved that, when Brownian motion is used as initial condition for the inviscid Burgers equation, the set of initial points, which have not fallen into a shock after a given positive time, has Hausdorff dimension 1/2. This result has been extended (in a partially heuristic way) to the fractional Brownian motion and has application to the mass–distribution of large–scale structures in the Universe (Vergassola *et al.* 1994).

During the Workshop some background material was presented (by U. Frisch and M. Vergassola) on how the study of large–scale structures can be reduced to analyzing solutions of the Burgers equation and key points in Sinai's proof were highlighted. The proof uses in particular two results in probability theory:

(i) The zero–crossings of fractional Brownian with scaling exponent h form a set of Hausdorff dimension $1 - h$; when an ultra–violet cutoff is introduced successive zero–crossings form Lévy flights.

(ii) A universality property of symmetrical random walks (see below).

Detail on item (i) will not be presented here; an elementary presentation may be found in Vergassola *et al.* (1994). Item (ii) has also important applications in the theory of radiative transfer in stellar atmospheres. In the latter, one studies the distribution of photons undergoing scatterings, absorptions and reemissions by different atomic processes. They often give rise to Lévy walks, for example when the mean–free–path of photons is controlled by the profile of a spectral line and the frequency can change at each scattering. Such material was presented at the Workshop by H. Frisch (details may be found in Frisch and Frisch (1977) and in Bell *et al.* (1978)). In radiative transfer, there is an important universality result, known as the $\sqrt{\epsilon}$–law, which relates, roughly, the properties of the radiation field at the surface of a star to the probability, denoted ϵ, of photon absorption without subsequent reemission. The short proof of this result given by Frisch and Frisch (1975), using a quadratic functional, can actually be recast in purely probabilistic random walk language and gives then a direct proof of the universality result for symmetrical random walks. The traditional proof, which may be found, e.g., in Feller (1966, pp. 393-397) uses somewhat intricate combinatorial arguments.

2 Universality for symmetrical random walks

Theorem *Let $S_1, S_2, \ldots$ be a sequence of independent random variables with the same even probability density $K(x)$, then*

$$p_n \equiv \text{Prob}\left\{S_1 > 0,\, S_1 + S_2 > 0, \ldots, S_1 + S_2 + \ldots + S_n > 0\right\}, \qquad (1)$$

is a universal function of n, independent of the specific functional form of $K(x)$.

Note that p_n is the probability that the random walk never visits the negative half-space during the n first steps. Note also that the universality result applies both to ordinary (finite variance) walks and to Lévy flights. For the proof of the theorem we shall make use of two lemmas.

Lemma 1 *Let $K(x)$ be an even positive function such that $\int_{-\infty}^{+\infty} K(x)dx = 1$. Let $f(x)$ and $g(x)$ be two functions, defined for $x \geq 0$, such that $\int_0^\infty |f(x)|\, dx$ and $\int_0^\infty |g(y)|\, dy$ are finite, and let*

$$Q(f,g) \equiv \int_0^\infty dx \int_0^x f(z)\, dz \int_0^\infty K(x-y)g(y)\, dy + f \leftrightarrow g. \qquad (2)$$

Then

$$Q(f,g) = \int_0^\infty f(x)\, dx \int_0^\infty g(y)\, dy. \qquad (3)$$

In (2), $f \leftrightarrow g$ means "interchange f and g".

Proof of Lemma 1
We write

$$Q(f,g) = \lim_{X \to \infty, Y \to \infty} Q_{XY}(f,g), \qquad (4)$$

where

$$Q_{XY}(f,g) \equiv \int_0^X dx \int_0^x f(z)dz \int_0^Y K(x-y)g(y)dy + f \leftrightarrow g. \qquad (5)$$

Note that the limit on Y must be taken first; without loss of generality we can assume $Y > X$. We introduce the function

$$K_1(x) \equiv \int_0^x K(y)dy. \qquad (6)$$

264

We can thus rewrite $Q_{XY}(f,g)$ as

$$Q_{XY}(f,g) = \int_0^X dx \int_0^x f(z)\,dz \frac{\partial}{\partial x}[\int_0^Y K_1(x-y)g(y)dy] + f \leftrightarrow g. \tag{7}$$

An integration by parts yields

$$Q_{XY}(f,g) = \int_0^X f(z)\,dz \int_0^Y K_1(X-y)g(y)\,dy$$
$$- \int_0^X \int_0^Y K_1(x-y)f(x)g(y)\,dx\,dy + f \leftrightarrow g. \tag{8}$$

Let us first consider the second integral in (8),

$$B_{XY}(f,g) \equiv \int_0^X \int_0^Y [f(x)g(y) + f(y)g(x)]K_1(x-y)dxdy. \tag{9}$$

We claim that

$$\lim_{X\to\infty, Y\to\infty} B_{XY} = 0. \tag{10}$$

Indeed, the integration domain is the union of a square $[0 \le x \le X, 0 < y \le X]$ and of a rectangle $[0 \le x \le X, X \le y \le Y]$. The integral over the square is zero for symmetry reasons, because $K_1(x)$ is odd. The integral over the rectangle is

$$\int_0^X \int_X^Y [f(x)g(y) + f(y)g(x)]K_1(x-y)\,dx\,dy. \tag{11}$$

From the fact that $f(x)$ and $g(x)$ are absolutely integrable and that $K_1(x)$ has a finite limit at infinity, it follows that the integral in (11) goes to zero in the limit $X \to \infty$, $Y \to \infty$, $Y > X$.

As for the first integral in (8), using the fact that $K_1(+\infty) = 1/2$, we find

$$\lim_{X\to\infty, Y\to\infty} \int_0^X f(z)\,dz \int_0^Y K_1(X-y)g(y)\,dy + f \leftrightarrow g = \int_0^\infty f(z)\,dz \int_0^\infty g(y)\,dy, \tag{12}$$

which concludes the proof. **QED**

Lemma 2 Let $K(x)$ be as in Lemma 1 and $f(x)$ such that $\int_0^\infty f(x)\,dx$ is finite, then

$$A \equiv \int_0^\infty K(x)\,dx \int_0^x f(z)\,dz + \int_0^\infty f(x)\,dx \int_0^\infty K(z-x)\,dz = \int_0^\infty f(x)\,dx. \tag{13}$$

Proof of Lemma 2. Using the same function $K_1(x)$ as in Lemma 1 we can write

$$A = \int_0^\infty [\frac{d}{dx}K_1(x)]\,dx \int_0^x f(z)\,dz + \int_0^\infty f(x)[K_1(\infty) - K_1(-x)]\,dx. \tag{14}$$

We now integrate the first term by parts. Using $K_1(-x) = -K_1(x)$, we observe that the term $\int_0^\infty K_1(x)f(x)\,dx$ appears twice with opposite signs. Using now $K_1(\infty) = 1/2$, we find

$$A = \int_0^\infty f(x)dx, \tag{15}$$

which concludes the proof. **QED**

Proof of the theorem
Let us define

$$P_n(x)dx \equiv \text{Prob}\{S_1 > 0, \ldots, S_1 + S_2 + \ldots + S_{n-1} > 0, \tag{16}$$
$$x \leq S_1 + S_2 + \ldots + S_n < x + dx\}.$$

That is, $P_n(x)$ is the probability density for the random walk $S_1 + S_2 + \ldots$ of being at x after n steps and having never visited negative (or zero) locations before. Obviously,

$$p_n = \int_0^\infty P_n(x)dx. \tag{17}$$

By definition the $P_n(x)$'s satisfy the recurrence relations

$$P_1(x) = K(x), \qquad P_n(x) = \int_0^\infty K(x - y)P_{n-1}(y)\,dy, \quad n > 1. \tag{18}$$

In the same spirit as in Frisch and Frisch (1975), we introduce the quadratic quantity

$$Q_n \equiv \sum_{\substack{r+s=n, \\ s \geq 1, r \geq 1}} \int_0^\infty P_s(x)\,dx \int_0^x P_r(z)\,dz, \quad n \geq 2. \tag{19}$$

By elementary manipulations we now show that the p_n's satisfy recurrence relations. First we remark that

$$\int_0^\infty P_r(x)dx \int_0^x P_s(z)\,dz + s \leftrightarrow r = \int_0^\infty P_r(x)\,dx \int_0^\infty P_s(z)\,dz, \tag{20}$$

where $s \leftrightarrow r$ means the same term with the interchange of s and r. Thus, we can write

$$Q_n = \sum_{\substack{r+s=n, \\ r \geq 1, s \geq 1}} \frac{1}{2} \int_0^\infty P_r(x)\,dx \int_0^\infty P_s(z)\,dz = \frac{1}{2} \sum_{\substack{r+s=n, \\ r \geq 1, s \geq 1}} p_r p_s. \tag{21}$$

Now we make use of the recurrence relation (18) to obtain a different expression for Q_n. Expressing $P_s(x)$ in terms of $P_{s-1}(x)$ and introducing $t = s - 1$, we can rewrite (19) as

$$Q_n = \sum_{\substack{r+t=n-1, \\ r\geq 1, t\geq 0}} \int_0^\infty dx \int_0^x P_r(z)\,dz \int_0^\infty K(x-y)P_t(y)\,dy. \qquad (22)$$

To deal with an expression symmetrical in r and t we take the term with $t = 0$ out of the summation and write

$$Q_n = A_n + B_n, \qquad (23)$$

with

$$A_n \equiv \sum_{\substack{r+t=n-1, \\ r\geq 1, t\geq 1}} \int_0^\infty dx \int_0^x P_r(z)\,dz \int_0^\infty K(x-y)P_t(y)\,dy, \qquad (24)$$

and

$$B_n \equiv \int_0^\infty K(x)\,dx \int_0^x P_{n-1}(z)\,dz. \qquad (25)$$

We can apply Lemma 1 to calculate A_n since the $P_n(x)$'s are positive and integrable. We thus find

$$A_n = \frac{1}{2} \sum_{\substack{r+t=n-1, \\ r\geq 1, t\geq 1}} \int_0^\infty P_r(x)\,dx \int_0^\infty P_t(y)\,dy = \frac{1}{2} \sum_{\substack{r+t=n-1, \\ r\geq 1, t\geq 1}} p_r p_t. \qquad (26)$$

We now consider B_n. From Lemma 2, we have

$$\int_0^\infty K(x)\,dx \int_0^x P_{n-1}(z)\,dz + \int_0^\infty P_{n-1}(x)\,dx \int_0^\infty K(z-x)\,dz = \int_0^\infty P_{n-1}(x)\,dx. \qquad (27)$$

According to the recurrence relation (18) we can rewrite the second term in the l.h.s. as $\int_0^\infty P_n(z)\,dz$. Introducing the probabilities p_n and p_{n-1} given by (17), we obtain

$$B_n = p_{n-1} - p_n. \qquad (28)$$

Regrouping (21), (26) and (28) we finally obtain the desired the recurrence relation,

$$p_n = p_{n-1} + \frac{1}{2} \sum_{\substack{r+t=n-1, \\ r\geq 1, t\geq 1}} p_r p_t - \frac{1}{2} \sum_{\substack{r+s=n, \\ r\geq 1, s\geq 1}} p_r p_s, \quad n \geq 2. \qquad (29)$$

Here, $p_1 = \int_0^\infty K(x)dx = 1/2$. This relation recursively determines all the p_n in a way which does not involve the functional form of the probability distribution function $K(x)$. Hence it proves the universality. **QED.**

The p_n can be calculated in explicit form using the generating function

$$p(\lambda) = \sum_{n=1}^{\infty} p_n \lambda^n. \tag{30}$$

Multiplying (29) by λ^n and summing over n from 1 to ∞ yields a quadratic equation for the generating function :

$$p(\lambda) = \lambda p(\lambda) + \lambda p_1 + \frac{1}{2}\lambda p^2(\lambda) - \frac{1}{2}p^2(\lambda). \tag{31}$$

Since $p_1 = 1/2$, this may also be written as

$$[p(\lambda) + 1]^2 = \frac{1}{1 - \lambda}. \tag{32}$$

Hence,

$$p(\lambda) = \frac{1}{\sqrt{1 - \lambda}} - 1, \tag{33}$$

which, expanded in powers of λ, gives :

$$p_n = \frac{(2n)!}{2^{2n}(n!)^2}, \tag{34}$$

an expression already obtained in a partially heuristic way by Landi Degl'Innocenti (1979). The first three coefficients are $p_1 = 1/2$, $p_2 = 3/8$ and $p_3 = 5/16$. The Stirling formula gives for the asymptotic behavior at large n :

$$p_n \sim 1/\sqrt{\pi n}. \tag{35}$$

We note that the (discrete) Bernoulli random walk starting from the origin with steps $+1$ or -1, although it does not satisfy the assumptions of the Theorem, has the same asymptotic behavior for large n. The p_n's for finite n are mostly different; for example $p_2 = 1/2$.

Let us finally observe that in radiative transfer, the parameter λ can be interpreted as the probability for a photon to survive in a scattering event. The generating function $p(\lambda)$ is then the mean number of scatterings undergone by photons created at the surface of a half–plane. We also note that the generating function $\sum_{n=1}^{\infty} \lambda^n P_n(x)$ satisfies a Wiener–Hopf integral equation (Vergassola *et al.* 1994). The latter can be solved explicitly by complex variable methods (Wiener and Hopf 1931; Frisch and Frisch 1982; Frisch 1988) and other techniques (Ivanov 1994).

References

[1] Frisch, U. and Frisch, H. 1975 Non–LTE transfer. $\sqrt{\epsilon}$ revisited, *Mon. Not. R. astr. Soc* **173**, 167-182.

[2] Bell, T.L., Frisch, U. and Frisch, H. 1978 Renormalization–group approach to noncoherent radiative transfer, *Phys. Rev. A* **17**, 1049-1057.

[3] Frisch, H. and Frisch U. 1982 A Method of Cauchy Integral Equation for non-coherent Transfer in half-space, *J. Quant. Spectrosc. Radiat. Transfer* **28**, 361-375.

[4] Frisch, H. 1988 A Cauchy Integral Equation Method for Analytic solutions of half-space Convolution Equations, *J. Quant. Spectrosc. Radiat. Transfer* **39**, 149-162.

[5] Ivanov, V.V. 1994 Resolvent method: exact solutions of half–space transport problems by elementary means, *Astron. Astrophys.* **286**, 328-337.

[6] Landi Degl'Innocenti, E. 1979 Non-LTE transfer. An alternative derivation for $\sqrt{\epsilon}$. *Mon. Not. R. astr. Soc* **186**, 369-375.

[7] Sinai, Ya.G. 1992a Statistics of shocks in solutions of inviscid Burgers equation, *Commun. Math. Phys.* **148**, 601-622.

[8] Sinai, Ya.G. 1992b Distribution of some functionals of the integral of the Brownian motion, *Theor. Math. Phys.* **90**, no. 3, 323-353.

[9] Vergassola, M., Dubrulle, B., Frisch, U. and Noullez, A. 1994 Burgers' equation, Devil's staircases and the mass distribution for large-scale structures, *Astron. Astrophys.* **289**, 325-356.

[10] Wiener, N. and Hopf, E. 1931 Über eine Klasse singulärer Integralgleichungen, *Sitzungsber. Preuss. Akad. Wiss. Berlin Phys-Math. Kl. 30/32*, 696-706.

Derivation of Lévy-Type Anomalous Superdiffusion from Generalized Statistical Mechanics

Constantino TSALLIS[1,2], André M. C. de SOUZA[2,3] and Roger MAYNARD[4]

[1] Department of Physics and Astronomy and Center for Fundamental Materials Research, Michigan State University, East Lansing, Michigan 48824-1116, USA
[2] Centro Brasileiro de Pesquisas Físicas/CNPq, Rua Xavier Sigaud 150, 22290-180, Rio de Janeiro-RJ, Brazil
[3] Departamento de Física, Universidade Federal de Sergipe, 49000-000, Aracaju-SE, Brazil
[4] Laboratoire d'Expérimentation Numérique, Maison des Magistères-CNRS, B.P. 166, 38042 Grenoble Cedex 9, France

Abstract. The robustness and ubiquity of the macroscopic normal diffusion is well known to be derivable within Boltzmann-Gibbs statistical mechanics. It is essentially founded on (i) a variational principle applied to $S = -\int dx p(x) ln[p(x)]$ with simple a priori constraints, and (ii) the central limit theorem. Its basic characterization consists in the time evolution $< x^2 > \propto t$. A recently generalized statistical mechanics enables the extension of the same program in order to also cover the long-tail Lévy-like anomalous superdiffusion, a phenomenon frequently encountered in Nature. By so doing, this formalism succeeds where standard statistical mechanics and thermodynamics are known to fail.

Keywords. Anomalous Diffusion, Lévy Distribution, Generalized Statistical Mechanics, Generalized Entropy

1 Introduction and Normal Diffusion

Modern understanding of what is now called *normal* diffusion started with Einstein's discussion [1] of the Brownian motion. Since then, the statistical-mechanic essence of this basic nonequilibrium process has been further and further clarified. In terms of a prototype model, it can be thought as a passive single particle which, as a consequence of its interaction with its environment, performs N successive jumps (N = 1,2,3,...) along a straight line ($d = 1$ model). These jumps are randomly and equally distributed on both right and left directions (of the x-axis), and their size follows a distribution law $p(x)$, *one and the same for all jumps, excepting for the fact that every new jump starts from the point where it arrived at the previous jump.* More precisely, the first jump occurs according to $p(x)$ and starts at $x = 0$, the second jump occurs according to the *same* distribution $p(x)$ but relocating $x = 0$ at the point attained after the first jump, and so on. In other words, the process has a *one-step* memory, i.e., it is Marcovian (*short* microscopic memory). Furthermore, each jump is assumed to be a *short-range* one (e.g., first-neighbor jumps if we are dealing with a $d = 1$ lattice, i.e., $p(x) = [\delta(x+1) + \delta(x-1)]/2$, where $\delta(x)$ is the Dirac distribution; or Gaussian-distributed jumps if we are dealing with a continuous space,etc). To be precise, by short-range we mean the class of $\{p(x)\}$ such that $< x^2 > \equiv \int_{-\infty}^{\infty} dx\, x^2\, p(x)$ is *finite* (besides, of course, the condition $p(x) = p(-x)$, hence $< x > \equiv \int_{-\infty}^{\infty} dx\, x\, p(x) = 0$, since we are assuming right-left symmetry on the axis).

To illustrate the statistical-mechanic grounds of this problem let us work out explicitly the $d = 1$ continuous case. The single-jump distribution $p(x)$ can be obtained by optimizing (maximizing, in fact) the Boltmann-Gibbs-Shannon entropy

$$S_1[p] = -\int_{-\infty}^{\infty} dx\, p(x)\, lnp(x) \tag{1}$$

(we choose units such as $k_B = 1$)
with the constraints

$$\int_{-\infty}^{\infty} dx\, p(x) = 1 \tag{2}$$

and

$$< x^2 > \equiv < x^2 >_1 \equiv \int_{-\infty}^{\infty} dx\, x^2\, p(x) = \sigma^2 \tag{3}$$

where the use of the subindex 1 will become clear later on, and $\sigma > 0$ is a *finite* characteristic length. If we introduce the Lagrange parameters α and β to take into account constraints (2) and (3) respectively, we obtain that $S_1[p]$ is optimal for

$$p_1(x) = \frac{e^{-\beta x^2}}{Z_1(\beta)} = \sqrt{\frac{\beta}{\pi}} e^{-\beta x^2} \tag{4}$$

where we have already eliminated α by using Eq. (2) and by introducing the partition function $Z_1 \equiv \int_{-\infty}^{\infty} dx e^{-\beta x^2} = \sqrt{\frac{\pi}{\beta}}$. The use of Eq. (3) of course yields the connection between $\beta \equiv 1/T$ and σ, namely

$$\sigma^2 = 1/2\beta = T/2. \tag{5}$$

After N such jumps (and since the final position is just the *sum* of the displacements attained after each jump) the distribution is given by

$$p_1(x, N) = p_1(x) * p_1(x) * ... * p_1(x) \quad (N factors) \tag{6}$$

where $*$ denotes the *convolution product* (of course $p_1(x, 1) = p_1(x)$). Since the Gaussian distribution (4) is *form-invariant* under convolution and since, according to the central limit theorem, x scales with $\sqrt{N}$, we have

$$p_1(x, N) = \frac{1}{\sqrt{N}} p_1(x/\sqrt{N}, 1) = \frac{1}{\sqrt{N}} p_1(x/\sqrt{N}) \tag{7}$$

$$= \sqrt{\frac{\beta}{\pi N}} e^{-\beta x^2/N} \quad (\forall N)$$

Now, the jumps are short-range ones, hence the average time required for them to be accomplished is *finite* and denoted by $\tau > 0$; τ incorporates *both* the *waiting time* (between the end of a jump and the beginning of the next one) and the *flight time* (time during which the particle is physically in motion, either ballistically or any other type). Therefore we have essentially

$$t = N\tau \qquad (8)$$

(see [2-7] for the subtleties related to the fundamental distinction that must, in general, be done between N and t). As a consequence of Eq. (8), and by choosing units such that $\tau = 1$, Eq. (7) can be rewritten as follows:

$$p_1(x,t) = \sqrt{\frac{\beta}{\pi t}} e^{-\beta x^2/t} \quad (\forall t) \qquad (9)$$

hence

$$< x^2 > (t) \equiv < x^2 >_1 (t) \equiv \int_{-\infty}^{\infty} dx \; x^2 \; p_1(x,t) \; = \frac{Tt}{2} \quad (\forall t) \qquad (10)$$

thus re-establishing the celebrated Einstein result (Fick's law; see [8] and references therein). If for any reason, the single jumps were not precisely Gaussian ones, but any other type within the class of the short-range jumps (i.e., *finite* $< x^2 >_1$), the $t \gg 1$ results would essentially be the same, because the Gaussian (as basically given by Eq. (9)) is the corresponding *attractor* in the distribution-space. So the fundamental idea which comes out from the present simple model is as follows: *normal* diffusion (i.e., $< x^2 >_1 \propto Tt$ for $t \gg 1$) (i) is *ubiquitous* because many physical systems exist for which the microscopic jumps have *short-memory* and are *short-ranged*; (ii) is *robust* because the central limit theorem guarantees that, after many jumps, the system is eventually driven onto the Gaussian distribution, with space x scaling as $\sqrt{t}$.

The aim of the present paper is to extend this analysis to the *long-range* jumps (i.e., whenever $< x^2 >_1 = \infty$). However, we shall restrict our discussion to the *short-memory* class (herein represented by the already described Marcovian process); indeed, the very interesting *long-memory* class (characterized by a "memory function" $M(t,t')$ such that $\int_{-\infty}^{t} dt' t' M(t,t')$ *diverges*) would deserve a specific analysis which is out of the scope of the present work.

2 Anomalous Diffusion

2.1 Introduction

Any diffusive process which violates $< x^2 >_1 (t) \propto t$ ($t \gg 1$) is frequently said to be *anomalous*, and we shall adopt here this terminology. By defining the *transport exponent* $2/d_W$ ($d_W \equiv$ random walk fractal dimensionality; see, for instance, [9] and references therein) through

$$< x^2 >_1 (t) \propto t^{2/d_W} \quad (t \gg 1) \qquad (11)$$

$d_W = 2$ corresponds to *normal* diffusion, $d_W > 2$ corresponds to *subdiffusion* (see, for instance, [10] and references therein), and $d_W < 2$ corresponds to *superdiffusion*. However, this terminology implicitly assumes that $\lim_{t\to\infty} < x^2 >_1 (t)/t^{2/d_W}$ is *finite*. In what concerns superdiffusion (main interest of the present work), the fact that $d_W < 2$ means that *some* fraction of the motion is of the long-range type (typically Lévy flights); some, *but not all*, otherwise the above limit would *diverge*. In fact, if the analysis of the data is done in such a way as to disconsider the sticking-like portions of the random motion and only retain the flights, then we should have $< x^2 > (t)$ diverging for *all* values of $t > 0$, and not only for $t \to \infty$. We shall refer to this type of behavior as *strong superdiffusion*; it can be considered, as will become clear later on, as particularly complex from the standpoint of Statistical Mechanics and it is the one that we focus in the present work

Let us mention a few examples of Lévy distributions in Physics, besides the very many (in both physical and human sciences) discussed by Montroll, Shlesinger and collaborators [11], as well as by Mandelbrot [12]. These distributions have been used to modelize incoherent radiative transfer conformation of a polymer just at adsorption threshold, diffusion in billiards, in chaotic maps (see, for instance, [2-5,14]), and very particularly for turbulent diffusion in an open system where energy is pumped in through mixing [15]. They have also been referred to in analyses of floating tracers in ocean and surface wave flows [16].

Before being more specific, let us recall that the d-dimensional Lévy distributions $L_\gamma(r)$ behave, at *long* distances (i.e., $r \equiv |\vec{r}| >> \sigma$), as $1/r^{d+\gamma}$ with $0 < \gamma < 2$. Their Fourier-transforms equal $exp\{-b|\vec{k}|^\gamma\}$ for *all* values of the wave-vector $\vec{k}$, where b is a positive prefactor and γ is the *fractal dimension* associated with random motion following $L_\gamma(r)$ at every jump. *All* distributions which share with $L_\gamma(r)$ the long-distance power-law $1/r^{d+\gamma}$, share with its Fourier-transform the *same* short wave-vector behavior. Any distribution whose long-distance behavior is a power-law $1/r^{d+\mu}$ with $\mu > 2$, or faster than that (e.g., a Gaussian distribution), has a Fourier transform $exp\{-b|\vec{k}|^2\}$ for small wave-vectors. The whole situation can be summarized by saying that $\gamma = \mu$ if $0 \leq \mu \leq 2$ and $\gamma = 2$ if $\mu \geq 2$. The long-distance behavior of these distributions is of course responsible for the fact that $< r^2 > (t)$ might diverge for *all* values of $t > 0$; indeed, $< r^2 > (1)$ is characterized by $\int_\sigma^\infty dr \, r^{d-1} \, r^2 \, \frac{1}{r^{d+\mu}}$, which diverges for all $\mu \leq 2$.

In order to have in mind some typical values of $\gamma = \mu < 2$ (i.e., neither Gaussian nor any other type of short-range distribution), let us mention some specific examples:

(i) The direct experimental observation of Lévy distributions has been recently possible [17] with amphiphilic molecules of CTAB (cetyl trimethyl ammonium bromide) dissolved in salted water. Indeed, these molecules tend to form elongated flexible breakable cylindrical micelles which, through a reptation mechanism [18], anomalously diffuse in a Lévy-like manner with a (CTBA-concentration, salinity, temperature) -dependent $\gamma = \mu \in (1.5, 2)$.

Micelles are systems quite close to membranes and soaps. So, the present discussion might also applicable to a great variety of systems, very particularly to biological or quasi-biological ones (see [19] for an illustration of the peculiarities that might occur in diffusion processes in such mesoscopic systems);

(ii) The analysis of measurements of heartbeat histograms in healthy individuals has recently provided [20] $\gamma = \mu \simeq 1.7$, whereas some deep cardiomyopathies yield a Gaussian behavior ($\gamma = 2$);

(iii) A recent computer simulation of a leaky faucet[21] (supported by experimental results) has provided $\gamma \in [1.66, 1.85]$;

(iv) Anomalous transport has been recently exhibited and discussed [6] in particle chaotic dynamics along the stochastic web associated with a $d = 3$ Hamiltonian flow with hexagonal symmetry in a plane. It was observed $\gamma = \mu \simeq 1/0.72 \simeq 1.39$;

(v) A computer simulation of a conservative motion in a two-dimensional periodic potential [22] yielded $\gamma = 1.25$;

(vi) Chaotic transport in a laminar fluid flow of a water-glycerol mixture in a rapidly rotating annulus was experimentally studied[23] by tracking large numbers of passive tracer particles for long times and $\gamma = 1.3 \pm 0.2 (\mu - 1$ of [23]) was obtained.

To close this section let us mention that Chandrasekhar has discussed [24] the Holtsmark distribution of forces for some astrophysical systems and power laws have once more emerged; finally, it has been recently argued [25] that bond percolation might correspond to the $\gamma = \mu \to 0$ limit.

2.2 Generalized Statistical Mechanics and Thermodynamics

The scope of Statistical Mechanics is to provide a bridge between microscopic and macroscopic behaviors, and this for *all* types of systems in Nature, *including those where long jumps frequently occur*! The inability of Boltzmann-Gibbs (BG) statistics to satisfactorily accomplish this task is since long well known [11]. The difficulty essentially comes from the impossibility of obtaining (say $d = 1$) Lévy distributions from $S[p] = -\int dx p(x) ln p(x)$ with acceptable *a priori* constraints [11]. Let us briefly review here the generalized thermostatistics which we refer to, and show in Section 2.3 how this extended formalism overcomes the BG difficulty related to the Lévy distributions.

BG Statistical Mechanics is essentially based on the usual (*extensive* and *concave*) entropy $-\sum_i p_i ln p_i$ (with $\sum_i p_i = 1$). It constitutes no doubt an impressive theoretical tool which (at least in principle!) enables the satisfactory discussion of an enormous amount of physical systems. However, *it is not universal*! Its basic restrictions are two-folded:

(i) the spatial range of the (effective) microscopic interactions must be *small* compared to the (linear) size of the macroscopic system (e.g., short-range forces);
(ii) the time range of the microscopic memory must be *small* compared to the observation time (e.g., Marcovian processes).

Whenever one and/or the other of these restrictions is not satisfied, BG statistics *fails* (in one way or another). These quite simple facts are rarely clearly stated or recalled. Fortunately, an everyday increasing literature on gravitational systems [24,26], long-range magnetic systems [27], Lévy-flight-like anomalous diffusion [2-7,11-17,20-23] and similar systems, provides strong evidence along the present lines. For example, in Kaburaki's dramatic style [28], "*Even as a thought experiment a black hole itself cannot be divided into two or more subsystems*". The way out from these difficulties quite clearly seems to be *nonextensive* Statistical mechanics and Thermodynamics. One of us proposed, in 1988 [29], a nonextensive entropy, namely

$$S_q = \frac{1 - \sum_i (p_i)^q}{q - 1} \quad (q \in \Re) \tag{12}$$

which recovers $-\sum_i p_i ln(p_i)$ in the $q \to 1$ limit. This entropy is concave if $q > 0$ and convex if $q < 0$. It enables the consistent generalization of Statistical Mechanics and Thermodynamics[30]. For this to be possible, the standard mean value associated with an arbitrary observable $\hat{O}$ is generalized into the *q-expectation value*

$$< O >_q \equiv \sum_i O_i \, p_i^q \equiv < O \, p^{q-1} >_1 \tag{13}$$

with $\sum_i p_i = 1$. The extended formalism based in Eqs. (12) and (13) has remarkable mathematical properties: among others, *it preserves the Legendre transformation structure* of Thermodynamics [30], and it leaves *form-invariant* the Ehrenfest theorem [31], the von Neumann equation [32] and the Onsager reciprocity theorem [33], *for all values of q*. Also, it has enabled the generalization of Boltzmann H-theorem [34], Shannon theorem [30], Langevin and Fokker-Planck equations [35], Bogolyubov inequality [36], fluctuation-dissipation theorem [37], classical equipartition principle [38], among others. Its effects have been illustrated in simple models such as the free particle [39], Ising model [40], Larmor precession [32]. It has been successfully applied in gravitational systems[41,42] (where it overcomes the already mentionned BG difficulties) and has been tackled in connection with self-organized biological systems [43] and with quantum grups [44]. For reviews, see [45].

A major contribution has been recently given by Alemany and Zanette [46] in connection with anomalous diffusion, which we focus next.

2.3 One-dimensional Lévy-like Anomalous Superdiffusion

We first discuss the *one-jump* problem (in which we extend along the lines of the recent work by Alemany and Zanette[46] and by ourselves[47]) and then address the *many-jumps* problem. Consistently with Eq. (12) we define

$$S_q[p] = \frac{1 - \int_{-\infty}^{\infty} d(x/\sigma)[\sigma p(x)]^q}{q - 1} \quad (q \in \Re) \tag{14}$$

We want to optimize $S_q[p]$ with respect to $p(x)$ with the norm-constraint as given by Eq. (2) as well as with the supplementary constraint

$$< x^2 >_q \equiv \int_{-\infty}^{\infty} dx \; x^2 \; [p(x)]^q = \sigma^{3-q}. \tag{15}$$

which generalizes Eq. (3) along the lines of Eq. (13). By introducing Lagrange parameters we straightfowardly obtain that the optimizing solution is given by

$$p_q(x) = \frac{[1 - \beta(1 - q)x^2]^{1/(1-q)}}{Z_q} \tag{16}$$

with

$$Z_q \equiv \int_{-\infty}^{\infty} dx[1 - \beta(1 - q)x^2]^{1/(1-q)}. \tag{17}$$

We notice that the $\lim_{q \to 1 \pm 0} p_q(x)$ reproduces Eq. (4), and that the norm-constraint can be satisfied for $q \in [-\infty, 3)$ whereas it cannot be if $q \geq 3$. The explicit calculation of Z_q yields, for $1 < q < 3$,

$$p_q(x) = \frac{1}{\sigma} \sqrt{\frac{\beta(q - 1)}{\pi}} \frac{\Gamma(\frac{1}{q-1})}{\Gamma(\frac{1}{q-1} - \frac{1}{2})} \frac{1}{[1 + \beta(q - 1)x^2]^{1/(q-1)}} \tag{18}$$

and, for $-\infty < q < 1$,

$$p_q(x) = \begin{cases} \sqrt{\frac{\beta(1-q)}{\pi}} \frac{\Gamma(\frac{1}{1-q} + \frac{3}{2})}{\Gamma(\frac{1}{1-q} + 1)}[1 - \beta(1 - q)x^2]^{1/(1-q)} & \text{if } |x| \leq \frac{1}{\sqrt{\beta(1-q)}} \\ 0 & \text{otherwise.} \end{cases} \tag{19}$$

These distributions are illustrated in Fig. 1, and it can be shown [48] that they precisely recover the Student's t-distribution for $1 \leq q \leq 3$, and the r-distributions for $-\infty \leq q \leq 1$.

Let us check what happens with the "width" of these (symmetric) distributions. Straightforward calculations yield the standard expectation value

$$< x^2 >_1 (1) \equiv \int dx \; x^2 \; p_q(x) = \frac{1}{\beta}A(q) \propto T \tag{20}$$

with

$$A(q) = \begin{cases} \frac{1}{5-3q} & \text{if } -\infty \leq q < 5/3; \\ \infty & \text{if } 5/3 \leq q \leq 3 \end{cases} \tag{21}$$

as well as the q-expectation value

$$< x^2 >_q (1) \equiv \int dx \; x^2 \; [p_q(x)]^q = \frac{1}{\beta^{\frac{3-q}{2}}}B(q) \propto T^{\frac{3-q}{2}} \tag{22}$$

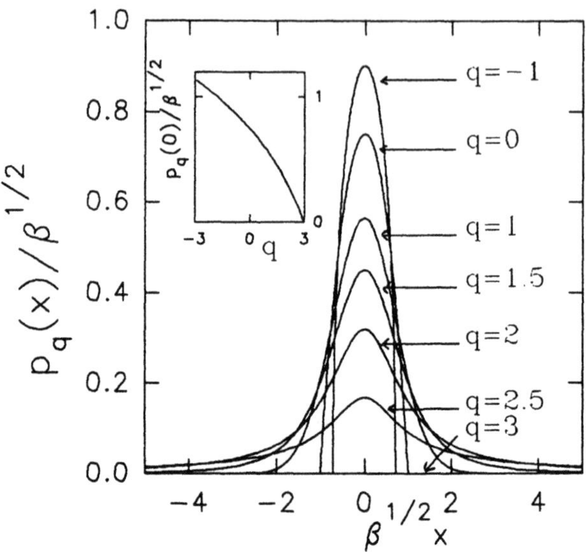

Fig. 1. $d = 1$ distributions $(p_q(x)/\sqrt{\beta})$ versus $(\sqrt{\beta}\,x)$ for typical values of q ($q = 1$: Gauss; $q = 2$: Lorentz; $q = 3$: asymptotically flat distribution; $q = -\infty$: Dirac's delta distribution). Inset: Values of $p_q(0)/\sqrt{\beta}$ as a function of q.

with

$$
B(q) = \begin{cases}
\frac{1}{2}\left[\sqrt{\frac{q-1}{\pi}}\,\frac{\Gamma(\frac{1}{q-1})}{\Gamma(\frac{1}{q-1}-\frac{1}{2})}\right]^{q-1} & \text{if } 1 < q < 3; \\
\frac{1}{2} & \text{if } q = 1; \\
\frac{1}{2}\left[\sqrt{\frac{1-q}{\pi}}\,\frac{\Gamma(\frac{1}{1-q}+\frac{3}{2})}{\Gamma(\frac{1}{1-q}+1)}\right]^{q-1} & \text{if } -\infty \leq q < 1;
\end{cases}
\tag{23}
$$

We of course verify that $A(1) = B(1) = 1/2$, thus recovering Eq. (5). Eqs. (21) and (23) are plotted in Fig. 2. We can see that $< x^2 >_1$ is *finite* only for $-\infty \leq q < 5/3$, and *diverges* otherwise, whereas $< x^2 >_q$ is *finite* for the entire region $-\infty \leq q \leq 3$. A beautiful illustration is the $q = 2$ case, which corresponds to the famous Lorentzian (or Cauchy) distribution $p_2(x) = (\sqrt{\beta}/\pi)(1+\beta x^2)^{-1}$; indeed, $< x^2 >_1 \propto \int_0^\infty dx\; x^2\,(1+\beta x^2)^{-1}$ *diverges*, whereas $< x^2 >_2 \propto \int_0^\infty dx\; x^2\,(1+\beta x^2)^{-2}$ *converges*. To say it in order words, the standard BG thermostatistics ($q = 1$) *cannot* provide the Lorentz distribution within a variational entropic formalism with simple a priori constraints (such as $< x^2 >_1$ known and finite), whereas the $q \neq 1$ formalism *can* (as Montroll and collaborators wanted!).

The use of constraint (15) yields the connection between β and σ, namely

$$
\beta = \frac{[B(q)]^{\frac{2}{3-q}}}{\sigma^2} \qquad (-\infty \leq q \leq 3)
\tag{24}
$$

This essentially closes the one-jump discussion. Let us now discuss the N-jumps

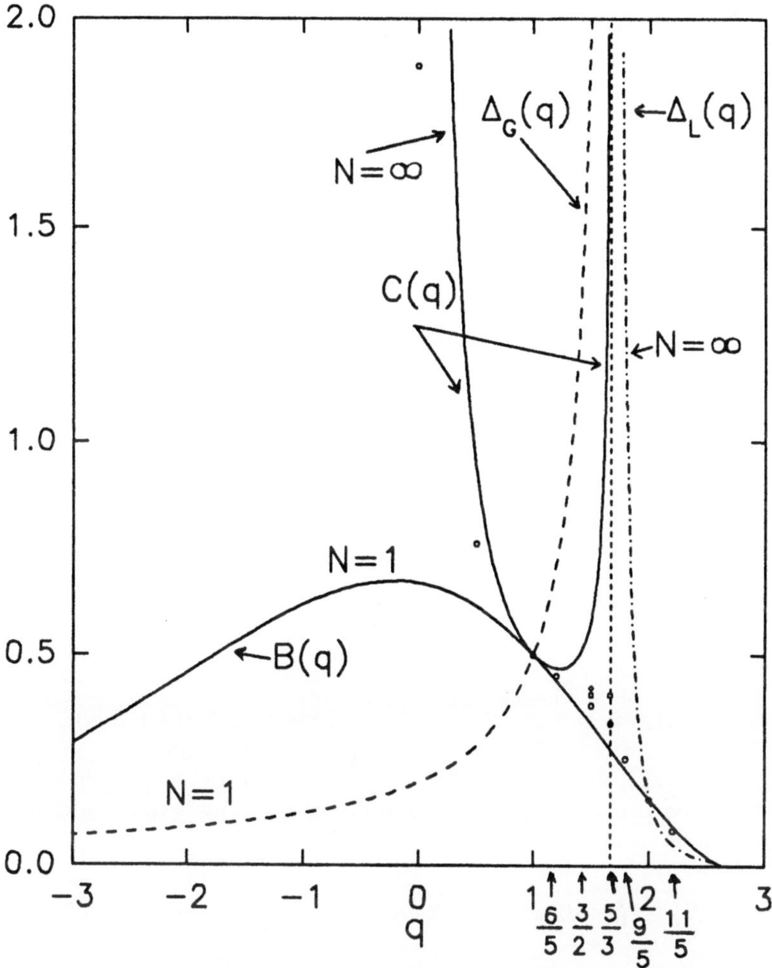

Fig. 2. $d = 1$ q-dependence of the various q-expectation values of x^2. $\Delta_G(q) \equiv A(q) \equiv \beta < x^2 >_1 (N)$ (dashed line; G stands for Gauss); $B(q) \equiv \beta^{\frac{3-q}{2}} < x^2 >_q$ (1) $(B(1) = 1/2, B(2) = 1/2\pi$ and $B(q)$ attains its maximum value 0.6713 for $q \simeq -0.21$); $C(q) \equiv \lim_{N\to\infty} (\beta/N)^{\frac{3-q}{2}} < x^2 >_q (N)$; $\Delta_L(q) \equiv D(q) \equiv \lim_{N\to\infty} (\beta^{\frac{3-q}{2}}/N^{q-1}) < x^2 >_q (N)$ (dot-dashed line; L stands for Lévy). Both $\Delta_G(q)$ and $\Delta_L(q)$ diverge at $q = 5/3$. Some typical points are also indicated: $(\beta/2)^{3/2} < x^2 >_0$ (2) for $q = 0$; $(\beta/2)^{5/4} < x^2 >_{1/2}$ (2) for $q = 1/2$; $(\beta/2)^{9/10} < x^2 >_{6/5}$ (2) for $q = 6/5$; $(\beta/2)^{3/4} < x^2 >_{3/2}$ (2), $(\beta/3)^{3/4} < x^2 >_{3/2}$ (3) and $(\beta/4)^{3/4} < x^2 >_{3/2}$ (4) for $q = 3/2$; $(\beta/2)^{2/3} < x^2 >_{5/3}$ (2) and $(\beta/3)^{2/3} < x^2 >_{5/3}$ (3) for $q = 5/3$; $(\beta^{3/5}/2^{4/5}) < x^2 >_{9/5}$ (2) for $q = 9/5$; $(\beta^{2/5}/2^{6/5}) < x^2 >_{11/5}$ (2) for $q = 11/5$. We recall that $(\beta/N) < x^2 >_1 (N) = 1/2$ and $(\beta^{1/2}/N) < x^2 >_2 (N) = 1/2\pi$ for *all* values of N, and also that $(\beta/N) < x^2 >_1 (N)$ diverges for $q \geq 5/3$ for all values of N and that $\lim_{N\to\infty} (\beta/N)^{\frac{3-q}{2}} < x^2 >_q (N)$ diverges for $q \leq 0$.

problem. The distribution in now given by

$$p_q(x, N) = p_q(x) * p_q(x) * \ldots * p_q(x) \quad \text{(N factors)} \tag{25}$$

The result is trivial for $q = 1$ and $q = 2$ because Gaussians and Lorentzians are form-invariant under convolution: for $q = 1$, $p_1(x, N)$ is given by Eq. (7) and x scales like $\sqrt{N}$, and, for $q = 2$, we have

$$p_2(x, N) = \frac{\sqrt{\beta}}{\pi} \frac{N}{N^2 + \beta x^2} \quad (\forall N) \tag{26}$$

and x scales like N.

If $q \neq 1, 2$, the result for $p_q(x, N)$ is not trivial for arbitrary N. Nevertheless, the $N >> 1$ asymptotic regime can be discussed since it is controlled by limit theorems. Indeed, if $q < 5/3$, $< x^2 >_1$ is *finite* hence the *central limit theorem* applies and therefore the asymptotic distribution is a *Gaussian* with the *same rescaled width* as $p_q(x)$, namely given by

$$p_q(x, N) \sim \sqrt{\frac{\beta}{N}} G_q(\sqrt{\frac{\beta}{N}} x) \quad (N \to \infty) \tag{27}$$

with

$$G_q(y) \equiv \sqrt{\frac{5 - 3q}{2\pi}} e^{-(5-3q)y^2/2} \quad (-\infty \leq q < 5/3; y \in \Re) \tag{28}$$

By using Eq. (20), we can verify that $\beta \int_{-\infty}^{\infty} dx \, x^2 \, p_q(x) = \int_{-\infty}^{\infty} dy \, y^2 \, G_q(y) = 1/(5-3q)$ $(-\infty \leq q < 5/3)$, and this is what we mean by "same rescaled width". In other words,

$$< x^2 >_1 (N) \equiv \int_{-\infty}^{\infty} dx \, x^2 \, p_q(x, N) = NT \frac{1}{5 - 3q} \quad (-\infty \leq q < 5/3; \forall N) \tag{29}$$

If, however, $q > 5/3$, $< x^2 >_1$ *diverges* and what now applies is the *general central limit theorem* [49]. Indeed, the long-distance behavior of $p_q(x)$ is given by $1/|x|^{2/(q-1)}$, consequently the $N \to \infty$ behavior of $p_q(x, N)$ is given by

$$p_q(x, N) \sim \frac{\sqrt{\beta}}{N^{1/\gamma}} L_\gamma((\sqrt{\beta}/N^{1/\gamma}) \, x) \quad (5/3 < q < 3) \tag{30}$$

where $L_\gamma(y)$ is the Lévy distribution whose long-distance behavior (given by $1/|x|^{1+\gamma}$) is the *same* as that of $p_q(x)$. Identification of the exponents yields $2/(q-1) = 1 + \gamma$, hence $\gamma = \frac{3-q}{q-1}$, which precisely is (*but only for $q \geq 5/3$*) the *one-jump* result of Alemany and Zanette [46]! The situation can be summarized by saying that, in the $N \to \infty$ limit, x scales like $N^{1/\gamma}$ where (see Fig. 3(a))

$$\gamma = \begin{cases} 2 & \text{if } -\infty \leq q \leq 5/3; \\ \frac{3-q}{q-1} & \text{if } 5/3 \leq q < 3 \end{cases} \tag{31}$$

It is worth noticing that, for $q \geq 5/3$, $2 - q \rightleftharpoons q - 2$ implies $\gamma \rightleftharpoons 1/\gamma$.

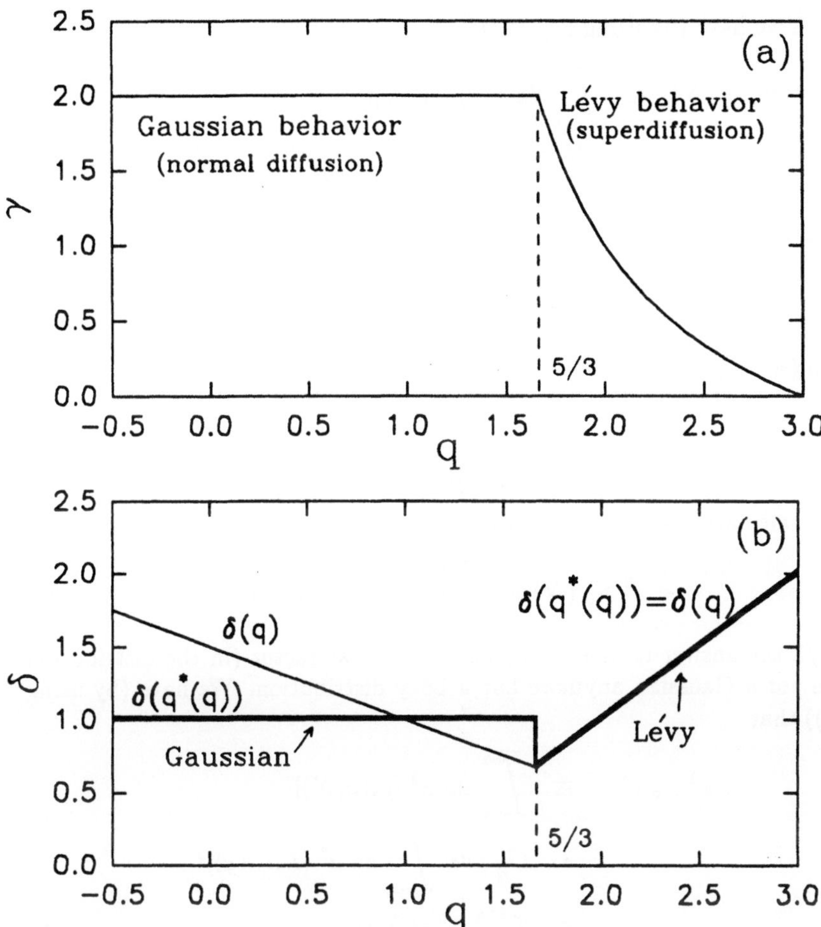

Fig. 3. $d = 1$ q-dependence of: (a) γ (x scales with $N^{1/\gamma(q)}$ for $N \gg 1$); (b) δ ($< x^2 >_q \propto T^{\frac{3-q}{2}} N^{\delta(q)}$, for $N \gg 1$; $\delta(q)$ equals $(3-q)/2$ if $q \leq 5/3$ and equals $(q-1)$ for $q \geq 5/3$; q^* denotes the *attractor* and equals 1 if $q < 5/3$ and q if $q > 5/3$).

To understand better this (fascinating!) anomalous diffusion problem let us now focus $< x^2 >_q (N)$. The quantity $< x^2 >_q (1)$ is given by Eqs. (22) and (23) and is depicted in Fig. 2. What happens for $N \gg 1$?

If $q < 5/3$ we have (by using Eq. (27))

$$< x^2 >_q (N) \equiv \int_{-\infty}^{\infty} dx \; x^2 \; [p_q(x, N)]^q \tag{32}$$

$$\sim \; (\frac{N}{\beta})^{\frac{3-q}{2}} \int_{-\infty}^{\infty} dy \; y^2 \; [G_q(y)]^q = (NT)^{\delta(q)} C(q)$$

with (see Fig. 3(b))

$$\delta(q) = \frac{3 - q}{2} \quad (-\infty \leq q \leq 5/3) \tag{33}$$

and with (see Fig. 1)

$$C(q) = \begin{cases} \infty & \text{if} \; -\infty \leq q \leq 0; \\ \dfrac{1}{\sqrt{(2\pi)^{q-1} q^3 (5-3q)^{3-q}}} & \text{if} \; 0 < q < 5/3 \end{cases} \tag{34}$$

A simple calculation shows that, for $q = 0$, the exact answer is given by

$$< x^2 >_0 (N) = \frac{2^{3N-2}}{3N^{3/2}} \quad (N = 1, 2, ...) \tag{35}$$

If $q > 5/3$ the answer is more complex since the attractor (in the distribution space) is not a Gaussian anymore but a Lévy distribution. We have (by using Eq. (30)) that

$$< x^2 >_q (N) \equiv \int_{-\infty}^{\infty} dx \; x^2 \; [p_q(x, N)]^q$$

$$\sim \; (\frac{N}{\beta})^{\frac{3-q}{\gamma(q)}} \int_{-\infty}^{\infty} dy \; y^2 \; [L_{\gamma(q)}(y)]^q$$

$$\equiv \; T^{\frac{3-q}{2}} N^{\delta(q)} D(q) \tag{36}$$

with (see Fig. 3(b))

$$\delta(q) = \frac{3 - q}{\gamma(q)} = q - 1 \quad (5/3 < q \leq 3) \tag{37}$$

and $D(q)$ a quantity whose general analytical expression we have been unable to calculate. Since $q = 2$ corresponds to the Lorentzian given by Eq. (29), we know that Eq. (36) is *exact for all values of N* and not only for $N \gg 1$, therefore

$$< x^2 >_2 (N) = \frac{NT^{1/2}}{2\pi} \quad (\forall N) \tag{38}$$

hence $D(2) = 1/2$. We have numerically studied the rest of the interval $q \in [5/3, 3]$ and our results suggest the curve indicated in Fig. 2; in particular, $D(5/3) = \infty$ and $D(3) = 0$. It is interesting also to remark that the limit $q = 3$ present a "ballistic" nature in the sense that $\delta(3) = 2$ (i.e., $< x^2 >_3 (N) \propto N^2$). The manner through which the distributions $p_q(x, N)$ approach, in the $N \to \infty$ limit, their attractor is illustrated in Fig. 4.

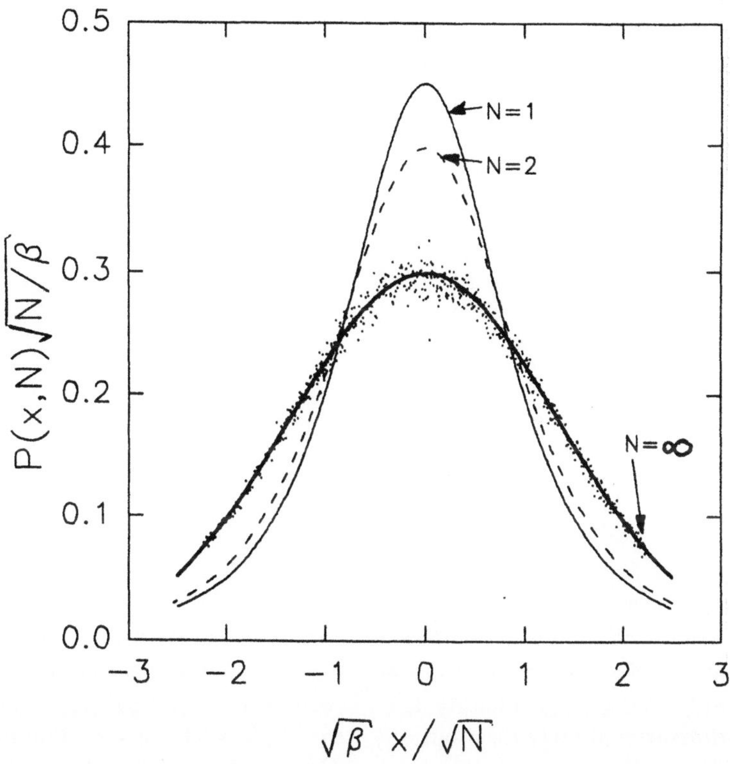

Fig. 4. N-scaling of $p_q(x, N)$ for q=3/2: $p_{3/2}(\sqrt{\beta}x, 1)/\sqrt{\beta} = \frac{4}{\pi}\frac{\sqrt{2}}{(2+\beta x^2)^2}$, $\sqrt{2}p_{3/2}(\sqrt{\beta}x, 2)/\sqrt{\beta} = \frac{64}{\pi}\frac{\beta x^2+80}{(\beta x^2+16)^3}$, $lim_{N\to\infty}\sqrt{N}p_{3/2}(\sqrt{\beta}x, N)/\sqrt{\beta} = \frac{1}{\sqrt{\pi}}e^{-\beta x^2}$ (for all $q < 5/3$, the rescaled *width* $< x^2 >_1$ is preserved through increasing N; the points have been obtained through a computational simulation for $N = 100$). For all $q > 5/3$, the *power* of the $|x| \to \infty$ asymptotic behavior is preserved through increasing N and, in variance with what happens for $q < 5/3$, remains in the $N \to \infty$ limit itself.

2.4 d-dimensional Lévy-like Anomalous Superdiffusion

We only consider here the isotropic case. The discussion essentially follows that of the $d = 1$ case. We consider the generalized entropy

$$S_q[p] = \frac{1 - \Omega_d \int_0^\infty d(r/\sigma)(r/\sigma)^{d-1}[\sigma^d p(r)]^q}{q - 1} \quad (q \in \Re) \tag{39}$$

where Ω_d is the hypersurface of the unitary d-dimensional sphere ($\Omega_1 = 2, \Omega_2 = 2\pi$, $\Omega_3 = 4\pi$). The first constraint is

$$\Omega_d \int_0^\infty dr \; r^{d-1} \; p(r) = 1 \qquad (40)$$

and the second one is

$$< r^2 >_q \equiv \Omega_d \int_0^\infty dr \; r^{d-1} \; r^2 \; [p(r)]^q = \sigma^{2+(1-q)d} \qquad (41)$$

The optimization yields

$$p_q(r) = \frac{[1 - \beta(1-q)r^2]^{1/(1-q)}}{Z_q} \qquad (42)$$

with

$$Z_q \equiv \Omega_d \int_0^\infty dr \; r^{d-1} \; [1 - \beta(1-q)r^2]^{1/(1-q)} \qquad (43)$$

The long-distance behavior of $p_q(r)$ is given by $1/r^{2/(q-1)}$ ($q > 1$), which implies that constraint (40) can be satisfied for all $q < q_{max} \equiv (2+d)/d$. Also, $< r^2 >_1$ is *finite* for all $q < q_c \equiv (4+d)/(2+d)$ and *diverges* for all $q \geq q_c$. As for the $d = 1$ case we must, for $q \geq q_c$, identify $1/r^{\frac{2}{q-1}}$ with $1/r^{d+\gamma}$ (long-distance behavior of the d-dimensional Lévy distribution), hence $2/(q-1) = d + \gamma$. Finally, if we consider $N >> 1$ jumps, the attractor is a Gaussian if $-\infty \leq q < q_c$ and a Lévy distribution if $q_c < q < q_{max}$; consequently, Eq. (31) is generalized as follows

$$\gamma = \begin{cases} 2 & \text{if } -\infty \leq q \leq q_c \equiv \frac{4+d}{2+d} ; \\ \frac{2}{q-1} - d & \text{if } q_c \leq q < q_{max} \equiv \frac{2+d}{d} \end{cases} \qquad (44)$$

which is depicted in Fig. 5.

The Lorentzian particular case corresponds to $\gamma = 1$, hence $q_L = (3+d)/(1+d)$.

2.5 Possible Connection to the Fractal Fokker-Planck-Kolmogorov Equation

A fractal-based generalization of the Fokker-Planck-Kolmogorov (or just Fokker-Planck, as frequently referred to) equation is available in the literature [7], and it is referred to as the *fractal* Fokker-Planck-Kolmogorov equation. Its form is essentially characterized by the prototype (see Eq. (5.17) of [7])

$$\frac{\partial^{\beta'} P(x,t)}{\partial t^{\beta'}} = \frac{\partial^{2\alpha}[BP(x,t)]}{\partial x^{2\alpha}} \qquad (45)$$

where the notation refers to *fractional derivatives* (with $0 < \beta' \leq 1$ and $0 < \alpha \leq 1$; see [5]), B is some function of x, and $P(x,t)$ is the space-time dependent

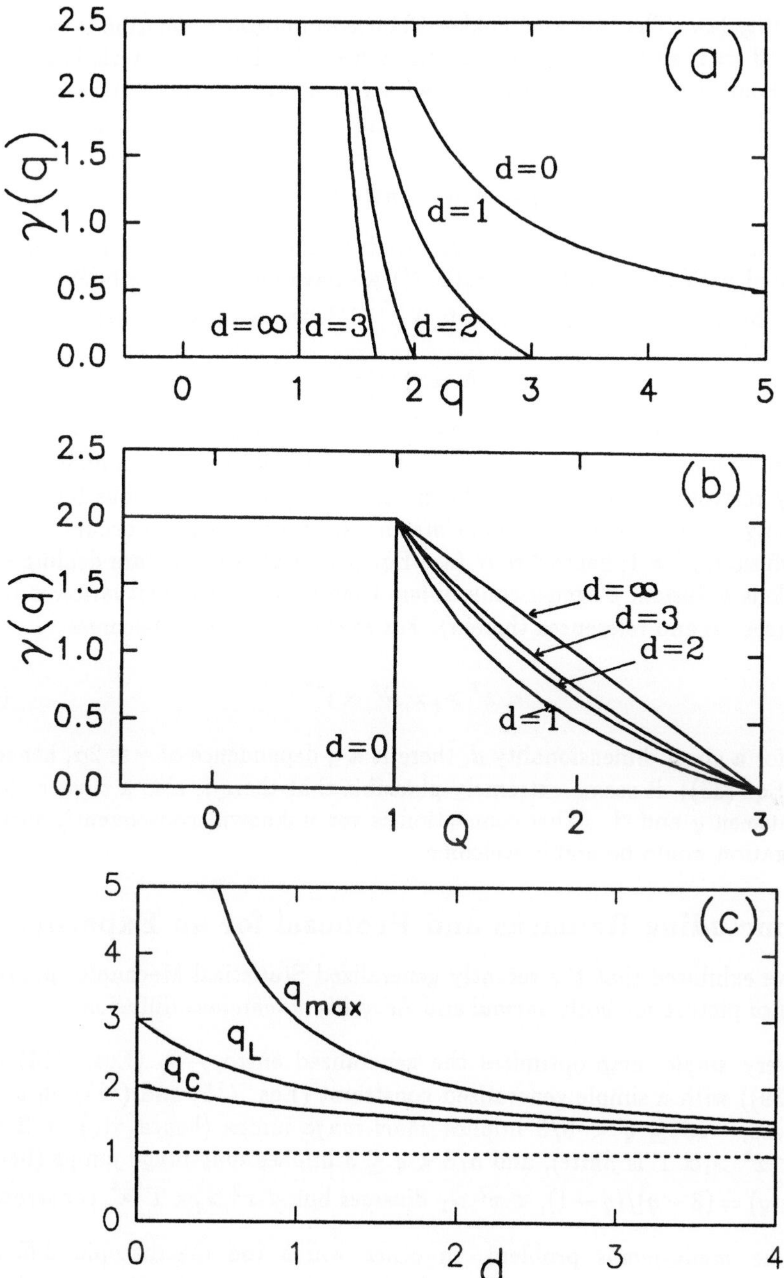

Fig. 5. (q,d)-dependence of γ: (a) γ versus q for typical values of d; (b) γ versus $Q \equiv 1 - \frac{d(4+d)}{2} + \frac{d(2+d)}{2} q$ (conveniente way for representing the $d \to \infty$ behavior, which then becomes $\gamma = 3 - Q$ for $Q > 1$); $Q = 2$ implies $\gamma = d/(1+d)$; (c) d-dependence of q_{max}, q_L and q_c (see the text).

probability law. The standard Fokker-Planck-Kolmogorov equation is recovered for $\alpha = \beta' = 1$, and its solution is given by Eq. (9). The characteristic function (space-Fourier transform) associated with $P(x,t)$ is essentially given (see Eqs. (7.4) and (7.5) of [7] where there is a misprint: one should read 2α instead of α) by

$$P(k,t) = exp\{-constant\ t^{\beta'}|\vec{k}|^{2\alpha}\} \qquad (46)$$

where $\vec{k}$ is the associated wave-vector variable. Since the characteristic function associated with the distribution $p_q(x,N)$ we have been working in Section 2.3 is given by $\tilde{p}_q(k,t) = exp\{-constant\ N|\vec{k}|^{\gamma(q)}\}$, we can readily identify $\gamma \equiv 2\alpha$ and

$$N = (t/\tau)^{\beta'} \qquad (47)$$

where we recall that $\tau > 0$ is a time-dimension characteristic constant of the problem. The physical interpretation of Eq. (47) is basically simple and extremely relevant. What we have been considering in the previous Sections is *how many jumps* (N), and *not the time ellapsed* (t) for them to occur. For normal diffusion, $\beta' = 1$, hence $t \propto N$ (see Eq. (8)). But when we are dealing with anomalous diffusion, β' generically differs from unity and the situation is more subtle (see [2] and references therein). For example, Eq. (32) becomes

$$< x^2 >_q \propto N^\delta \propto t^{\delta\beta'} \qquad (48)$$

Since, for a given dimensionality d, there is a q-dependence of $\gamma \equiv 2\alpha$, hence of δ (see Eq. (33)), it seems extremely plausible that there is also a close connection between q and β'. This connection is yet unknown, consequently further investigation would be highly welcome.

3 Concluding Remarks and Proposal for an Experiment

We have exhibited that the recently generalized Statistical Mechanics provides an *unified* picture for both *normal* and *Lévy-like anomalous* diffusions:

(i) Every *single jump* optimizes the generalized entropy S_q (Eqs. (14) and (39)) with a simple generalized constraint (Eqs. (15) and (41)); in $d = 1$ case, $-\infty \le q < 5/3$ implies *short-range* jumps (hence $\gamma(q) = 2$ and $< x^2 >_1 \propto T$ is *finite*), and $5/3 < q \le 3$ implies *long-range* jumps (hence, $\gamma(q) = (3-q)/(q-1)$, $< x^2 >_1$ *diverges* but $< x^2 >_q \propto T^{\frac{3-q}{2}}$ *converges*);

(ii) The *many-jumps* problem, in other words the macroscopic diffusion ($N \to \infty$, hence $t \to \infty$), is *robust* and determined by the *general central limit theorem*; for $-\infty \le q < 5/3$, the attractor in the distribution space is a *Gaussian*, x scales like $\sqrt{N} \propto \sqrt{t}$, and $\lim_{N\to\infty} \frac{<x^2>_1(N)}{N} \propto \lim_{t\to\infty} \frac{<x^2>_1(N(t))}{t} \propto T$ is *finite* and monotonically increases from zero to infinity when q increases from $-\infty$ to 5/3; for $5/3 < q < 3$, the attractor is the *Lévy distribution* $L_\gamma(y)$, x scales like $N^{1/\gamma(q)} \propto t^{\beta'/\gamma(q)}$, and

285

$\lim_{N\to\infty} \frac{<x^2>_q(N)}{N^{1/\gamma(q)}} \propto \lim_{t\to\infty} \frac{<x^2>_q(N(t))}{t^{\beta'/\gamma(q)}} \propto T^{\frac{3-q}{2}}$ is *finite* and seems to monotonically decrease from infinity to zero when q increases from 5/3 to 3.

These facts provide a beautiful analogy with standard phase transitions where the relevant susceptibility diverges at the critical point (from both sides). Here, the quantity $\lim_{N\to\infty} \frac{<x^2>_{q^*}(N)}{N^{1/\gamma(q^*)}}$ (q^* refers to the attractor, i.e., $q^* = 1$ when it is Gaussian, and $q^* = q$ when it is a Lévy distribution) presents exactly the same behavior as a function of $q \in [-\infty, 3]$, the critical divergence ocurring at $q = 5/3$ (where the single jump $< x^2 >_1 (1)$ diverges). Summarizing, for $N >> 1$ and all values of T,

$$N^{1/\gamma(q^*)}p_q(x,N)/\sqrt{\beta} \sim L_{\gamma(q^*)}(\frac{x\sqrt{\beta}}{N^{1/\gamma(q^*)}}) \tag{49}$$

(with the notation $L_{\gamma(1)}(y) \equiv G_q(y)$) and

$$< x^2 >_{q^*} (N) \sim \Delta(q)T^{\frac{3-q^*}{2}} N^{\delta(q)} \tag{50}$$

with $\delta(q) = 1$ for $q^* = 1$ and $\delta(q) = q - 1$ for $q^* = q$ and where $\Delta(q)$ is a pure number (see Fig. 2) diverging at $q = 5/3$ (from both sides). The Gauss-Lévy crossover can be followed by defining $q_{eff}(q, N)$ through

$$\frac{< x^2 >_{q_{eff}(q,N)} (N)}{T^{\frac{3-q_{eff}(q,N)}{2}}} = \frac{< x^2 >_{q^*(q)} (N)}{T^{\frac{3-q^*(q)}{2}}} \quad (N \geq 1) \tag{51}$$

The result is indicated in Fig. 6. We verify that $\lim_{N\to\infty} q_{eff}(q, N) = q^*(q)$. The present theoretical considerations could, in principle, be easily checked experimentally. Indeed, let s be an external parameter which might drive, in a given physical system, the diffusion from a normal regime to a Lévy-like superdiffusion regime (e.g., in the micelles experiment described in [17], s could be the CTBA concentration). Let us assume that for s below (above) a critical value s_c the regime is a Gaussian (Lévy) one. Let us also assume that we are able to experimentally follow, during long times, many passive tracers and digitalize their trajectories (this was achieved, for instance, in the experiment described in [23]). Then we can computationally erase the "stickings" that might be present in the trajectories of the tracers, and only leave the "flights" (as an illustration of such procedure, an operational distinction between "stickings" and "flights" has been worked out by the authors of [23] for their particular experiment). Finally, the appropriate histograms of what is left will provide approximations for the probability law $p_s(x, N)$, i.e., the "pure" Lévy-like distribution. We should then check that the data can be made essentially N-invariant by representing them in a plot $N^{1/\gamma(s)}p_s(x, N)$ versus $x/N^{1/\gamma(s)}$ where the function $\gamma(s)$ is obtained through the fitting. One expects $\gamma = 2$ for all $s < s_c$, and then γ should decrease below 2 for further increasing s above s_c. The relation $q = (3+\gamma(s))/(1+\gamma(s))$, or the analogous one if $d > 1$, will provide $q(s)$ for $s > s_c$; we expect $q^*(s)$ to exhibit a jump (from 1 to 5/3 if $d = 1$) at s_c. We can then evelute

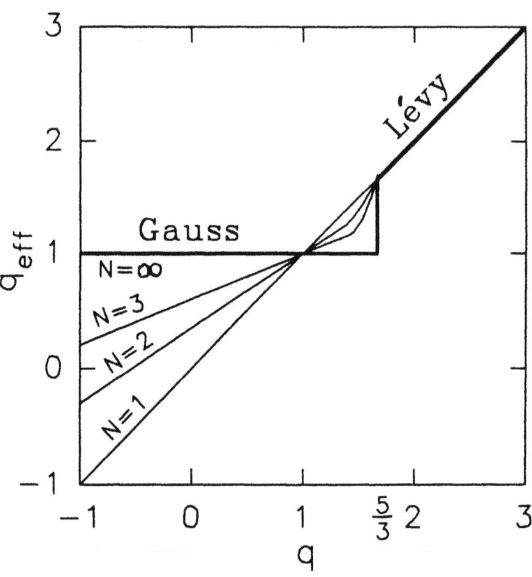

Fig. 6. q-dependence of q_{eff} for typical values of N (the $N = 1$, ∞ curves are exact; the $N = 2, 3$ curves are indicative); $\lim_{N \to \infty} q_{eff}(q, N) = q^*(q)$ equals 1 if $q < 5/3$ and equals q if $q > 5/3$.

$< x^2 >_{q^*} (N) \equiv \int dx \, x^2 \, [p_s(x, N)]^{q^*(s)}$ versus N and plot this in a log-log representation. The slope, as a function of s, should be 1 for $s < s_c$, and given by $q(s) - 1$ for $s > s_c$; the ordinate at the origin, as a function of s, should approach a divergence (from both sides) at s_c. Though slightly more complex, similar analyses can in principle be done without erasing the "stickings" (we should then use one more exponent, namely $\beta'(s)$, which makes the connection between N and t (see section 2.5).

Let us conclude by adding that the whole picture suggests interesting dynamical effects of the manner with which the rescaled distributions approach their attractor as N increases: there is probably a q-dependent relaxation time which, for $d = 1$, diverges at $q = 5/3$ (from both sides), but this study remains to be done.

We acknowledge very useful remarks from A. Araújo, R. R. da Silva and M. E. Vares. One of us (CT) also acknowledges the benefit of stimulating discussions with P. G. de Gennes, A. Libchaber, M. F. Shlesinger, D. Prato, G. M. Zaslavsky, O. N. de Mesquita, S. D. Mahanti, T. A. Kaplan and P. M. Duxbury, as well as warm hospitality received from R. Maynard (at the Laboratoire d'Expérimentation Numérique, Université Joseph Fourier/Grenoble, where the present work started) and from M. F. Thorpe (at the Department of Physics and Astronomy, Michigan State University, where it was concluded).

References

[1] A. Einstein, Ann. Phys. **17**, 549(1905), **33**, 1275(1910), and "Investigations on the Theory of Brownian Movement" (Methuen and Co, London, 1926).

[2] M. F. Shlesinger, J. Klafter and Y. M. Wong, J. Stat. Phys. **27**, 499(1982); M. F. Shlesinger, Rev. Phys. Chem. **39**, 269(1988); H. Scher, M. F. Shlesinger and J. T. Bendler, Physics Today **1**, 26(1991).

[3] M. F. Shlesinger, G. M. Zaslavsky and J. Klafter, Nature **363**, 31(1993).

[4] V. V. Afanasiev, R. Z. Sagdeev and G. M. Zaslavsky, in "Topological Aspects of the Dynamics of Fluids and Plasmas", Eds. H. K. Moffatt, G. M. Zaslavsky, P. Comte and M. Tabor (Kluwer, Dordrecht, 1992), page 481.

[5] G. M. Zaslavsky, Chaos **4**, 25(1994).

[6] G. M. Zaslavsky, D. Stevens and H. Weitzner, Phys. Rev. E **48**, 1683(1993).

[7] G. M. Zaslavsky, Physica D **76**, 110 (1994).

[8] S. Havlin and D. Ben-Abraham, Adv. Phys. **36**, 695(1987); S. N. Majumdar and M. Barma, Physica A **177**, 367(1991) and Phys. Rev. B **44**, 5306(1991).

[9] J. S. Helman, A. Coniglio and C. Tsallis, Phys. Rev. Lett. **53**, 1195(1984) and **54**, 1735(1985).

[10] P. M. Richards, Phys. Rev. B **16**, 1393(1977); A. Bhattacharya, Phys. Rev. E **49**, 4946(1994).

[11] B. D. Hughes, M. F. Shlesinger and E. W. Montroll, Proc. Natl. Acad. Sci. USA **78**, 3287(1981); M. F. Shlesinger and B. D. Hughes, Physica A **109**, 597(1981); E. W. Montroll and M. F. Shlesinger, J. Stat. Phys. **32**, 209(1983); E. W. Montroll and M. F. Shlesinger, in *Nonequilibrium Phenomena II: From Stochastic to Hydrodynamics*, Eds. J. L. Lebowitz and E. W. Montroll (North-Holland, Amsterdam, 1984).

[12] B. B. Mandelbrot, *The Fractal Geometry of Nature* (Freeman, San Francisco, 1982).

[13] T. L. Bell, U. Frisch and H. Frisch, Phys. Rev. A **17**, 1049(1978).

[14] J. P. Bouchaud and A. Georges, Phys. Rep. **195**, 127(1990).

[15] M. F. Shlesinger, B. J. West and J. Klafter, Phys. Rev. Lett. **58**, 1100(1987).

[16] M. G. Brown and K. B. Smith, Phys. Fluids A **3**, 1186(1991); A. R. Osborne, A. D. Kirwan, A. Provenzale and L. Bergamasco, Tellus **41A**, 416(1989); R. Ramshankar and J. P. Gollub, Phys. Fluids A **3**, 1344(1991).

[17] A. Ott, J. P. Bouchaud, D. Langevin and W. Urbach, Phys. Rev. Lett. **65**, 2201(1990); J. P. Bouchaud, A. Ott, D. Langevin and W. Urbach, J. Phys. II France **1**, 1465(1991).

[18] S. F. Edwards, Proc. Phys. Soc. **92**, 9(1967); P. G. de Gennes, J. Chem. Phys. **55**, 572(1971); see also W. Hess, Macromolecules **19**, 1395(1986), **20**, 2589(1987) and **21**, 2670(1988), M. Doi, J. Polym. Sci., Polym. Phys. Ed. **21**, 667(1983), J. des Cloizeaux, Europhys. Lett. **5**, 437(1988) and **6**, 475(1988) and G. T. Barkema, J. F. Marko and B. Widom, Phys. Rev. E **49**, 5303(1994).

[19] M. O. Magnasco, Phys. Rev. Lett. **71**, 1477(1993); A. Libchaber, talk presented at the International Conference on "Scaling Concepts and Complex Fluids" (4-8 July 1994, Catanzaro-Italy).

[20] C. -K. Peng, J. Mietus, J. M. Hausdorff, S. Havlin, H. E. Stanley and A. L. Goldberger, Phys. Rev. Lett. **70**, 1343(1993).

[21] T. J. P. Penna, P. M. C. de Oliveira, J. C. Sartorelli, W. M. Gonçalves and R. D. Pinto, *Long range anticorrelations: is the healthy heart a leaky faucet?*, preprint (1994).

[22] J. Klafter and G. Zumofen, Phys. Rev. E **49**, 4873(1994).

[23] T. H. Solomon, E. R. Weeks and H. L. Swinney, Phys. Rev. Lett. **71**, 3975(1993).

[24] S. Chandrasekhar, Rev. Mod. Phys. **15**, 1(1943); see also P. H. Coleman and L. Pietronero, Phys. Rep. **213**, 311(1992).

[25] S. Roux, A. Hansen, L. R. da Silva, L. S. Lucena and R. B. Pandey, J. Stat. Phys. **65**, 183(1991).

[26] L. G. Taff, "Celestial Mechanics" (John Wiley and Sons, New York, 1985) pages 437 and 438; W. C. Saslaw, " Gravitational Physics of Stellar and Galactic Systems" (Cambridge University Press, Cambridge, 1985), pages 217 and 218; R. Balian, "From Microphysics to Macrophysics", Vol. 1, (Springer-Verlag, Berlin, 1991), page 134; A. M. Salzberg, J. Math. Phys. **6**, 158(1965); H. E. Kandrup, Phys. Rev. A **40**, 7265(1989); J. Binney and S. Tremaine, "Galactic Dynamics" (Princeton University Press, Princeton, 1987), page 268; P. T. Landsberg, J. Stat. Phys. **35**, 159(1984); D. Pavon, Gen. Rel. and Grav. **19**, 375(1987).

[27] B. J. Hiley and G. S. Joyce, Proc. Phys. Soc. **85**, 493(1965).

[28] O. Kaburaki, Phys. Lett. A **185**, 21(1994).

[29] C. Tsallis, J. Stat. Phys. **52**, 479(1988).

[30] E. M. F. Curado and C. Tsallis, J. Phys. A **24**, L69(1991); Corrigenda: J. Phys. A **24**, 3187(1991) and **25**, 1019(1992).

[31] A. R. Plastino and A. Plastino, Phys. Lett. A **177**, 177(1993).

[32] A. R. Plastino and A. Plastino, Physica A **202**, 438(1994).

[33] A. Chame and E. V. L. de Mello, *The Onsager reciprocity relations within Tsallis statistics*, preprint (1994); M. O. Caceres, *Irreversible thermodinamics in the framework of Tsallis entropy*, preprint (1994).

[34] A. M. Mariz, Phys. Lett. A **165**, 409(1992); J. D. Ramshaw, Phys. Lett. A **175**, 169 and 171(1993).

[35] D. A. Stariolo, Phys. Lett. A **185**, 262(1994).

[36] A. Plastino and C. Tsallis, J. Phys. A **26**, L893(1993).

[37] A. Chame and E. V. L. de Mello, J. Phys. A **27**, 3663(1994).

[38] A. R. Plastino and A. Plastino and C. Tsallis, J. Phys. A **27**, 5707(1994).

[39] E. P. da Silva, C. Tsallis and E. M. F. Curado, Physica A **199**, 137(1993); Erratum: Physica A **203**, 160(1994).

[40] R. F. S. Andrade, Physica A **175**, 285(1991) and **203**, 486(1994).

[41] A. R. Plastino and A. Plastino, Phys. Lett. A **174**, 384(1993); J. J. Aly, in "N-body Problems and Gravitational Dynamics", Proceedings of the Meeting held at Aussois-France (21-25 March 1993), Eds. F. Combes and E. Athanassoula (Publications de l'Observatoire de Paris, 1993), page 19.

[42] A. R. Plastino and A. Plastino, Phys. Lett. A **193**, 251(1994).

[43] P. T. Landsberg, in "On Self-Organization", Springer Series on Synergetics **61**, 157(1994).

[44] C. Tsallis, *Non Extensive Physics: A Connection between Generalized Statistical Mechanics and Quantum Groups*, Phys. Lett. A (1994), in press.

[45] C. Tsallis, in "New Trends in Magnetism, Magnetic Materials and their Applications", Eds. J. L. Morán-López and J. M. Sánchez (Plenum Press, New York, 1994), page 451; in "Chaos, Solitons and Fractals", Ed. G. Marshall (Pergamon Press, 1994), in press; Quimica Nova (1994), in press.

[46] P. A. Alemany and D. H. Zanette, Phys. Rev. E **49**, 956(1994).

[47] C. Tsallis, R. Maynard and A. M. C. de Souza, *Statistical-mechanic Foundation of the Robustness of the Lévy-type Anomalous Diffusion*, preprint (1994).

[48] A. M. C. de Souza and C. Tsallis, *Student's t- and r-Distributions: Unified Derivation From an Entropic Variational Principle*, preprint (1994).

[49] A. Araújo and E. Giné, *The Central Limit Theorem for Real and Banach Valued Random Variables*(J. Wiley, New York, 1980), chapter 2.

A dynamical model leading to the breakdown of the Green-Kubo predictions

[1] Elena Floriani, [2] György Trefán,
[1, 2, 3] Paolo Grigolini and [2] Bruce J. West

[1]Dipartimento di Fisica dell'Università di Pisa,
Piazza Torricelli 2, 56100 Pisa, Italy
[2]Department of Physics of the University of North Texas,
P.O. Box 5368, Denton, Texas 76203
[3]Istituto di Biofisica del Consiglio Nazionale delle Ricerche,
Via San Lorenzo 26, 56127 Pisa, Italy

Abstract: We study a dynamical realization of a Lévy diffusion process and the response of the diffusing system to an external perturbation. We find that the response is characterized by two time regimes, an early regime, still fitting the Green-Kubo prediction, albeit implying a conductivity which would diverge if this regime were unlimited, and a later regime, characterized by a constant conductivity but departing from the Green-Kubo prediction. We focus on the shape of the resulting distribution of particles, and we show that the anomalous properties of the response depend on one distribution tail growing much faster than the other.

Keywords. Green-Kubo, Lévy statistics, conductivity, mappings

I. Introduction

The basis of our understanding of statistical physics is the phenomenological theory of Brownian motion and stochastic processes [1] together with the chaotic behavior of nonintegrable Hamiltonian systems [2]. The latter motion is an intrinsic property of nonlinear dynamical systems [3], whereas the former motion results from the system of interest being coupled to the environment. If a dynamical system such as the standard map is fully chaotic, meaning that all the KAM tori have dissolved in a chaotic sea, then the average energy of the system increases linearly with time and the system is diffusive [4]. Even more interesting are certain non-diffusive systems that are described by Lévy statistics rather than those of Brownian motion [5,6]. The connection between Lévy statistics and dynamical systems is through weak chaos, which means stable islands persisting in the chaotic sea. A number of investigators maintain that this coexistence of stable islands, cantori and the chaotic sea produces waiting time distributions with an inverse power law due to the chaotic orbits sticking to the cantori at the phase space boundary betwen the two types of motion [7-13], and hence results in Lévy statistics [10,11,13,14]. Of some interest is the fact that in a preceding paper [14] we have developed a dynamical approach to Lévy diffusion processes based on a Master Equation (ME) method tailored for this specific purpose.

It has recently been stressed that there is a close connection between the microscopic derivation of the conventional linear response and the microscopic derivation of ordinary statistical mechanics [15]. The conventional theory of conductivity is a significant expression of ordinary statistical mechanics and essentially rests on a clearcut separation between the microscopic regime of the processes responsible for the statistical nature of the process under study and the macroscopic process of transport which has to be compared with experimental observation.

The phenomenon of anomalous diffusion resulting from the lack of a microscopic time scale is very interesting since the system response to external perturbation might be suggestive of the behavior of real systems departing from ordinary statistical mechanics. Let us see in detail this aspect. From a dynamical point of view, all diffusion processes, anomalous as well as normal, result from the integration of the equation of motion

$$\dot{x}(t) = \xi , \tag{1.1}$$

where ξ is a statistical variable characterized by the equilibrium autocorrelation function

$$\Phi_\xi(t) \equiv \frac{<\xi(0)\xi(t)>}{<\xi^2>} . \tag{1.2}$$

Under the stationarity assumption, it can be shown [16] that

$$<x^2(t)> = <x^2(0)> + 2 <\xi^2>_{eq} \int_0^t dt' \int_0^{t'} dt'' \, \Phi_\xi(t'') . \tag{1.3}$$

The phenomenon of anomalous diffusion is triggered by the fact that the correlation function $\Phi_\xi(t)$ has a long-time behavior characterized by an inverse power law,

$$\lim_{t\to\infty} \Phi_\xi(t) \equiv \pm \frac{k}{t^\beta} . \tag{1.4}$$

It can be shown that the long-time behavior of $< x^2(t) >$ is described by a generalized version of ordinary diffusion,

$$<x^2(t)> = K t^{2H} , \tag{1.5}$$

with $H=1-\beta/2$ as can be seen by twice differentiating both (1.5) and (1.3) with respect to time. In this paper we shall focus on the case $1/2 < H < 1$, namely the case of superdiffusion, or diffusion faster than standard, corresponding to the following values of β: $0 < \beta < 1$.

According to the standard Green-Kubo method [17] the response of a normal diffusion process to a perturbation resulting in a bias is given by

$$\frac{d}{dt} <x(t)> = K \int_0^t <\xi(0)\,\xi(t')>_{eq} dt' \;, \tag{1.6}$$

where K is a constant depending on the physical details of the system considered. If we imagine (1.6) to be applicable also to a case of anomalous diffusion, we are led to conclude that a stationary current, namely a constant velocity $<\dot{x}>$, is not allowed. The inverse power-law dependence (1.4) with the condition $0 < \beta < 1$ suggests that in the case of anomalous diffusion conductivity might become time dependent and increase as a function of time.

To assess the real behavior of the system it is convenient to study a dynamical model characterized by the property (1.4). Thus we decided to study the maps originally proposed by Geisel *et al.* [7], which have more recently been the subject of the research work of Zumofen and Klafter [10,11]. These are essentially dynamical models making the variable ξ in (1.1) fluctuate between the two values 1 and -1, with the asymptotic form of the waiting time distribution $\psi(t)$ in each of these two states satisfying

$$\lim_{t \to \infty} \psi(t) = \frac{1}{t^\mu} \;. \tag{1.7}$$

Notice that the waiting time distribution $\psi(t)$ and the correlation function $\Phi_\xi(t)$ are related [8] to one another by

$$\Phi_\xi(t) = \frac{1}{<\xi^2>} \frac{1}{<t>} \int_t^\infty dt' \, (t'-t)\, \psi(t') \;. \tag{1.8}$$

Thus it is immediately seen that (1.4) is recovered, with the following connection between μ and β, $\beta = \mu\text{-}2$. In accordance with $0 < \beta < 1$ our investigation is limited to the case of maps leading to $2 < \mu < 3$.

In an earlier paper [14] the present authors have studied the response of this dynamical system to perturbations. It was noticed that in the first part of the response process the mean value of the displacement $<x(t)>$ is proportional to the unperturbed mean squared displacement $<x^2(t)>$, thereby leading us to the conclusion that in this case the conductivity, rather than quickly converging to a stationary value, would instead tend to infinity. In other words, the early part of the response process adheres to the Green-Kubo-like prediction (1.6). In the latter part of the process, however, the conductivity becomes constant, albeit with an absolute value which does not have anything to do with the Green-Kubo prediction. Similar results were found years ago in the case of anomalous diffusion originated by diffusion in random environments [18]. At short times the first moment of the distribution $P(x,t)$ responding to the perturbation turned out to be proportional to the second moment (1.5), and consequently, in the case $H > 1/2$, this is equivalent to a conductivity increasing in time. The long-time regime of $<x(t)>$ was proved [18] to be linear in time, thereby implying a steady conductivity.

We are therefore led to believe that these might be general properties of the response to perturbations of the systems whose free diffusion would be anomalous. To complete our understanding of this interesting behavior, in this paper we determine the distribution $P(x,t)$ corresponding to these two regimes.

II. The Dynamical Model And Its Expected Conductivity

The discrete time dynamical model studied herein is one of the two discussed in the recent paper by Zumofen and Klafter [10], originally introduced by Geisel *et al.* [7]. We focus on the map $g(x)$ they used to discuss anomalous diffusion that evolves faster than normal. Assuming that $g(x)$ is antisymmetric by reflection around $x = 0$ and invariant by translation of a unit distance, they express this map in the reduced range $0 \leq x \leq 1/2$, where

$$x_{n+1} = g(x_n) ,$$

$$(2.1)$$

with

$$g(x) = x + ax^z , \qquad 0 \leq x \leq 1/2 .$$

$$(2.2)$$

They choose the constant to be $a = 2^z$.

We define the reduced map [10]

$$\tilde{x}_{n+1} = \tilde{g}(\tilde{x}_n) , \quad 0 < \tilde{x} < 1 ,$$

$$N_{n+1} = \hat{g}(\tilde{x}_n) + N_n .$$

$$(2.3)$$

The coordinate x of the trajectory has been decomposed into the box number N and the position $\tilde{x}$ within a box $(x_n = N_n + \tilde{x}_n)$. $\tilde{g}(\tilde{x})$ is the reduced map for the reduced coordinate illustrated in Fig. 1 and $\hat{g}(\tilde{x})$ is used to increase or decrease the box number N by unity. We are now in a position to express the variable velocity ξ in terms of the mapping variable. This is given by $\xi = 2\tilde{x} - 1$.

By adopting this procedure to make the numerical calculations we allow the velocity of the particle to also attain values distinct from 1 and - 1. However, we note that the closer the particle is to the left border of the left laminar region or to the right border of the right laminar region, the longer is the time spent in the corresponding laminar region. Consequently, the resulting dynamics is sligthly different from the case where the variable ξ has only the states 1 and - 1, with the same inverse power-law distribution $\psi(t)$ in each of them.

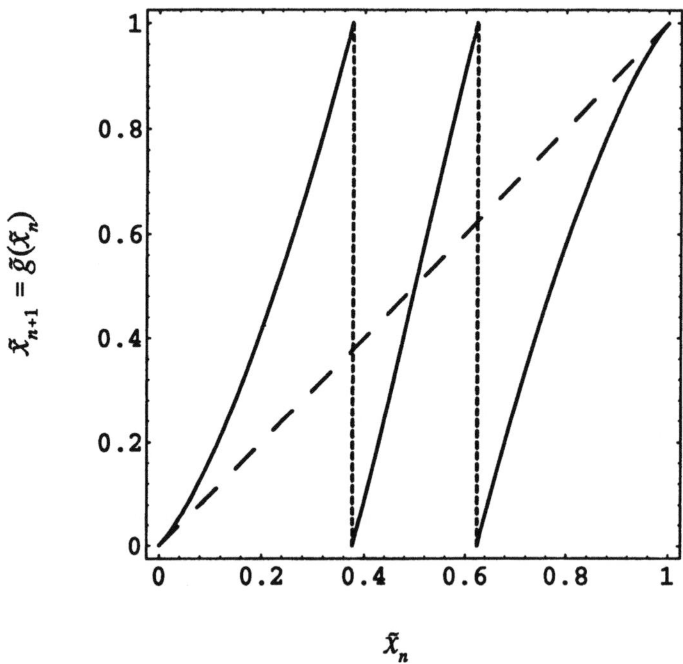

Fig. 1. The reduced map defined by (2.3) with $z=5/3$.

The numerical calculations are made by direct numerical implementations of the Zumofen-Klafter map, see equations (2.1) to (2.3). The displacement is the sum of the "kicks" ξ_n, i.e., $x_{n+1} = x_n + \xi_n$ with the initial condition $x_0 = 0$. The initial values of the reduced map are chosen randomly in the interval $[0,1]$ and the iteration is repeated 10,000 times. The average is taken over 10,000 trajectories.

The waiting time distribution in each of the two laminar regions of the map turns out to be [10]

$$\psi(t) = \frac{A}{(B+t)^\mu} \, ,$$

$$(2.4)$$

where the exponents μ and z are related by $\mu = z/(z-1)$ and

$$B = \frac{\mu-1}{2} \ ,$$
(2.5)

$$A = (\mu-1) B^{\mu-1} \ .$$
(2.6)

A dynamical bias is obtained by changing the left laminar region, namely (2.2), into

$$g(x) = (1+\lambda) x + ax^z$$
(2.7)

with $0 < \lambda \ll 1$ and keeping the right laminar region unperturbed. Thus for the left laminar region the waiting time distribution is

$$\psi_\epsilon(t) = \frac{A_\epsilon}{(B+t)^\mu} \exp(-\epsilon t) \ ,$$
(2.8)

where $\epsilon = (\mu - 2) \lambda/[2(\mu - 1)]$ and the constant A_ϵ is chosen to ensure normalization of $\psi_\epsilon(t)$.

The quantities of interest, *i.e.* the probability distribution $P(x,t)$ and its moments $<x^n(t)>$ can be calculated with Continuous Time Random Walk (CTRW) methods as in ref. [10,11], or through the ME approach developed in [14]. In this paper we focus on the ME approach, which consists in solving an evolution equation for $P(x,t)$ with the following structure

$$\frac{\partial}{\partial t} P(x,t) = \int_0^t dt' \int_{-\infty}^{\infty} dx' \, K(x-x',t-t') \, P(x',t') \ .$$
(2.9)

The memory kernel $K(x,t)$ is derived [14] using physical considerations and reads

$$K(x,t) = \frac{1}{<t>} \left[\psi(x,t) - \delta(x) \int_{-\infty}^{\infty} \psi(x',t) \, dx' \right] \ ,$$
(2.10)

where $< t >$ is the first moment of $\psi(t)$

$$<t> = \frac{B}{\mu-2}$$
(2.11)

and

$$\psi(x,t) = \frac{1}{2}\,\delta(x-t)\,\psi(t) + \frac{1}{2}\,\delta(x+t)\,\psi_\epsilon(t)\ , \tag{2.12}$$

with $\psi(t)$ and $\psi_\epsilon(t)$ given by (2.4), (2.8) respectively.

In the non-stationary case, the ME approach leads to [14]

$$<x(t)> \ = \frac{\epsilon}{2}\,\frac{(\mu-1)(\mu-2)}{(4-\mu)(3-\mu)}\,B^{\mu-2}\,t^{4-\mu}\ , \tag{2.13}$$

whereas in the stationary case it leads to [14]

$$<x(t)> \ = \epsilon^{\mu-2}\,\frac{(\mu-1)\Gamma(3-\mu)}{2}\,B^{\mu-2}\,t\ . \tag{2.14}$$

III. Numerical Results
And Theoretical Interpretation

This Section is devoted to illustrating the numerical results concerning the shape of the distribution $P(x,t)$. We set the system in an initial condition corresponding to a flat distribution of the reduced map, and we switch on the perturbation. In Fig. 2a we illustrate by means of black dots the distribution in the early part of the response process, namely in the time region characterized by $t \ll 1/\epsilon$, and corresponding to the first moment of x having the form (2.13). We see that at this stage the distribution is still almost symmetric. With increasing time, while the maximum of the distribution does not essentially move from the initial position, the right tail becomes more and more extended, and its broadening is much faster than that of the left tail. At times of the order of $1/\epsilon$ there is a crossover to a new regime, where the response is expressed by the first moment having the form (2.14). Fig. 2a depicts the distribution in this regime.

Using the ME theory [14] it is possible, via an inverse Fourier-Laplace analysis, to determine either the distribution $P(x,t)$ corresponding to (2.13) or that corresponding to (2.14). This is done by considering the Fourier-Laplace transform $\tilde{P}(k,s)$ of the distribution $P(x,t)$ satisfying (2.9):

$$\hat{P}(k,s) = \frac{1}{s - \hat{K}(k,s)} \tag{3.1}$$

where $\hat{K}(k,s)$ is the Fourier-Laplace transform of $K(x,t)$ defined by (2.10),

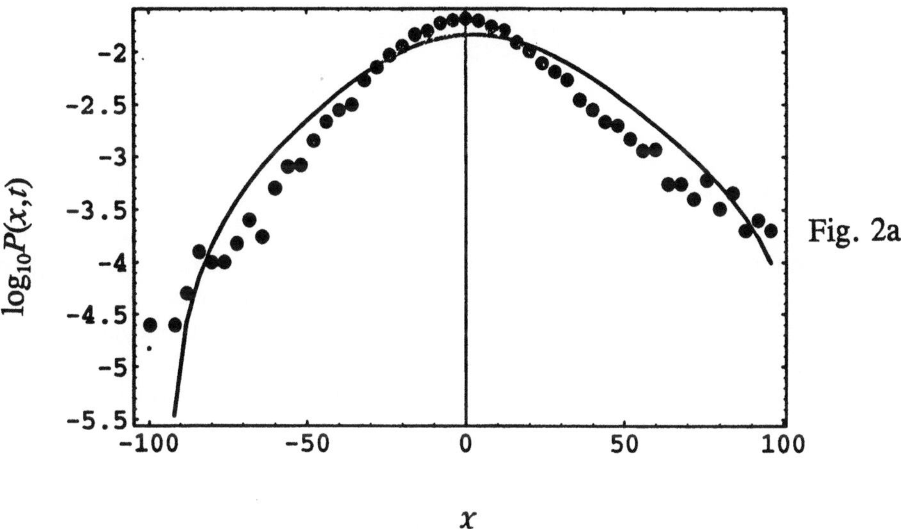

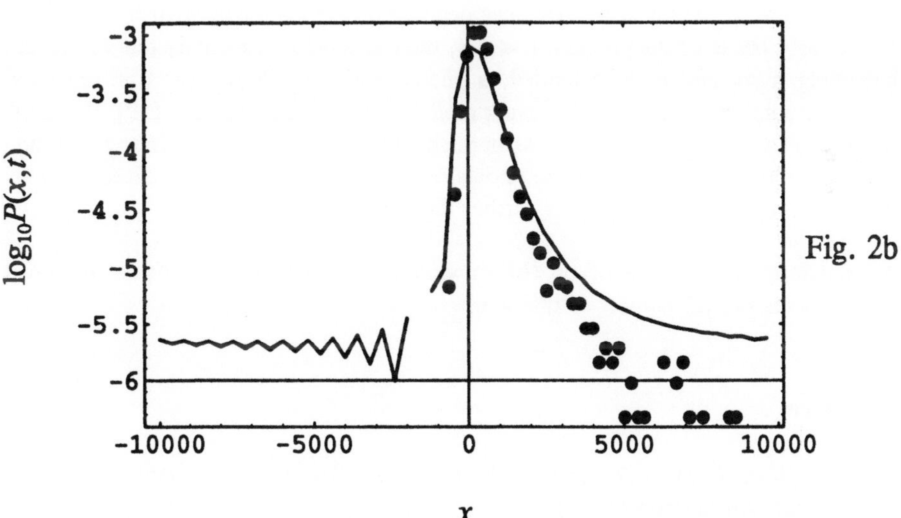

Fig. 2. The distribution $P(x,t)$ of the dynamical process (2.1-3) with $z=5/3$ (a) at time $t = 100$, in the nonstationary regime; (b) at time $t = 10,000$ in the stationary regime. The dots show the results of direct numerical calculations and the solid lines illustrate the results of the inverse Fourier transformation, according to the prescriptions of the text. The value adopted for the perturbation parameter is $\lambda = 0.05$, and the initial conditions $\hat{x}_0$ are random numbers with uniform distribution in the interval [0,1].

$$\hat{K}(k,s) = \frac{1}{<t>} \, [\hat{\psi}(k,s) - \hat{\psi}(0,s)] \, , \qquad\qquad (3.2)$$

and $\hat{\psi}(k,s)$ is the Laplace-Fourier transform of $\psi(x,t)$ of equation (2.12). We are interested in the evaluation of $\hat{P}(k,s)$ in the limiting case of s, k and ϵ very small. From the analysis carried out in [14] it appears that the non stationary situation corresponding to (2.13) is obtained by first sending s to zero, then ϵ (which corresponds to exploring the time regime $t \ll 1/\epsilon$), and making s tend to zero faster than k. The stationary situation corresponding to equation (2.14) is obtained by first sending s to zero, and then ϵ (which corresponds to taking $t \gg 1/\epsilon$), and again by making s tend to zero faster than k. This procedure results in expressions for $\hat{P}(k,s)$ that are independent of s, and which then are used to derive $P(x,t)$ by means of a numerical inverse Fourier transform method.

To make a fair comparison between the numerical "experiment" and the ME theory, we compare the numerical results of Fig. (2a) to the theoretical prediction for $P(x,t)$ obtained using the first of the above methods, and the numerical results of Fig. (2b) to the theoretical prediction for $P(x,t)$ obtained using the second of the above methods. The agreement is satisfactory and supports our main conclusion that the anomalous behavior of the response depends on the fast increase of the right tail. In the early phase of the process, $t \ll 1/\epsilon$, the transport is essentially due to this tail becoming more and more extended, a phenomenon similar to that first observed using the continuous time random walk model by Scher and Montroll [19]. We think that this interesting connection between the breakdown of the Green-Kubo prediction and a response reminiscent of the anomalous transit-time dispersion discovered by the authors of Ref. [19] deserves further theoretical and numerical investigation.

Acknowledgment. One of us (BJW) would like to thank the office of Naval Research for partial support of this work.

References

[1] N.G. Van Kampen, *Stochastic Processes in Physics and Chemistry*, North-Holland, Amsterdam (1981).

[2] H. Mori, H. Hata, T. Horita and T. Kobayashi, Suppl. Prog. Theor. Phys. **99**, 1 (1989).

[3] E. Ott, *Chaos in Dynamical Systems*, Cambridge University Press, Cambridge (1993).

[4] B.V. Chirikov, Prog. Rep. **52**, 265 (1979).

[5] E.W. Montroll and B.J. West, *Studies in Statistical Mechanics*, Volume VII, *Fluctuation Phenomena*, Editors E.W. Montroll, J. L. Lebowitz, North-Holland (1979); 2nd Edition North-Holland, Amsterdam (1987).

[6] E.W. Montroll and M.F. Shlesinger, in *From Stochastic to Hydrodynamics*, edited by J.L. Lebowitz and E.W. Montroll, North-Holland, Amsterdam (1984).

[7] T. Geisel and S. Thomae, Phys. Rev. Lett. **52**, 1936 (1984).

[8] T. Geisel, J. Nierwelberg and A. Zacherl, Phys. Rev. Lett. **54**, 616 (1985).

[9] J. D. Hanson, J. R. Cary and J. D. Meiss, J. Stat. Phys. **39**, 327 (1985); J. D. Meiss and E. Ott, Physica **20D**, 387 (1986).

[10] G. Zumofen and J. Klafter, Phys. Rev. E **47**, 851 (1993).

[11] G. Zumofen and J. Klafter, Physica A **196**, 102 (1993); G. Zumofen and J. Klafter, Physica D **69**, 436 (1993).

[12] R. Ishizaki, H. Hata, T. Horita and H. Mori, Prog. Theor. Phys. **84**, 179 (1990); R. Ishizaki, T. Horita, T. Kobayashi and H. Mori, Prog. Theor. Phys. **85**, 1013 (1991).

[13] J. Klafter, G. Zumofen, M. F. Shlesinger, Fractals, **1**, No. 3 (1993); G. Zumofen and J. Klafter, submited to Europhys. Lett.

[14] G. Trefán, E. Floriani, B. J. West and P. Grigolini, Phys. Rev. E, in press (1994).

[15] M. Bianucci, R. Mannella, B. J. West and P. Grigolini, Phys. Rev. E, in press (1994).

[16] R. Mannella, P. Grigolini and B.J. West, Fractals **2**, 81 (1994).

[17] R. Kubo, M. Toda, N. Hashitsume, *Statistical Physics II, Nonequilibrium Statistical Mechanics*, Second Edition, Springer-Verlag Berlin (1991).

[18] J.-P. Bouchaud and A. Georges, Phys. Rep. **195**, 127 (1990).

[19] E. W. Montroll and H. Scher, Phys. Rev. B **12**, 2455 (1975); E. W. Montroll and H. Scher, J. Stat. Phys. **9**, 101 (1973).

Ultra-Slow Convergence to a Gaussian:
The Truncated Lévy Flight

Rosario N. Mantegna[1] and H. Eugene Stanley

Center for Polymer Studies and Department of Physics
Boston University, 590 Commonwealth Av., Boston, MA 02215, USA

Abstract. We introduce a class of *quasi*-stable stochastic process, the *truncated* Lévy Flight (TLF). A TLF is a stochastic process with finite variance. We show theoretically and numerically that the convergence of the sum of n independent TLF to a Gaussian process is usually extremely slow. In fact a remarkably large value of n can be required to ensure the convergence to a Gaussian process. We also investigate the statistical properties of the S&P 500 (a financial index) and we show that they are qualitatively in agreement with the one of a TLF.

Lévy flights, i.e. stochastic processes with jumps being Lévy stable non-Gaussian stochastic variables [1,2,3], have infinite variance. An infinite variance is not expected in physical systems. In spite of this Lévy, or power-law, density functions have been observed in physical [4,5] and biological systems [6,7]. The observation of Lévy-like density functions in physical or biological systems is then paradoxical.

One resolution of the paradox of infinite variance is provided by a stochastic process called a Lévy walk [8]. A Lévy walk is a random walk performed by visiting the same sites of a Lévy flight. However in a Lévy walk instantaneous jumps, which are responsible for the infinite variance, are not allowed and a time cost is introduced so that long steps are penalized. In other words, a spatio-temporal coupling memory is present in a Lévy walk. It is worth to point out that the space-time coupling memory is mathematically essential to avoid the divergence of the moments in the Lévy walk, divergence which is present in the underlying Lévy flight.

In this paper we present an alternative resolution of the paradox of infinite variance in physical systems which is also valid in the absence of a spatio-temporal coupling. We introduce a *quasi*-stable stochastic process with a *finite* variance. We define the the *Truncated Lévy Flight* (TLF), a stochastic process $\{x\}$ characterized by the probability density

$$T(x) \equiv \begin{cases} 0 & x > \ell \\ c_1 L(x) & -\ell \leq x \leq \ell \\ 0 & x < -\ell \end{cases}, \tag{1}$$

---

[1]Present address: Dipartimento di Energetica ed Applicazioni di Fisica, Viale delle Scienze, I-90128 Palermo, Italia

where c_1 is a normalizing constant, ℓ is the cutoff length and

$$L(x) \equiv \frac{1}{\pi} \int_0^{+\infty} \exp(-\gamma q^\alpha) \cos(qx) dq \tag{2}$$

is the symmetrical Lévy stable distribution of index α ($0 < \alpha \le 2$) and scale factor γ ($\gamma > 0$).

We study the dynamics of a discrete random walk $\{z\}$ in which the successive jumps are independent stochastic processes $\{x\}$. We show that by analyzing the stochastic process $\{z\}$ a Lévy regime can be observed for a huge number of steps n. The crossover between the Lévy regime and the Gaussian regime of the stochastic process

$$z_n \equiv \sum_{i=1}^{n} x_i \tag{3}$$

is a function of the cut-off length ℓ. In Fig. 1a, we plot the probability density function (PDF) of a TLF characterized by $\alpha = 1.5$, $\gamma = 1$ and $\ell = 20$. In Fig. 1a the difference between the TLF and the Lévy stable distribution of the same index cannot be noticed. However the truncation of the PDF becames evident when we plot the density function in a semilogarithmic plot (Fig. 1b). In a TLF the "rare events" ($|x| > \ell$) are forbidden.

We study the stochastic process z_n as a function of n. In Eq. (3) x_i is a TLF and

$$< x_i \, x_j >= k \, \delta_{ij} \tag{4}$$

Since z_n is by definition a sum of n independent stochastic variables with finite variance, the central limit theorem implies that for $n \approx \infty$, z_n is a Gaussian stochastic process. Hence two distinct regimes are expected for the stochastic process z_n: (i)A regime where the stochastic process is *quasi*-stable and the PDF is

$$P(z_n) \approx L_\alpha(z) \qquad n \approx 1 \tag{5}$$

and (ii) a Gaussian regime observed for high values of n,

$$P(z_n) \approx G(z) \qquad n \gg 1 \tag{6}$$

The key question is: how fast is the convergence to the Gaussian? Or, in other words, for how long a *quasi*-stable stochastic process is observed?

We answer this question by studying the probability of return to the origin $P(z_n = 0)$ of the stochastic process z_n as a function of n. Our choice is motivated by two observations: (i) the maximal distance between $P(z_n)$ and the Gaussian distribution with the same variance σ_n is detected at $z_n = 0$ for any n. (ii) an analytical relation between $P(z_n)$ and n is known for Lévy stable processes when $z_n = 0$.

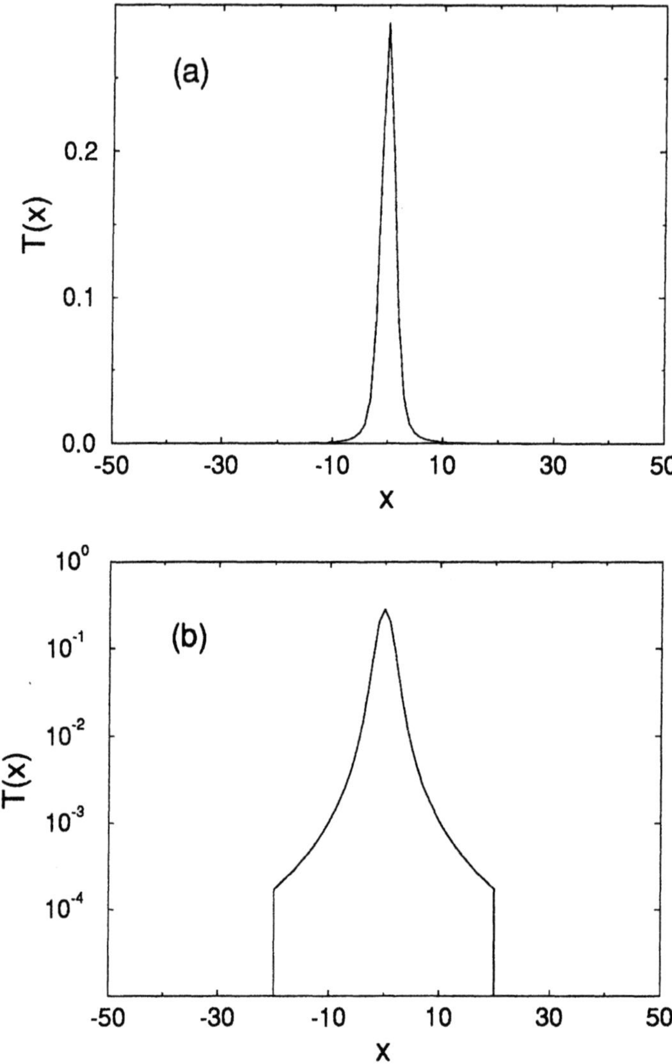

Figure 1: (a) Probability density function of a TLF characterized by $\alpha = 1.5$, $\gamma = 1.0$ and cut-off length $\ell = 20$; (b) the same PDF as in (a) in a semilogarithmic plot. The truncation of jumps longer than ℓ is evident in the semilogarithmic plot.

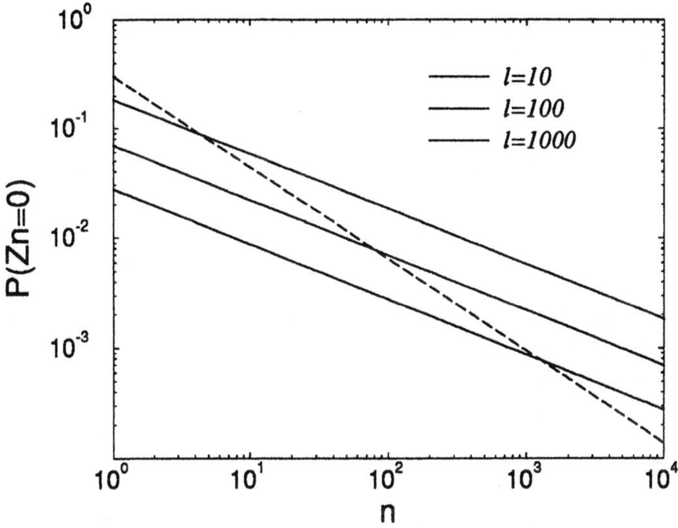

Figure 2: Logarithm of the probability of return to the origin for a Lévy flight of index $\alpha = 1.2$ (dashed line) and for three different Gaussian processes having the same standard deviation as a TLF of the same index and cut-off length $\ell = 10, 100$ and 1000 (solid lines from top to the bottom respectively).

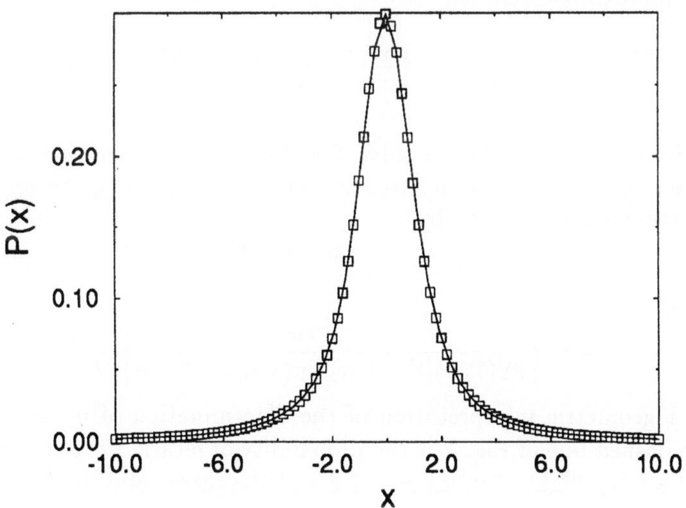

Figure 3: (a) Probability density function of the jumps of a 1-dimensional Lévy flight of index $\alpha = 1.20$ and $\gamma = 1.0$. The PDF is obtained by simulating the process (10^5 realizations) with the algorithm of ref. [9]. The symmetrical Lévy stable distribution of the same index and scale factor is shown for comparison (solid line).

The probability of return to the origin in the stable non-Gaussian regime is given by

$$P(z_n = 0) \simeq L(z_n = 0) = \frac{\Gamma(1/\alpha)}{\pi \alpha n^{1/\alpha}}. \tag{7}$$

whereas in the Gaussian regime the expected behavior is

$$P(z_n = 0) \simeq G(z_n = 0) = \frac{1}{\sqrt{2\pi}\sigma_o(\alpha,\ell)n^{1/2}}, \tag{8}$$

where $\sigma_o(\alpha,\ell)$ is the standard deviation of the TLF $\{x\}$ and we set $\gamma = 1$ for the sake of simplicity.

The crossover between the two regimes occurs for the value $n_\times$, which is the solution solution of the equation

$$L(z_n = 0) = G(z_n = 0) \tag{9}$$

The only implicit term in Eq.(9) is the TLFs standard deviation. By using the first term of the series expansion

$$L(z) \simeq -\frac{1}{\pi} \sum_{k=1}^{m} \frac{(-1)^k}{k!} \frac{\Gamma(\alpha k + 1)}{z^{\alpha k + 1}} \sin\left(\frac{k\pi\alpha}{2}\right) + R(z), \tag{10}$$

valid for a symmetrical Lévy distribution in the interval $1 < \alpha < 2$, we write the approximate relation

$$\sigma_o(\alpha,\ell) \simeq \left[\frac{2\Gamma(1+\alpha)\sin(\pi\alpha/2)}{\pi(2-\alpha)}\right]^{1/2} \ell^{(2-\alpha)/2}. \tag{11}$$

By using Eq. (11), we solve Eq. (9) and we determine the crossover $n_\times$. Under the approximations used to determine Eq. (11), the crossover between the two regimes is given by

$$n_\times \simeq A\,\ell^\alpha, \tag{12}$$

where

$$A = \left[\frac{\pi\alpha}{2\Gamma(1/\alpha)[\Gamma(1+\alpha)\sin(\pi\alpha/2)/(2-\alpha)]^{1/2}}\right]^{2\alpha/(\alpha-2)}. \tag{13}$$

A geometric interpretation of the determination of $n_\times$ is given in Fig. 2. The dashed line of Fig. 2 is the probability of return $P(z_n = 0)$ as a function of n of a Lévy flight of index $\alpha = 1.2$ and the three solid lines are the $P(z_n = 0)$ calculated for a Gaussian process having the same standard deviation of a TLF with the same index and cut-off length $\ell = 10, 100$ and 1000 (top to bottom respectively). For each value of ℓ, a crossover between the two regimes occurs when the probability of return to the origin of the associated Gaussian process exceeded the probability of return expected for the Lévy flight. In addition to the

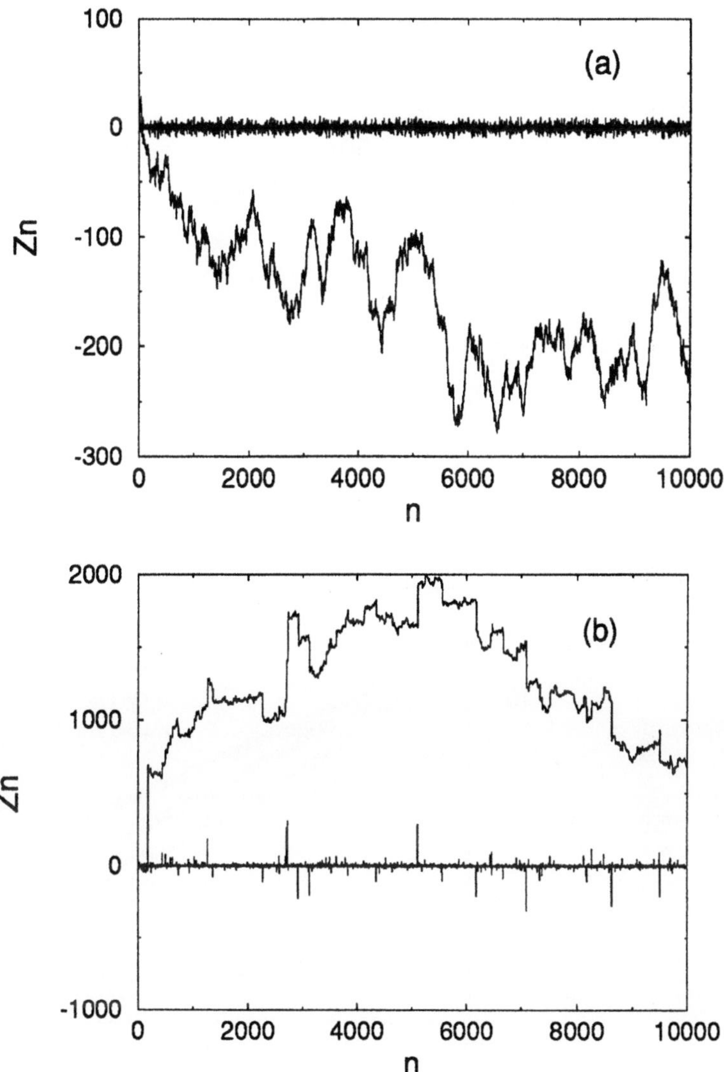

Figure 4: (a) z_n is the sum of n independent jumps distributed as a TLF of index $\alpha = 1.2$ and $\gamma = 1.0$ and cut-off length $\ell = 10$ (also in the figure as noise close the origin $z_n \approx 0$). The profile of the z_n walk is analogous to a Brownian random walk. From [14] .(b) Same as in (a) but with cut-off length $\ell = 1000$. The profile of the z_n walk is now close to the one observed for a Lévy flight of the same index (abrupt jumps are observed quite frequently). From [14].

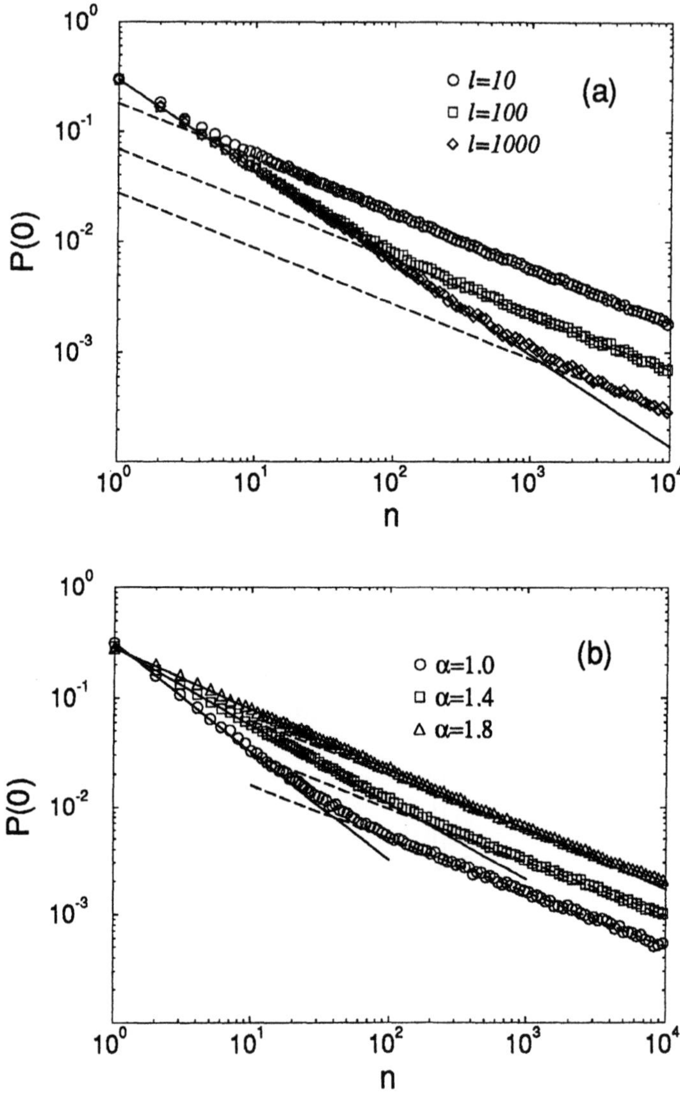

Figure 5: (a) Probability of return to the origin of z_n as a function of n for $\alpha = 1.2$ and $\ell = 10, 100$ and 1000. The simulations (circles, squares and diamonds) are compared with the Lévy regime (solid line) and the Gaussian regimes calculated for $\ell = 10, 100$ and 1000 (dotted lines from top to bottom respectively). From [14]. (b) Same as in (a) but for a fixed value of the cut-off length $\ell = 100$ and different values of α (1.0,1.4 and 1.8). The simulations (circles squares and triangles) are compared with the corresponding Lévy (solid lines) and Gaussian (dotted lines) regimes.

crossover $n_\times$ another interesting quantity is the "distance" between the probability of return to the origin of a TLF and the rescaled asymptotic Gaussian process

$$G_s(z_n = 0) \equiv G(z_n = 0)n^{1/2} \tag{14}$$

To be precise, we define

$$\Delta \equiv \log_{10} \frac{T(0)}{G_s(0)}, \tag{15}$$

to be the "distance" between the probability of return to the origin of a TLF and the Gaussian process with the same standard deviation $\sigma_o(\alpha, \ell)$. In Fig. 2 the "distance" is the separation between the two regimes observed at $n = 1$. Under the same assumptions used to determine the crossover $n_\times$, we obtain

$$\Delta \simeq \log_{10} \frac{2\Gamma(1/\alpha)}{\pi\alpha} \;+\; \frac{1}{2} \log_{10} \left(\frac{\Gamma(1 + \alpha)\sin(\pi\alpha/2)}{2 - \alpha} \right)$$
$$+ \; \frac{2 - \alpha}{2} \log_{10} \ell. \tag{16}$$

We test the accuracy of the results of Eqs. (12) and (16) by performing several numerical simulations of TLFs. In particular, we investigate the probability of return to the origin of the stochastic process z_n as a function of n. Since we investigate $P(z_n = 0)$, we need an algorithm which is accurate all over the definition range, including the origin. In our simulations we use a simple, fast and accurate algorithm proposed recently [9]; other algorithms can be found in the mathematical literature [10]. The algorithm is accurate over the entire range of the stochastic variable $\{x\}$. As an example, in Fig. 3 we show the PDF of jumps of a simulated Lévy flight together with the theoretical Lévy stable distribution of the same index. The index α is 1.2 and the scale factor γ is 1.0 . The PDF (black boxes) is measured by analyzing an ensemble of 10^5 independent realizations. It is in very good agreement with the theoretical distribution of the same index and scale factor (solid line).

In Fig. 4 we show single realizations of a 1-dimensional TLF random walk of index $\alpha = 1.2$ and cut-off length $\ell = 10$ and 1000 (Figs. 4a and 4b). In each figure, z_n is the sum of n independent jumps $\{x\}$ characterized by the PDF of Eq. (1). The variables x_n are also shown; they are the noise plotted around the origin ($z_n \approx 0$). In the Figure we can observe the role played by the cut-off length ℓ. In fact, the pattern of Fig. 4a is similar to the pattern observed in a Brownian random walk. In Fig. 4b a Lévy like pattern is observed due to the fact that a significant number of rare events are allowed.

We perform a quantitative analysis of the statistical properties of z_n as a function of n by investigating the probability of return to the origin ($P(z_n = 0)$). In Fig. 5 we show typical results of our simulations. In Fig. 5a we show $P(z_n = 0)$ when $\alpha = 1.2$ and $\ell = 10, 100$ and 1000. The solid line is the

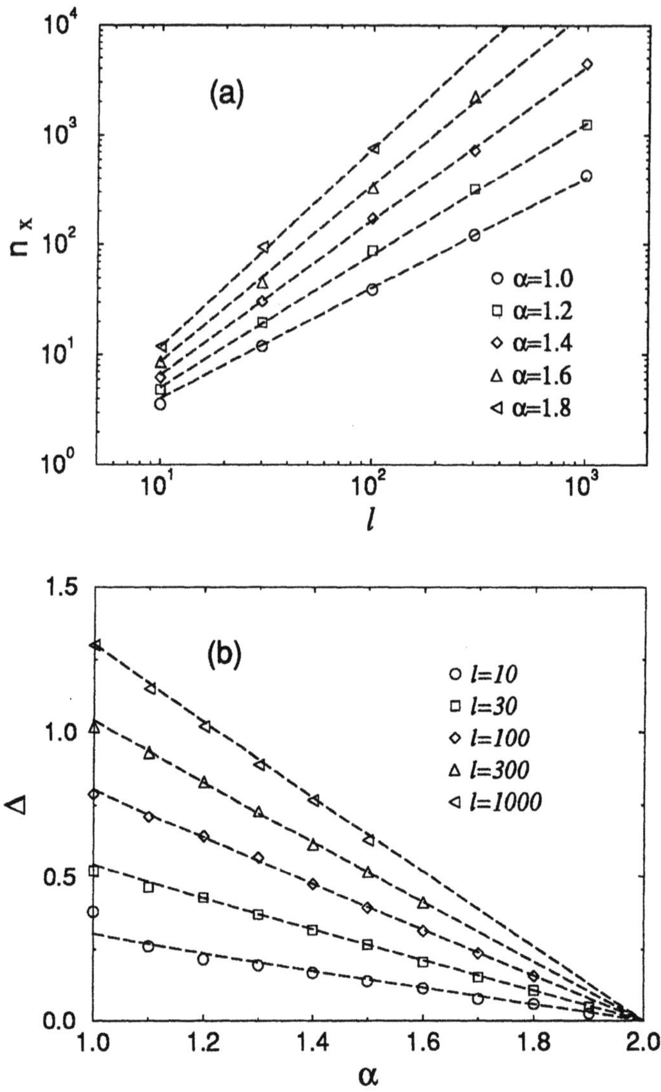

Figure 6: (a) Crossover between the Lévy and the Gaussian regime obtained from numerical simulations of z_n. The dotted lines are the theoretical predictions of Eqs. (12) and (13). From [14]. (b) *Distance* between the associated Lévy and the rescaled Gaussian process measured in numerical simulations of z_n. The dotted lines are the theoretical predictions of Eq. (16). From [14].

theoretical prediction for a Lévy flight of index $\alpha = 1.2$ (Eq. (7)) and the dotted lines are the predicted asymptotic Gaussian regimes of Eq. (8) calculated for the three different values of ℓ. In Fig. 5b we show the simulated $P(z_n = 0)$ obtained by setting $\ell = 100$ and $\alpha = 1.0$, 1.4 and 1.8. In the same Figure we show the theoretical prediction for the Lévy flights of the same indices and the associated asymptotic Gaussian regimes. We note that the crossover between the two regimes is observed for larger values of n when α increases at fixed values of ℓ, in spite of the fact that the distance Δ between the two regimes decreases. The agreement between numerical simulations and the theoretical prediction for the crossover between the Lévy and Gaussian regimes is very good for all the investigated values of α and ℓ.

We summarize the results of our numerical simulations in Fig. 6. In Fig. 6a we show the measured crossover between the Lévy and Gaussian regimes as a function of the cut-off length ℓ for different values of the index α. The dotted lines are obtained by using Eqs (12) and (13) with the same value of α used in numerical simulations. In Fig. 6b we show the "distance" Δ measured in numerical simulations, and we compare the values obtained from simulations with the theoretical prediction of Eq.(16) (dotted lines). The agreement between simulations and the theoretical predictions of Eqs. (12) (13) and (16) is quite good.

Before concluding, we show that a stochastic process having statistical properties similar to that of a TLF is the time evolution of the S&P 500. The study of economic time series as stochastic processes was pioneered by L.Bachelier at the beginning of this century [11]. The relation between economic series end Lévy (or Pareto) distribution was pointed out for the first time by Mandelbrot in the sixties [12]. Here we analize the time series as a common random process and study the successive differences of the time series without performing a nonlinear transformation.

The S&P 500 is one of the most important financial indices of the *New York Stock Exchange*, and has been recorded for between time intervals as short as 15 seconds. We analyze the dynamics of the S&P 500 with a very high temporal resolution (1 minute) in a time interval of 6 years. By studying the PDF of successive non-overlapping differences of the index $V(\Delta t)$, we measure a non-Gaussian scaling of the probability of return to the origin. In Fig. 7 we show the probability of return to the origin as a function of the time interval between the data. The investigated period is Jan 84 - Dec 89. The number of non-overlapping successive differences used in this analysis decreases from nearly 500,000 ($\Delta t = 1$ minute) to nearly 500 ($\Delta t = 1000$ minutes). In Fig. 7 we also show the best linear fit of the data performed in a double logarithmic plot. The functional form of the probability of return is a power-law $P(V = 0) = k/(\Delta t)^{0.712}$. The value of the exponent is different from the one expected for a Gaussian process, 0.5, so that experimental data are compatible with a Lévy walk or flight of index $\alpha = 1/0.712 = 1.40$. We check the hypothesis of Lévy stable PDF for the successive variations of the S&P 500 by comparing

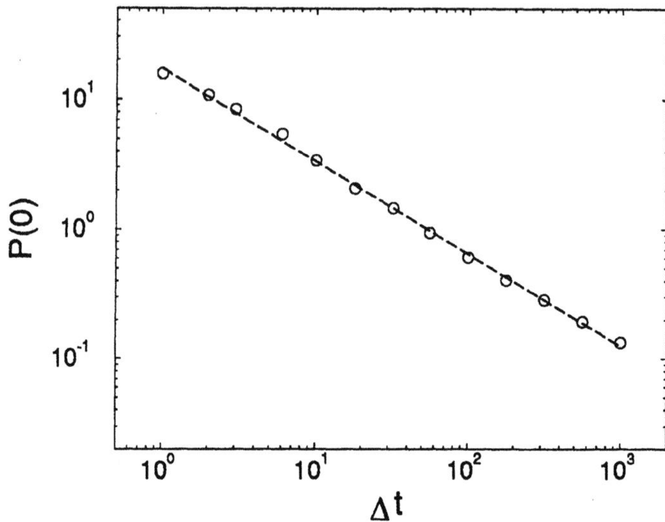

Figure 7: Probability of return to the origin of the successive changes of the S&P 500 measured between non-overlapping intervals ranging from 1 minute to 1000 minutes. The data are collected during the period Jan 84-Dec 89. The dotted line is the best linear fit of the experimental results. The slope of the fit is -0.712 a value different from the one expected for a normal diffusion (-0.5).

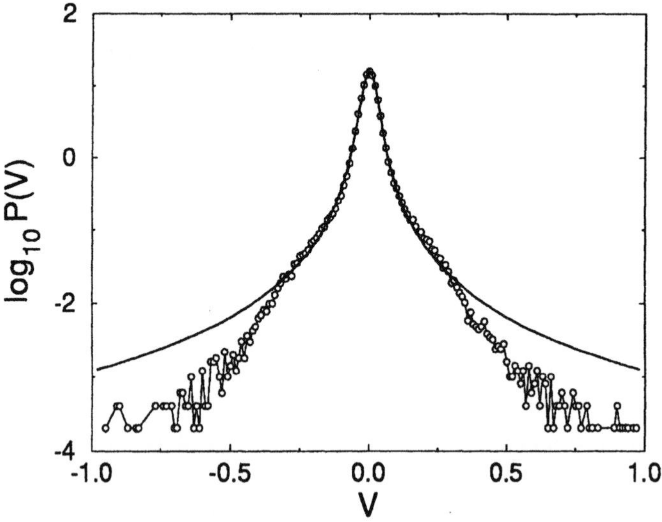

Figure 8: (a) Probability density function of the 1 minute changes of the S&P500 index measured in the period Jan 84-Dec 89. The solid line is a symmetrical Lévy stable distribution of index $\alpha = 1.40$ and $\gamma = 0.00375$. The agreement between the experimental PDF (circles) and the Lévy distribution is very good for $|V| \leq 6\sigma$. The observed exponential (or stretched exponential) wings ensure a finite variance for the observed stochastic process.

the PDF measured when $\Delta t = 1$ minute with the Lévy stable PDF of index $\alpha = 1.40$ and scale factor $\gamma = 0.00375$. The scale factor is determined by using the experimental values $P(0) = 15.66$ and $\alpha = 1.40$ and the theoretical relation $P(0) = \Gamma(1/\alpha)/(\pi\alpha\gamma^{1/\alpha})$ valid for a Lévy flight of scale factor γ ($\Delta t = 1$). In Fig. 8 we show the experimental PDF (circles) together with the Lévy stable distribution of the determined index and scale factor (solid line). The agreement between the experimental and Lévy distribution is excellent when the absolute value of the jumps V is less then 6 standard deviations σ ($|V| < 6\sigma$). Conversely, when $|V| > 6\sigma$ an approximately exponential fall-off of the wings is observed. Although the cut-off is not abrupt as in a TLF, exponential fall-off ensures a finite variance for the investigated stochastic process. Another similarity with the TLF is that the stochastic process $\{V\}$ is diffusive. In fact in the interval $\Delta t = 10,, 1000$ minutes, the measured variance $< V^2 >$ is fitted by the relation $< V^2 >= k_1 \Delta t^{1.08}$. The measured diffusion exponent, 1.08, is very close to the value (1) observed in normal diffusive processes. The absence of a super diffusive behavior of the variance of V rules out the possibility that a Lévy walk can describe the dynamics of the S&P 500.

In conclusion, our theoretical and numerical results show that the TLF, a *quasi*-stable stochastic process, can show a ultra-slow convergence to the asymptotic associated Gaussian process. In the interval $n \ll n_\times$ a TLF is a stochastic process well described by Lévy stable distribution, with the exception of the most rare events, but having a *finite* variance. This is a *practical* violation of the central limit theorem. With the previous statement we mean one can expect to observe experimentally a stochastic process with the characteristic of a Lévy flight for a long time (for example $n \approx 10^4$ time steps), even if the stochastic process has indeed a finite variance. A TLF provides a resolution of the paradox of the experimental observation of a Lévy-like stochastic processes in physical systems. This resolution is different and complementary to the resolution provided by Lévy walks. It is worthwhile to compare the general properties of Lévy walks and TLFs. Similarities: Lévy walks and TLFs have PDFs which are Lévy distributions, with the exception of the most rare events and finite variance, moreover Lévy walks and TLFs (in the Lévy regime) show a non-Gaussian scaling of the probability of return to the origin. Differences between the two processes are observed in the asymptotic basin of attraction and in the time evolution of the variance of the process. Lévy walks converge to Lévy flights whereas TLFs converge to Gaussian processes. Moreover, Lévy walks are superdiffusive whereas the diffusion of a TLF walk is normal. The empirical observation of a Lévy profile of the PDF together with a non-Gaussian scaling of the probability of return to the origin is not sufficient to discriminate between a Lévy walk and a TLF. Additional information about the properties of the spectral density (or the time evolution of the variance) is needed to discriminate between the two modeling processes. We provide as a possible example of TLF the time evolution of the S&P 500.

We thank S. V. Buldyrev, C. K. Peng, and F. Sciortino for helpful discussions and NSF for financial support. After this work was completed, the result (12) was derived by M. F. Shlesinger [13] using the Berry-Esseen theorem, which applies to the convergence to a Gaussian for a symmetric random walk whose jump probabilities have a finite third moment.

References

1. P. Lévy, *Théorie de l'Addition des Variables Aléatoires* (Gauthier-Villars, Paris, 1937).

2. W. Feller, *An Introduction to Probability Theory and Its Applications* (Wiley, New York, 1971).

3. B. B. Mandelbrot, *The Fractal Geometry of Nature* (Freeman, San Francisco, 1982).

4. T. H. Solomon, E. R. Weeks, and H. L. Swinney, Phys. Rev. Lett. **71**, 3975 (1993).

5. F. Bardou, J.-P. Bouchaud, O. Emile, A Aspect, and C. Cohen-Tannoudji, Phys. Rev. Lett. **72**, 203 (1994).

6. A. Ott, J.-P. Bouchaud, D. Langevin, and W. Urbach, Phys. Rev. Lett. **65**, 2201 (1990).

7. C.-K. Peng, J. Mietus, J.M. Hausdorff, S. Havlin, H.E. Stanley, and A.L. Goldberger, Phys. Rev. Lett. **70**, 1343 (1993).

8. M.F. Shlesinger, G.M. Zaslavsky, and J.Klafter, Nature **363**, 31 (1993), and references therein.

9. R. N. Mantegna, Phys. Rev. E **49**, 4677 (1994).

10. G. Samorodnitsky and M.S. Taqqu, *Stable Non-Gaussian Random Processes: Stochastic Models with Infinite Variance* (Chapman and Hall, NY, 1994).

11. L. J. B. Bachelier, *Théorie de la Speculation* (Gauthier-Villars, Paris, 1900); this article is reproduced in P. H. Cootner (ed.), *The Random Character of Stock Market Prices* (MIT Press, Cambridge MA, 1964).

12. B. Mandelbrot, J. Business **36**, 394 (1963).

13. M. F. Shlesinger, preprint

14. R. N. Mantegna and H. E. Stanley, Phys. Rev. Letters **73**, 2946 (1994).

PART 5:

LÉVY FLIGHTS IN BIOLOGY

Fractals in Physiological Control: From Heart Beat to Gait

C.-K. Peng,[1,2] J. M. Hausdorff,[1,3] J. E. Mietus,[1] S. Havlin,[2,4] H. E. Stanley[2] and A. L. Goldberger[1,3]

[1] Cardiovascular Division, Harvard Medical School, Beth Israel Hospital, Boston, MA 02215
[2] Center for Polymer Studies and Department of Physics, Boston University, Boston, MA 02215
[3] Department of Biomedical Engineering, Boston University, Boston, MA 02215
[4] Department of Physics, Bar Ilan University, Ramat Gan, ISRAEL

1 Introduction

Scale-invariant properties in biological systems have received much attention recently [1, 2]. The absence of characteristic length (or time) scales may confer important biological advantages, related to adaptability of response [2, 3]. In this paper, we present some recent progress in applying scale-invariant (fractal) analysis to physiological time series. We will concentrate on output from two physiological systems: (1) human heartbeat time series under neuroautonomic control; and (2) human gait time series under the control of central nervous system. We will emphasize the difficulties of analyzing physiological time series that arise mainly from their nonstationarity and sometimes short data length.

2 Human Heartbeat Dynamics

Clinicians often describe the normal activity of the heart as "regular sinus rhythm." But in fact cardiac interbeat intervals normally fluctuate in a complex, apparently erratic manner [2, 4] (Fig. 1). This highly irregular behavior has recently motivated researchers [5, 6] to apply time series analyses that derive from statistical physics, especially methods for the study of critical phenomena where fluctuations at all length (time) scales occur. These studies show that under healthy conditions, interbeat interval time series exhibit long-range power-law correlations reminiscent of physical systems near a critical point [7, 8]. Furthermore, certain disease states may be accompanied by alterations in this scale-invariant (fractal) correlation property. Here we explore the potential utility of such scaling alterations in the detection of pathological states.

Our analysis in this paper is based on the beat-to-beat heart rate fluctuations of digitized electrocardiograms recorded with an ambulatory (Holter) monitor. The time series obtained by plotting the sequential intervals between beat i and beat $i+1$, denoted by $B(i)$, typically reveals a complex type of variability (Fig. 1).

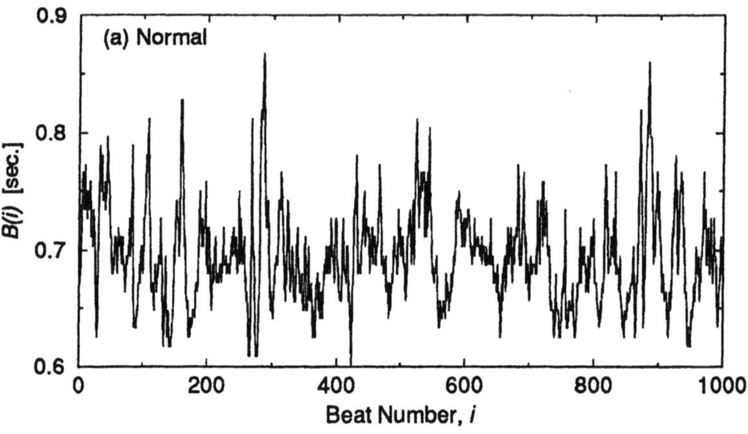

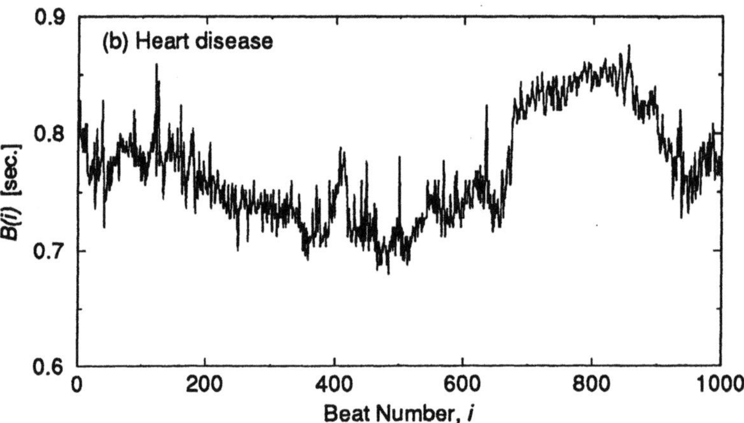

Figure 1: The interbeat interval time series $B(i)$ of 1000 beats for (a) a healthy subject and (b) a patient with severe cardiac disease (dilated cardiomyopathy). The healthy heartbeat time series shows more complex fluctuations compared to the diseased heart rate fluctuation pattern.

The mechanism underlying such fluctuations appears to be related primarily to countervailing neuroautonomic inputs. Parasympathetic stimulation decreases the firing rate of pacemaker cells in the heart's sinus node. Sympathetic stimulation has the opposite effect. The nonlinear interaction (competition) between the two branches of the autonomic nervous system is the postulated mechanism for the type of erratic heart rate variability recorded in healthy subjects [4, 9].

An immediate problem facing researchers applying time series analysis to interbeat interval data is that the heartbeat time series is often highly nonstationary. Two approaches can be taken to reduce these nonstationary effects. We will discuss those methods and their physiological implications separately in the following sections.

2.1 Increment of Interbeat Interval

A standard treatment in time series analysis is to remove nonstationarities by taking the first difference of the original time series. In our case, to study the increment of interbeat interval, $I(i) \equiv B(i+1) - B(i)$, might actually be physiologically reasonable. One important mechanism for regulating heart rate is the baroreflex, which keeps the arterial blood pressure in proper range. This reflex operates by arterial pressure sensors collecting information about the blood pressure and possibly its rate of change which, in turn, is related to the increment of heart rate. This feedback information can then be used to regulate the volume of blood in the arteries by adjusting the heart rate.

Unlike the original time series $B(i)$, $I(i)$ is *stationary*, i.e., the average and the variance are independent of the sampling position. We have verified this stationarity property experimentally by measuring the stability of the average and variance in a moving window. One nice advantage of the stationarity property of $I(i)$ is that, unlike the original interbeat intervals, the comparison of histograms between data from healthy and diseased subjects becomes feasible.

We find that $I(i)$ for the two time series in Figs. 1a and 1b have virtually identical histograms, and can be well described by a Lévy stable distribution (see Fig. 2):

$$P(I, \psi, \gamma) = \frac{1}{\pi} \int_0^\infty \exp(-\gamma q^\psi) \cos(qI) dq, \tag{1}$$

with $\psi = 1.7$ and $\gamma > 0$ [10]. Since the histograms of the increments are the same for both normal and diseased conditions, *the different fluctuation patterns observed in health and disease in this example must relate to the ordering of these increments*, i.e., to the correlations between the length of successive increments produced by the underlying dynamics of the heartbeat.

To investigate these dynamical differences that we observed visually in Fig. 1, it is helpful to study further the correlation properties of the time series. To this end, we choose to study $I(i)$ because of the aforementioned reason. Since $I(i)$ is stationary, we can apply standard spectral analysis techniques [12]. Figures 3a and 3b show the power spectra $S_I(f)$, calculated as the square of the Fourier transform amplitudes for $I(i)$, derived from the same data sets used in Fig. 1. The fact that the log-log plot of $S_I(f)$ vs f is linear implies

$$S_I(f) \sim f^\beta. \tag{2}$$

Furthermore, β can serve as an indicator of the presence and type of correlations: (i) If $\beta = 0$, there is no correlation in the time series $I(i)$ ("white noise"). (ii) If $-1 < \beta < 0$, then $I(i)$ is correlated such that positive values of I are likely to be close (in time) to each other, and the same is true for negative I values. (iii) If $0 < \beta < 1$, then $I(i)$ is also correlated; however, the values of I are organized such that positive and negative values are more likely to alternate in time ("anti-correlation") [13].

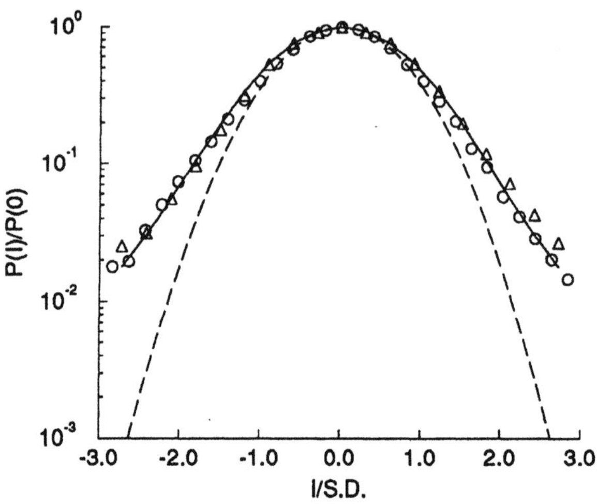

Figure 2: The histogram of $I(i)$ for the healthy (circles) and diseased (triangles) subjects shown in Fig. 1. $P(I)$ is the probability of finding an interbeat increment in the range $[I - \Delta I/2,\ I + \Delta I/2]$. To facilitate comparison, we divide the variable I by the standard deviation (S.D.) of the increment data and rescale P by $P(0)$. In Lévy stable distributions, ψ is related to the power law exponent describing the distribution for large values of the variable, while the width of the distribution is characterized by γ. Since we have rescaled I by the width, ψ is the only relevant parameter. Both histograms are indistinguishable and are well-fit by a Lévy stable distribution with $\psi = 1.7$ (solid line). To find the best fit of our data with a Lévy stable distribution, we have followed the procedure described in Ref. [11]. The dashed line is a Gaussian distribution, which is a special case of a Lévy stable distribution with $\psi = 2$. Although the second moment diverges for a Lévy stable distribution, for a finite sample the second moment remains finite. Similar fits were obtained for 8 of the 10 normal subjects and all 10 subjects with heart disease we studied [5]. The slow decay of Lévy stable distributions for large increment values may be of adaptive importance and may relate to the plasticity of response of the cardiovascular system.

For the diseased data set, we observe a flat spectrum ($\beta \simeq 0$) in the low frequency region (Fig. 3b) confirming that $I(i)$ are not correlated over long time scales (low frequencies). Therefore, $I(i)$, the first derivative of $B(i)$, can be interpreted as being analogous to the *velocity* of a random walker, which is uncorrelated on long time scales, while the values of $B(i)$—corresponding to the *position* of the random walker—are correlated. However, this correlation is of a trivial nature since it is simply due to the summation of uncorrelated random variables.

In contrast, for the data set from the healthy subject (Fig. 3a), we obtain $\beta \simeq 1$, indicating *non-trivial* long-range correlations in $B(i)$—these correlations are not the consequence of summation over random variables or artifacts of non-stationarity. Furthermore, the "anti-correlation" properties of $I(i)$ indicated

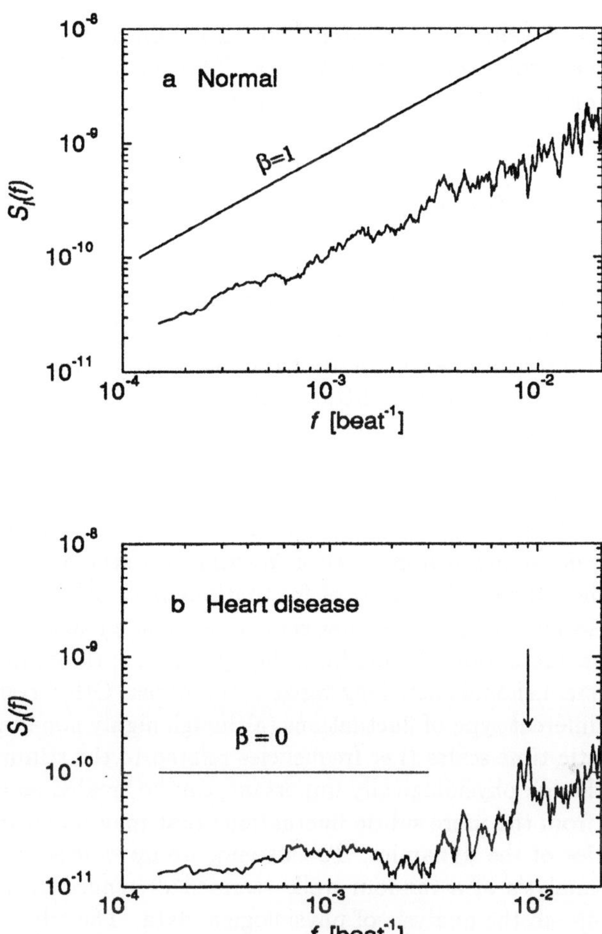

Figure 3: The power spectrum $S_I(f)$ for the interbeat interval increment sequences over $\sim$ 24 hours for the same subjects in Fig. 1. (a) Data from a healthy adult. The best-fit line for the low frequency region has a slope $\beta = 0.93$. The heart rate spectrum is plotted as a function of "inverse beat number" (beat⁻¹) rather than frequency (time⁻¹) to obviate the need to interpolate data points. The spectral data are smoothed by averaging over 50 values. (b) Data from a patient with severe heart failure. The best-fit line has slope 0.14 for the low frequency region, $f < f_c = 0.005$ beat⁻¹. The appearance of a pathologic, characteristic time scale is associated with a spectral peak (arrow) at about 10^{-2} beat⁻¹ (corresponding to Cheyne-Stokes respiration, an abnormal type of periodic breathing commonly associated with low cardiac output in heart failure). After [5].

by the positive β value are consistent with a nonlinear feedback system that "kicks" the heart rate away from extremes. This tendency, however, does not only operate on a beat-to-beat basis (local effect) but on a wide range of time scales.

2.2 Interbeat Interval Time Series: Detrended Fluctuation Analysis

In this section we discuss an alternative approach of analyzing the original interbeat interval time series by detrending the data. One question is whether this observed heterogeneous time series structure arises trivially from changes in environmental conditions having little to do with the intrinsic dynamics of the heart rate itself. Alternatively, these heart rate fluctuations may arise from a complex non-linear dynamical system rather than being an epiphenomenon of environmental stimuli.

From a practical point of view, if the fluctuations driven by uncorrelated stimuli can be decomposed from intrinsic fluctuations generated by the dynamical system, then these two classes of fluctuations may be shown to have very different correlation properties. If that is the case, then a plausible consideration is that only the fluctuations arising from the dynamics of the complex, multiple-component system should show long-range correlations. Other responses should give rise to a different type of fluctuations (although highly non-stationary) having characteristic time scales (i.e. frequencies related to the stimuli). This type of "noise," although physiologically important, can be treated as a "trend" and distinguished from the more subtle fluctuations that may reveal intrinsic correlation properties of the dynamics. To this end, we introduced a modified root mean square analysis of a random walk—termed *detrended fluctuation analysis* (DFA)[1] [14]—to the analysis of physiological data. The advantages of DFA over conventional methods (e.g. spectral analysis and Hurst analysis) are that it permits the detection of long-range correlations embedded in a seemingly nonstationary time series, and also avoids the spurious detection of apparent long-range correlations that are an artifact of non-stationarities. This method has been validated on control time series that consist of long-range correlations with the superposition of a non-stationary external trend [14]. The DFA method has also been successfully applied to detect long-range correlations in highly heterogeneous DNA sequences [14, 15, 16], and other complex physiological signals [17, 18, 19].

To illustrate the DFA algorithm, we use the interbeat time series shown in Fig. 1a as an example. Briefly, the interbeat interval time series (of total length N) is first integrated, $y(k) = \sum_{i=1}^{k}[B(i) - B_{\mathrm{ave}}]$, where $B(i)$ is the i-th interbeat interval and B_{ave} is the average interbeat interval. Next the integrated time series is divided into boxes of equal length, n. In each box of length n, a least squares line is fit to the data (representing the *trend* in that box) (Fig. 4). The

---

[1]Computer software of DFA algorithm is available upon request; contact C.-K. Peng (e-mail: peng@chaos.bih.harvard.edu).

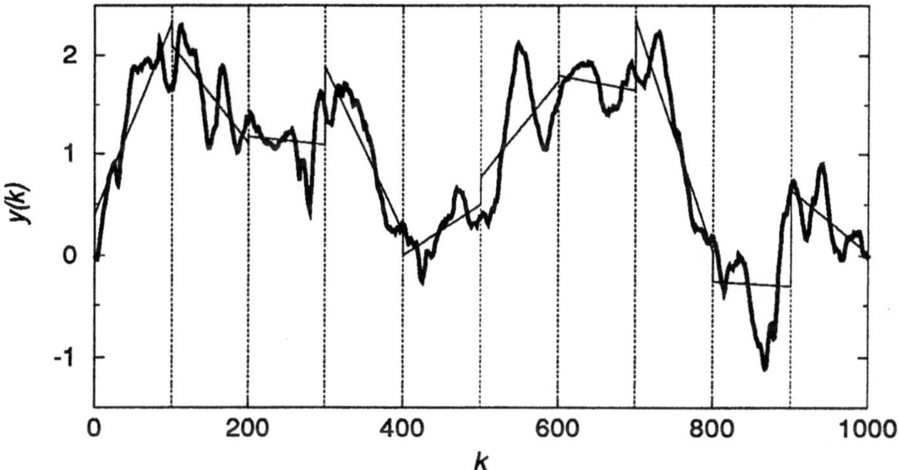

Figure 4: The integrated time series: $y(k) = \sum_{i=1}^{k}[B(i) - B_{\text{ave}}]$, where $B(i)$ is the interbeat interval shown in Fig. 1(a). The vertical dotted lines indicate box of size $n = 100$, the solid straight line segments represent the "trend" estimated in each box by a linear least-squares fit.

y coordinate of the straight line segments is denoted by $y_n(k)$. Next we detrend the integrated time series, $y(k)$, by subtracting the local trend, $y_n(k)$, in each box. The root-mean-square fluctuation of this integrated and detrended time series is calculated by

$$F(n) = \sqrt{\frac{1}{N}\sum_{k=1}^{N}[y(k) - y_n(k)]^2}. \tag{3}$$

This computation is repeated over all time scales (box sizes) to provide a relationship between $F(n)$, the average fluctuation as a function of box size, and the box size n (i.e. the number of beats in a box which is the size of the window of observation). Typically, $F(n)$ will increase with box size n. A linear relationship on a double log graph indicates the presence of scaling. Under such conditions, the fluctuations can be characterized by a scaling exponent α, the slope of the line relating $\log F(n)$ to $\log n$. Consider first a process where the value at one interbeat interval is completely uncorrelated from any previous values, e.g. white noise. This can be achieved by using a time series for which the order of the points has been shuffled (a so called "surrogate" data set). For this type of uncorrelated data, the integrated value, $y(k)$, corresponds to a random walk, and therefore $\alpha = 0.5$ [20]. If there are only short-term correlations, the initial slope may be different from 0.5, but α will approach 0.5 for large window sizes. An α greater than 0.5 and less than or equal to 1.0 indicates persistent long-range power-law correlations such that a large (compared to the average) interbeat interval is more likely to be followed by a large interval and vice versa. In contrast, $0 < \alpha < 0.5$ indicates a different type of power-law correlations such

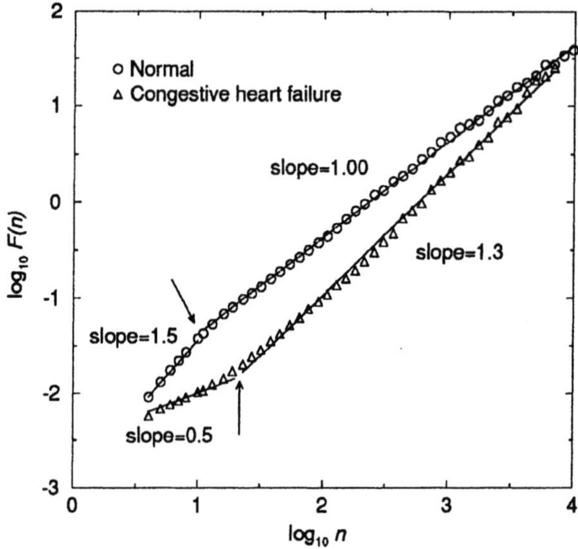

Figure 5: Plot of $\log F(n)$ vs $\log n$ (see description of DFA computation in text) for two very long interbeat interval time series ($\sim$ 24 hours). The circles are from a healthy subject while the triangles are from a subject with congestive heart failure. Arrows indicate "crossover" points in scaling. After [17].

that large and small value of the time series are more likely to alternate [13]. A special case of $\alpha = 1$ corresponds to $1/f$ noise [8, 21]. For $\alpha \geq 1$, correlations exist but cease to be of a power-law form; $\alpha = 1.5$ indicates *Brown* noise, the integration of white noise. The α exponent can also be viewed as an indicator that describes the "roughness" of the original time series: the larger the value of α, the smoother the time series. In this context, $1/f$ noise can be interpreted as a "compromise" between the complete unpredictability of white noise (very rough "landscape") and the relatively smooth "landscape" of Brownian noise [22, 23].

Figure 5 compares the DFA analysis of representative 24 hour interbeat interval time series of a healthy subject ($\bigcirc$) and a patient with congestive heart failure ($\triangle$). Notice that for large time scales (asymptotic behavior), the healthy subject interbeat interval time series shows almost perfect power-law scaling over two decades ($20 \leq n \leq 10000$) with $\alpha = 1$ (i.e., $1/f$ noise) while for the pathologic data set $\alpha \approx 1.3$ (closer to Brownian noise). This result is consistent with our previous finding that there is a significant difference in the long-range scaling behavior between healthy and diseased states (Fig. 3) [5, 6].

2.3 Normal vs. Pathologic Time Series

To test for statistical significance using the DFA method, we re-analyzed cardiac interbeat data from two different groups of subjects reported in our previous work [5]: 12 healthy adults without clinical evidence of heart disease (age range: 29–64 years, mean 44) and 15 adults with severe heart failure[2] (age range: 22–71 years; mean 56). Data from each subject consist of approximately 24 hours of ECG recording. Data from patients with heart failure due to severe left ventricular dysfunction are likely to be particularly informative in analyzing correlations under pathologic conditions since these individuals have abnormalities in both the sympathetic and parasympathetic control mechanisms [25] that regulate beat-to-beat variability. Previous studies have demonstrated marked changes in short-range heart rate dynamics in heart failure compared to healthy function, including the emergence of intermittent relatively low frequency (~ 1 cycle/minute) heart rate oscillations associated with the well-recognized syndrome of periodic (Cheyne-Stokes) respiration, an abnormal breathing pattern often associated with low cardiac output (see Fig. 3) [25].

We observe the following scaling exponents[3] (for time scales $10^2 \sim 10^4$ beats) for the group of healthy cardiac interbeat interval time series (mean value $\pm$ S.D.): $\alpha = 1.00 \pm 0.11$. This result is consistent with previous reports of $1/f$ fluctuations in healthy heart rate (by spectral analysis) [3, 26]. The pathologic group shows a significant ($p < 0.01$ by Student's t-test) deviation of the long-range correlation exponent from normal. For the group of heart failure subjects, we find that $\alpha = 1.24 \pm 0.22$. Of interest, some of the heart failure subjects show an α exponent very close to 1.5 (Brownian noise), indicating random walk-like fluctuations, also consistent with our previous findings in this group. The group-averaged exponent α is less than 1.5 for the heart failure patients, suggesting that pathologic dynamics may only transiently operate in the random walk regime or may only approach this extreme state as a limiting case. We obtained similar results when we divided the time series into three consecutive subsets (of ~ 8 hours each) and repeated the above analysis. Therefore our findings are not simply attributable to different levels of daily activities.

2.4 Crossover Phenomena

Although this asymptotic scaling exponent may serve as a useful index for selected diagnostic purposes, a drawback is that very long data sets are required (at least 24 hours) for statistically robust results. For practical purposes, clinical investigators are often interested in the possibility of using substantially shorter time series. In this regard, we note that for short time scales, there is an apparent *crossover* exhibited for the scaling behavior of both data sets (arrows in Fig. 5). For the healthy subject, the α exponent estimated from very

---

[2]ECG recordings of Holter monitor tapes were processed both manually and in a fully automated manner using our computerized beat recognition algorithm (Aristotle). Abnormal beats were deleted from each data set. The deletion has practically no effect on the DFA analysis since less than 1% of total beats were removed. Patients in the heart failure group were receiving conventional medical therapy prior to receiving an investigational cardiotonic drug; see Ref. [24].

[3]Typical regression fit shows excellent linearity of double log graph (indicated by correlation coefficient $r > 0.97$) for both groups. However, usually data from healthy subjects show even better linearity on log-log plots than data from subjects with heart disease.

small n (< 10 beats) is larger than that calculated from large n (> 10 beats). This is probably due to the fact that on very short time scales (a few beats to ten beats), the physiologic interbeat interval fluctuation is dominated by the relatively smooth heartbeat oscillation associated with respiration, thus giving rise to a large α value. For longer scales, the interbeat fluctuation, reflecting the intrinsic dynamics of a complex system, approaches that of $1/f$ behavior as previously noted. *In contrast, the pathologic data set shows a very different crossover pattern* (Fig. 5). For very short time scales, the fluctuation is quite random (close to white noise, $\alpha \approx 0.5$). As the time scale becomes larger, the fluctuation becomes smoother (asymptotically approaching Brownian noise, $\alpha \approx 1.5$). These findings are consistent with our previous report of altered correlation properties under pathologic conditions [5, 6].

2.5 Clinical Application: Preliminary Results

The above observation of a differential crossover pattern for healthy versus pathologic data motivated us to extract two parameters from each data set by fitting the scaling exponent α over two different time scales: one short, the other long. To be more precise, for each data set we calculated an exponent α_1 by making a least squares fit of $\log F(n)$ vs $\log n$ for $4 \leq n \leq 16$. Similarly, an exponent α_2 was obtained from $16 \leq n \leq 64$. Since these two exponents are not extracted from the asymptotic region, relatively short data sets are sufficient, thereby making this technique applicable to "real world" clinical data.

We applied this quantitative fluctuation analysis to the two different groups of subjects mentioned above to measure the two scaling exponents α_1 and α_2. All data set records were divided into multiple sub-sets (each with $N = 8192$ beats ~ 2 hours) and the two exponents were calculated for each subset. For healthy subjects, we find the following exponents (mean value $\pm$ S.D.) for the cardiac interbeat interval time series: $\alpha_1 = 1.201 \pm 0.178$ and $\alpha_2 = 0.998 \pm 0.124$. For the group of congestive heart failure subjects, we find that $\alpha_1 = 0.803 \pm 0.259$ and $\alpha_2 = 1.125 \pm 0.216$, both significantly ($p < 0.0001$ for both α_1 and α_2) different from normal. Furthermore, we show in Fig. 6 that fairly good discrimination between these two groups can be achieved by using these two scaling exponents. We note that not all subjects in our preliminary study show an obvious crossover in their scaling behavior. Only 8 out of 12 healthy subjects exhibited this crossover, while 11 out of 15 pathologic subjects exhibited a "reverse" crossover. However, the two scaling exponents (α_1 and α_2) measured from relatively short data sets can still be potentially useful indicators to distinguish normal from pathologic time series.

To test the effect of data length on these calculations, we repeated the same DFA measurements for longer data sets ($N = 16384$) and also for shorter data sets ($N = 4096$). As expected, the results for shorter data sets are less reliable (more overlap between two groups) due to anticipated statistical error related to finite sample size [27]. On the other hand, longer data sets result in little improvement for the distinction between groups. Therefore, the data length of

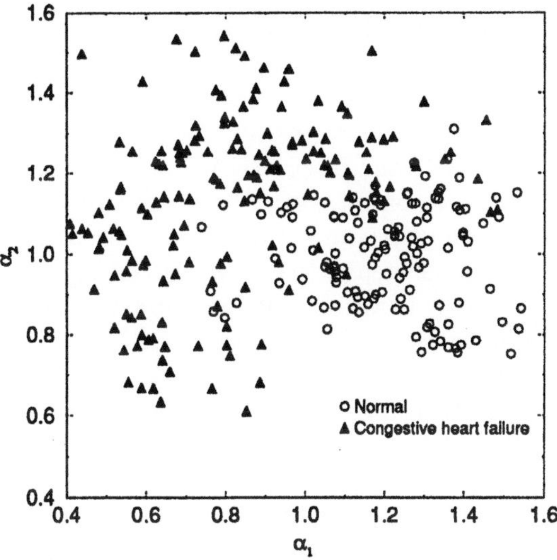

Figure 6: Scatter plot of scaling exponents α_1 vs α_2 for the healthy subjects (O) and subjects with congestive heart failure ($\triangle$). The α's were calculated from interbeat interval data sets of length 8192 beats. Longer data set records were divided into multiple data sets (each with 8192 beats). Note good separation between healthy and heart disease subjects, with clustering of points in two distinct "clouds." After [17].

8192 seems to be a statistically reasonable choice[4].

Furthermore, we note that data from normal interbeat interval time series are tightly clustered suggesting that there may exist a "universal" scaling behavior for physiologic interbeat time series. In contrast, the pathologic data show more variation, a finding which may be related to different clinical conditions and varying severity of the pathologic states.

The potential *practical* applications of DFA analysis to the assessment of patient survival using Holter monitor data was supported by a preliminary report of a prospective, population based study [28]. We found that the DFA analysis confirmed a breakdown of long-range correlations in subjects with heart failure vs. age and sex-matched controls. Furthermore, the DFA analysis appeared to add prognostic information about mortality not extractable from traditional methods of heart rate variability analysis.

---

[4] We also tested these calculations by varying the fitting range for α_2. We find that the results are very similar when we measure α_2 from 16 beats to 128 beats. However, when we move the upper fitting range for α_2 from 128 beats to 256 beats or more, the pathologic data sets show larger variation of α_2 leading to less obvious separation from normal subjects. This is partly due to the fact that, for finite length data sets, the calculation error of $F(n)$ increases with n [27]. Therefore, scaling exponents obtained over larger values of n will have greater uncertainty.

3   The Dynamics of Human Walking

Human gait is a complex process. The locomotor system incorporates input from the cerebellum, the motor cortex and the basal ganglia, as well as feedback from visual, vestibular and proprioceptive sensors [29, 30]. Under healthy conditions, this multi-level control system produces a remarkably stable walking pattern; the kinetics, kinematics and muscular activity of gait appear to remain relatively constant from one step to the next even during unconstrained walking [31, 32, 33, 34]. Actually, however, closer examination reveals fluctuations in the gait pattern, even under stationary conditions [31, 35, 36]. The origin and the implications of these fluctuations are unknown. In this section, we analyze the step-to-step fluctuations in gait in order to gain insight into locomotor function and its control mechanisms. To this end, we use the same DFA method we developed for studying the dynamics of heartbeat time series. Ultimately, these insights should increase the understanding of neurophysiological control of normal and pathological walking and might also prove useful clinically in the diagnosis and prognosis of gait disorders.

A representative stride interval time series is shown in Figure 7 (top). First, note the stability of the stride interval; during a nine minute walk the coefficient of variation was only 4 %. Thus, a good first approximation of the dynamics of the stride interval would be a constant. However, fluctuations occur about the mean. The stride interval varies irregularly with some underlying complex "structure." This structure changes after random shuffling, as seen in Figure 7, demonstrating that the original structure is a result of the sequential ordering of the stride interval and not a result of the stride interval distribution. Figure 7 (bottom) shows $F(n)$ versus n plotted on a double log graph, for the original time series and the shuffled time series. The slope of the line relating $\log F(n)$ to $\log n$ is 0.83 for the original time series and 0.50 after random shuffling. Thus, fluctuations in the stride interval scale as $F(n) \approx n^{0.83}$ exhibiting long-range correlations, while the shuffled data set behaves as uncorrelated white noise; $\alpha = 0.50$. Figure 7 (bottom) also displays the power spectrum of the original time series. The spectrum is broad band and scales as $f^{-0.92}$. The two scaling exponents are consistent with each other within statistical error due to finite data length [27] and both indicate the presence of long-range correlations.

For a group of ten healthy young (age range: 20–30 years) subjects, $\alpha = 0.76 \pm 0.11$ (mean $\pm$ standard deviation) for the original stride interval time series (range: 0.56 to 0.91) and, after random shuffling, $\alpha = 0.50 \pm 0.03$ (range: 0.45 to 0.55). The possibility that the long-range correlations were somehow due to instrumentation and measurement errors was excluded by 1) using an electro-mechanical load cell to periodically activate a footswitch, 2) computing the differences between the footswitch-based estimates of the period and the actual input period, and 3) re-calculating α from each subject's stride interval time series after adding Gaussian noise equal in magnitude to that of the measurement system noise. This procedure did not significantly change the results; re-calculated α's changed towards 0.5 by less than 5 %. More detailed discussions and model can be found in Refs. [18, 19]. Thus, similar to the interbeat

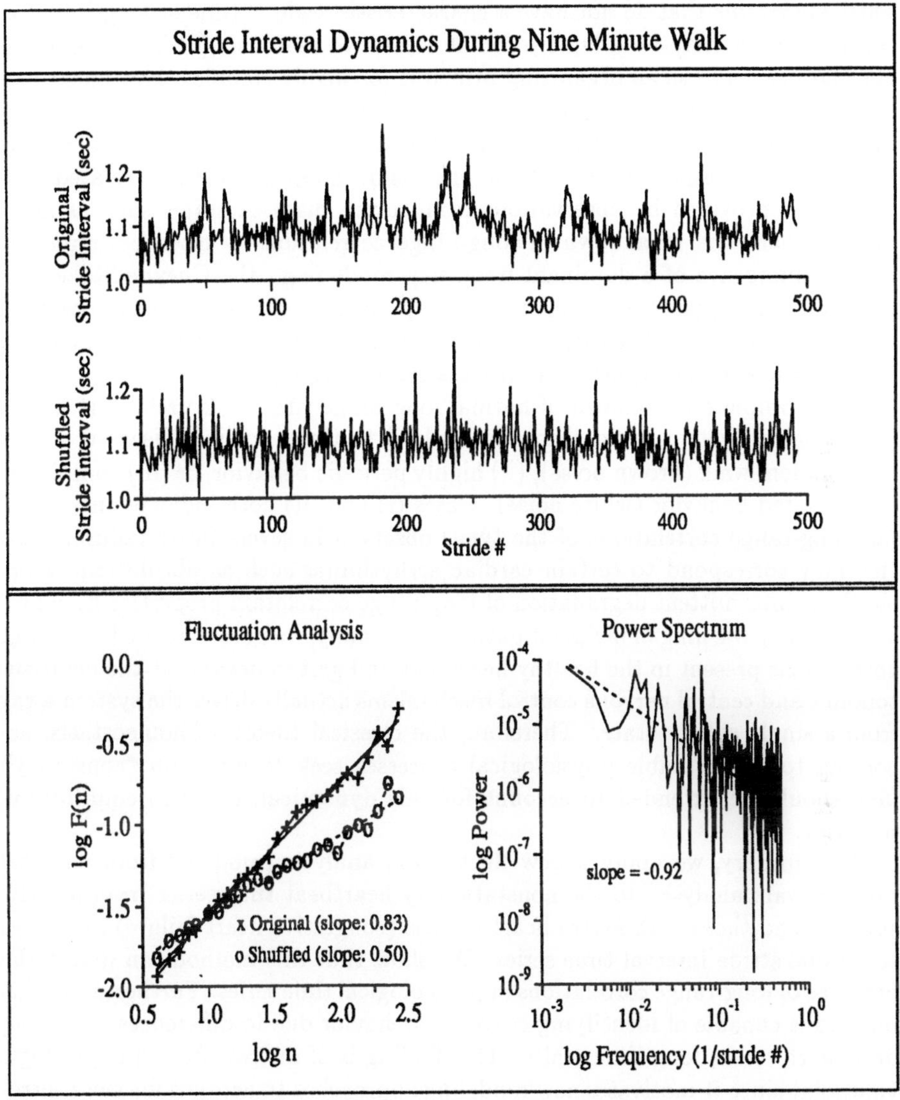

Figure 7: Representative stride interval time series before and after shuffling (above) and the fluctuation and power spectrum analysis (below). After [18].

interval time series, human interstride interval time series also display fractal long-range correlations.

4   Conclusion

Our finding of non-trivial long-range correlations in healthy heart rate and gait dynamics is consistent with the observation of long-range correlations in other biological systems that do not have a characteristic scale of time or length [2, 37]. Such behavior may be adaptive for at least two reasons. (i) The long-range correlations serve as an organizing principle for highly complex, non-linear processes that generate fluctuations on a wide range of time scales. (ii) The lack of a characteristic scale helps prevent excessive *mode-locking* that would restrict the functional responsiveness of the organism. Support for these related conjectures is provided by observations from severe diseased states such as heart failure where the breakdown of long-range correlations is often accompanied by the emergence of a dominant frequency mode (e.g., the Cheyne-Stokes frequency). Analogous transitions to highly periodic regimes have been observed in a wide range of other disease states including certain malignancies, sudden cardiac death, epilepsy and fetal distress syndromes [3].

The complete breakdown of normal long-range (fractal) correlations in any physiological system could theoretically lead to three possible diseased states: (i) a random walk (brown noise), (ii) highly periodic behavior, or (iii) completely uncorrelated behavior (white noise). Cases (i) and (ii) both indicate only "trivial" long-range correlations of the types observed in severe heart failure. Case (iii) may correspond to certain cardiac arrhythmias such as fibrillation. More subtle or intermittent degradation of long-range correlation properties may provide an early warning of incipient pathology. Finally, we note that the long-range correlations present in the healthy heartbeat and gait indicate that the neuroautonomic and central nervous control mechanisms actually drives the system away from a single steady state. Therefore, the classical theory of homeostasis, according to which stable physiological processes seek to maintain "constancy" [38], should be extended to account for this dynamical, far from equilibrium, behavior.

In summary, we apply a new fluctuation analysis (modified from classical random walk analysis) to the nonstationary heartbeat time series from healthy subjects and those with severe heart disease (congestive heart failure) as well as to normal stride interval time series. We show that this method can detect the presence of long-range correlations in physiological time series. Furthermore, this method is capable of identifying crossover behavior due to differences in scaling over short versus long time scales. This finding is of interest from a physiologic viewpoint since it motivates new modeling approaches to account for the control mechanisms regulating cardiac and neuromuscular dynamics on different time scales. From a practical point of view, quantification of these scaling exponents may have potential applications for bedside and ambulatory monitoring.

Acknowledgments

We are grateful to S. V. Buldyrev, Z. Ladin, G. B. Moody, P. Purdon, D. R. Rigney, M. Rosenblum and J. Y. Wei for valuable discussions. Partial support was provided to CKP by NIH/NIMH NRSA Postdoctoral Fellowship, to JMH by NIA, to HES and SH by NSF and to ALG by the G. Harold and Leila Y. Mathers Charitable Foundation, NIDA, and NASA. This review is mainly based on the Refs. [5, 17, 18].

References

[1] A. Bunde, S. Havlin, eds.: *Fractals in Science* (Springer-Verlag, Berlin 1994)

[2] A. L. Goldberger, D. R. Rigney and B. J. West, *Sci. Am.* **262**, 42 (1990).

[3] A. L. Goldberger and B. J. West, *Yale J. Biol. Med.* **60**, 421(1987)

[4] R. I. Kitney and O. Rompelman, *The Study of Heart-Rate Variability*, Oxford U. P., Oxford, 1980;
S. Akselrod, D. Gordon, F. A. Ubel, D. C. Shannon, A. C. Barger and R. J. Cohen, *Science* **213**, 220 (1981).

[5] C.-K. Peng, J. Mietus, J. M. Hausdorff, S. Havlin, H. E. Stanley, and A. L. Goldberger, *Phys. Rev. Lett.* **70**, 1343–1346 (1993).

[6] C.-K. Peng, S. V. Buldyrev, J. M. Hausdorff, S. Havlin, J. E. Mietus, M. Simons, H. E. Stanley, and A. L. Goldberger, in *Fractals in Biology and Medicine*, T. F. Nonnenmacher, G. A. Losa and E. R. Weibel, eds. (Birkhaüser Verlag, Basel, 1994), pp. 55–65.

[7] H. E. Stanley, *Introduction to Phase Transitions and Critical Phenomena*, (Oxford University Press, London, 1971).

[8] P. Bak, C. Tang & K. Wiesenfeld, *Phys. Rev. Lett.* **59**, 381-384 (1987).

[9] M. N. Levy, *Circ. Res.*, **29**, 437-445 (1971).

[10] P. Lévy, *Théorie de l'Addition des Variables Aléatoires* (Gauthier-Villars, Paris, 1937).

[11] R. N. Mantegna, *Physica A* **179**, 232 (1991).

[12] A. M. Yaglom, *Correlation Theory of Stationary and Related Random Functions, Vol. I & II*, (Springer-Verlag, New York, 1987).

[13] S. Havlin, R. B. Selinger, M. Schwartz, H. E. Stanley and A. Bunde, *Phys. Rev. Lett.* **61**, 1438 (1988).

[14] C.-K. Peng, S. V. Buldyrev, S. Havlin, M. Simons, H. E. Stanley and A. L. Goldberger, *Phys. Rev. E* **49**, 1691–1695 (1994).

[15] S. V. Buldyrev, A. L. Goldberger, S. Havlin, C.-K. Peng, H. E. Stanley and M. Simons, *Biophysical Journal* **65**, 2675–2681 (1993).

[16] S. M. Ossadnik, S. V. Buldyrev, A. L. Goldberger, S. Havlin, R. N. Mantegna, C.-K. Peng, M. Simons, and H. E. Stanley, *Biophysical Journal* **67**, 64–70 (1994).

[17] C.-K. Peng, S. Havlin, H. E. Stanley, and A. L. Goldberger, *Chaos* (in press).

[18] J. M. Hausdorff, C.-K. Peng, Z. Ladin, J. Y. Wei and A. L. Goldberger, *J. Appl. Physiol.* **78**, 349–358 (1995).

[19] J. M. Hausdorff, P. Purdon, C.-K. Peng, Z. Ladin, J. Y. Wei and A. L. Goldberger (preprint).

[20] E. W. Montroll and M. F. Shlesinger, in *Nonequilibrium Phenomena II. From Stochastics to Hydrodynamics*, eds. J. L. Lebowitz and E. W. Montroll, pp. 1-121 (North-Holland, Amsterdam, 1984).

[21] W. H. Press, *Comments Astrophys.*, **7**, 103–119 (1978).

[22] C.-K. Peng, S. Buldyrev, A. L. Goldberger, S. Havlin, F. Sciortino, M. Simons and H. E. Stanley, *Nature* **356**, 168-170 (1992).

[23] S. V. Buldyrev, A. L. Goldberger, S. Havlin, C.-K. Peng and H. E. Stanley, in *Fractals in Science*, A. Bunde and S. Havlin, eds (Springer-Verlag, Berlin, 1994), 49–83.

[24] D. S. Baim et al., *J. Am. Coll. Cardiol.* **7**, 661–670 (1986).

[25] A. L. Goldberger, D. R. Rigney, J. Mietus, E. M. Antman & S. Greenwald, *Experientia* **44**, 983–987 (1988).

[26] M. Kobayashi and T. Musha, *IEEE Trans. Biomed. Eng.* **29**, 456 (1982).

[27] C.-K. Peng, S. V. Buldyrev, A. L. Goldberger, S. Havlin, M. Simons, and H. E. Stanley, *Phys. Rev. E* **47**, 3730–3733 (1993).

[28] K. K. L. Ho, G. B. Moody, C.-K. Peng, J. E. Mietus, M. G. Larson, A. L. Goldberger, and D. Levy, *Circulation* **90**, I-330 (1994).

[29] T. A. McMahon, *Muscles, reflexes, and locomotion*, Princeton: Princeton University Press, 1984.

[30] D. A. Winter, *The biomechanics and motor control of human gait*, Waterloo, Canada: University of Waterloo Press, 1987.

[31] J. Pailhous and M. Bonnard, *Behav Brain Res* **47**, 181-190 (1992).

[32] D. A. Winter, *Hum. Movement Sci.* **3**, 51-76 (1984).

[33] A. E. Palta, *J Motor Behav* **17**, 443-461 (1985).

[34] M. P. Kadaba, H. K. Ramakrishnan, M. E. Wootten, J. Gainey, G. Gorton and G. V. B. Cochran, *J Orthop Res* **7**, 849-860 (1989).

[35] M. Yamasaki, T. Sasaki and M, Torii, *Eur J Appl Phys* **62**, 99-103 (1991).

[36] M. Yamasaki, T. Sasaki, S. Tsuzki and M. Torii, *Ann Physiol Anthrop* **3**, 291-296 (1984).

[37] E. R. Weibel, *Am. J. Physiol.* **261**, L361 (1991); J. B. Bassingthwaighte and R. P. Beyer, *Physica D* **53**, 71 (1991).

[38] W. B. Cannon, *Physiol. Rev.* **9**, 399 (1929).

Long-Range Correlations and Generalized Lévy Walks in DNA Sequences

H. E. Stanley,[1] S. V. Buldyrev,[1] A. L. Goldberger,[3,4] S. Havlin,[1,2]
R. N. Mantegna,[1,5] C.-K. Peng,[1,3] M. Simons[3] and M. H. R. Stanley[1]

[1]Center for Polymer Studies and Department of Physics, Boston University, Boston, MA

[2]Department of Physics, Bar Ilan University, Ramat Gan, ISRAEL

[3]Cardiovascular Div., Harvard Medical School, Beth Israel Hospital, Boston, MA

[4]Department of Biomedical Engineering, Boston University, Boston, MA

[5]Dipartimento di Energetica ed Applicazioni di Fisica, Palermo University, Palermo, I-90128, Italy

1 Long-Range Power-Law Correlations

In recent years long-range power-law correlations have been discovered in a remarkably wide variety of systems. Such long-range power-law correlations are a physical fact that in turn gives rise to the increasingly appreciated "fractal geometry of nature" [1–12]. So if fractals are indeed so widespread, it makes sense to anticipate that long-range power-law correlations may be similarly widespread. Indeed, recognizing the ubiquity of long-range power-law correlations can help us in our efforts to understand nature, since as soon as we find power-law correlations we can quantify them with a critical exponent. Quantification of this kind of scaling behavior for apparently unrelated systems allows us to recognize similarities between different systems, leading to underlying unifications that might otherwise have gone unnoticed.

Traditionally, investigators in many fields characterize processes by assuming that correlations decay exponentially. However, there is one major exception: at the critical point, the exponential decay turns into to a power law decay [13]

$$C_r \sim (1/r)^{d-2+\eta}. \tag{1}$$

Many systems drive themselves spontaneously toward critical points [2, 14]. One of the simplest models exhibiting such "self-organized criticality" is invasion percolation, a generic model that has recently found applicability to describing anomalous behavior of rough interfaces.

In the following sections we will attempt to summarize some recent findings [15–35] concerning the possibility that—under suitable conditions—the sequence of base pairs or "nucleotides" in DNA also displays power-law correlations. The underlying basis of such power law correlations is not understood at present, but this discovery has intriguing implications for molecular evolution [32], as well as potential practical applications for distinguishing coding and noncoding regions in long nucleotide chains [34]. It also may be related to the presence of a language in noncoding DNA [36].

2 DNA

The role of genomic DNA sequences in coding for protein structure is well known [37]. The human genome contains information for approximately 100,000 different proteins, which define all inheritable features of an individual. The genomic sequence is likely the most sophisticated information database created by nature through the dynamic process of evolution. Equally remarkable is the precise transformation of information (duplication, decoding, etc) that occurs in a relatively short time interval.

The building blocks for coding this information are called *nucleotides*. Each nucleotide contains a phosphate group, a deoxyribose sugar moiety and either a *purine* or a *pyrimidine base*. Two purines and two pyrimidines are found in DNA. The two purines are adenine (A) and guanine (G); the two pyrimidines are cytosine (C) and thymine (T). The nucleotides are linked end to end, by chemical bonds from the phosphate group of one nucleotide to the deoxyribose sugar group of the adjacent nucleotide, forming a long polymer (*polynucleotide*) chain. The information content is encoded in the sequential order of the bases on this chain. Therefore, as far as the information content is concerned, a DNA sequence can be most simply represented as a symbolic sequence of four letters: A, C, G and T.

In the genomes of high eukaryotic organisms only a small portion of the total genome length is used for protein coding (as low as 3% in the human genome). The segments of the chromosomal DNA that are spliced out during the formation of a mature mRNA are called *introns* (for intervening sequences). The coding sequences are called *exons* (for expressive sequences).

The role of introns and intergenomic sequences constituting large portions of the genome remains unknown. Furthermore, only a few quantitative methods are currently available for analyzing information which is possibly encrypted in the noncoding part of the genome.

3 The "DNA Walk"

One interesting question that may be asked by statistical physicists would be whether the sequence of the nucleotides A,C,G, and T behaves like a one-dimensional "ideal gas", where the fluctuations of density of certain particles obey Gaussian law, or if there exist long range correlations in nucleotide content (as in the vicinity of a critical point). These result in domains of all size with different nucleotide concentrations. Such domains of various sizes were known for a long time but their origin and statistical properties remain unexplained. A natural language to describe heterogeneous DNA structure is long-range correlation analysis, borrowed from the theory of critical phenomena [13].

In order to study the scale-invariant long-range correlations of a DNA sequence, we first introduced a graphical representation of DNA sequences, which we term a *fractal landscape* or *DNA walk* [15]. For the conventional one-dimensional random walk model [38, 39], a walker moves either "up" [$u(i) = +1$]

or "down" $[u(i) = -1]$ one unit length for each step i of the walk. For the case of an uncorrelated walk, the direction of each step is independent of the previous steps. For the case of a correlated random walk, the direction of each step depends on the history ("memory") of the walker [40–42].

One definition of the DNA walk is that the walker steps "up" if a pyrimidine (C or T) occurs at position i along the DNA chain, while the walker steps "down" if a purine (A or G) occurs at position i. The question we asked was whether such a walk displays only short-range correlations (as in an n-step Markov chain) or long-range correlations (as in critical phenomena and other scale-free "fractal" phenomena).

There have also been attempts to map DNA sequence onto multi-dimensional DNA walks [16, 43]. However, recent work [34] indicates that the original purine-pyrimidine rule provides the most robust results, probably due to the purine-pyrimidine chemical complementarity.

The DNA walk allows one to visualize directly the fluctuations of the purine-pyrimidine content in DNA sequences: Positive slopes correspond to high concentration of pyrimidines, while negative slopes correspond to high concentration of purines. Visual observation of DNA walks suggests that the coding sequences and intron-containing noncoding sequences have quite different landscapes.

4 Correlations and Fluctuations

An important statistical quantity characterizing any walk [38, 39] is the root mean square fluctuation $F(\ell)$ about the average of the displacement of a quantity $\Delta y(\ell)$ defined by $\Delta y(\ell) \equiv y(\ell_0+\ell)-y(\ell_0)$. If there is no characteristic length (i.e., if the correlation were "infinite-range"), then fluctuations will also be described by a power law

$$F(\ell) \sim \ell^\alpha \tag{2}$$

with $\alpha \neq 1/2$.

Figure 1a shows a typical example of a gene that contains a significant fraction of base pairs that do *not* code for amino acids. It is immediately apparent that the DNA walk has an extremely jagged contour which corresponds to long-range correlations.

The fact that data for intron-containing and intergenic (i.e., noncoding) sequences are linear on this double logarithmic plot confirms that $F(\ell) \sim \ell^\alpha$. A least-squares fit produces a straight line with slope α substantially larger than the prediction for an uncorrelated walk, $\alpha = 1/2$, thus providing direct experimental evidence for the presence of long-range correlations.

On the other hand, the dependence of $F(\ell)$ for coding sequences is not linear on the log-log plot: its slope undergoes a crossover from 0.5 for small ℓ to 1 for large ℓ. However, if a single patch is analyzed separately, the log-log plot of $F(\ell)$ is again a straight line with the slope close to 0.5. This suggests that within a large patch the coding sequence is almost uncorrelated.

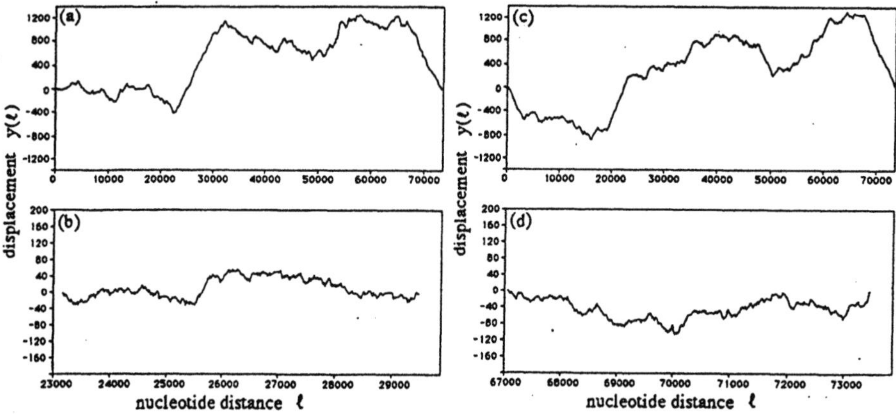

Figure 1: DNA walk displacement $y(\ell)$ (excess of purines over pyrimidines) vs nucleotide distance ℓ for (a) HUMHBB (human beta globin chromosomal region of the total length $L = 73,239$); (b) the LINE1c region of HUMHBB starting from 23,137 to 29,515; (c) the generalized Lévy walk model of length 73,326 with $\mu = 2.45$, $l_c = 10$, $\alpha_o = 0.6$, and $\epsilon = 0.2$; and (d) a segment of a Lévy walk of exactly the same length as the LINE1c sequence from step 67,048 to the end of the sequence. This sub-segment is a Markovian random walk. Note that in all cases the overall bias was subtracted from the graph such that the beginning and ending points have the same vertical displacement ($y = 0$). This was done to make the graphs clearer and does not affect the quantitative analysis of the data.

5 Lévy Walk Model and its Generalization

Although the correlation is long-range in the non-coding sequences, there seems to be a paradox: long *uncorrelated* regions of up to thousands of base-pairs can be found in such sequences as well. For example, consider the human beta-globin intergenomic sequence of length $L = 73,326$ (GenBank name: HUMHBB). This long non-coding sequence has 50% purines (no *overall* strand bias) and $\alpha = 0.7$ (see Fig. 1(a). However, from nucleotide #67,089 to #73,228, there occurs the LINE-1 region (defined in Ref. [44]. In this region of length 6139 base pairs, there is a strong strand bias with 59% *purines*. In this non-coding sub-region, we find power-law scaling of F, with $F \sim l^{\alpha}$, with $\alpha = 0.55$, quite close to that of a random walk.

Even more striking is another region of 6378 base pairs, from nucleotide #23,137 to #29,515, which has 59% *pyrimidines* and is *uncorrelated*, with remarkably good power-law scaling and correlation exponent $\alpha = 0.49$ (Fig. 1(b)). This region actually consists of three sub-sequences, complementary to shorter parts of the LINE-1 sequence.

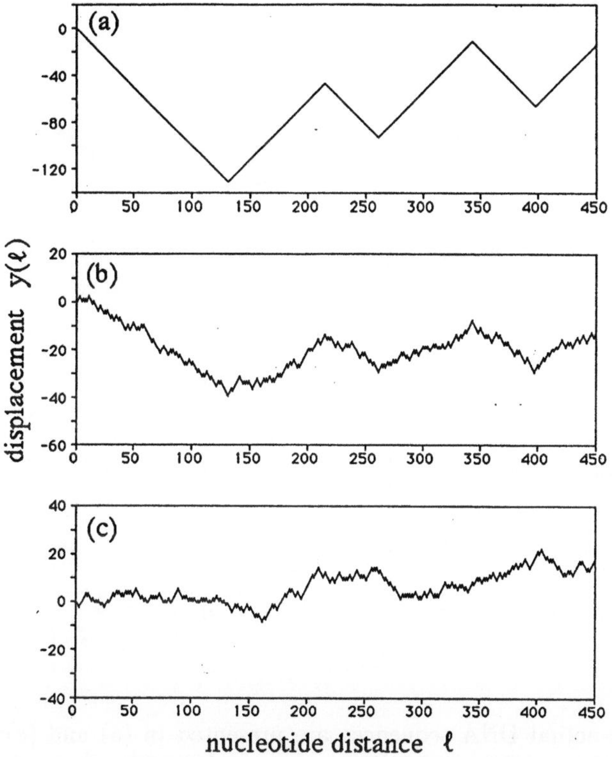

Figure 2: Displacement $y(\ell)$ vs number of steps for (a) the classical Lévy walk model consisting of 6 strings of l_j steps, each taken in alternating directions; (b) the generalized Lévy walk model consisting of 6 biased random walks of the same length with a probability of p_+ that it will go up equal to $(1 \pm \epsilon)/2$ [$\epsilon = 0.2$]; and (c) the unbiased uncorrelated random walk. Note that the vertical scale in (b) and (c) is twice that in (a).

These features motivated us to apply a generalized Lévy walk model (see Figs. 1c, 1d and 2) for the non-coding regions of DNA sequences [30]. We will show in the next section how this model can explain the long-range correlation properties, since there is no characteristic scale "built into" this generalized Lévy walk. In addition, the model simultaneously accounts for the observed large subregions of non-correlated sequences within these non-coding DNA chains.

The classic Lévy walk model describes a wide variety of diverse phenomena that exhibit long-range correlations [45–48]. The model is defined schematically in Fig. 2a: A random walker takes not one but l_1 steps in a given direction. Then the walker takes l_2 steps in a new randomly-chosen direction, and so forth. The lengths l_j of each string are chosen from a probability distribution, with

$$P(l_j) \propto (1/l_j)^\mu, \tag{3}$$

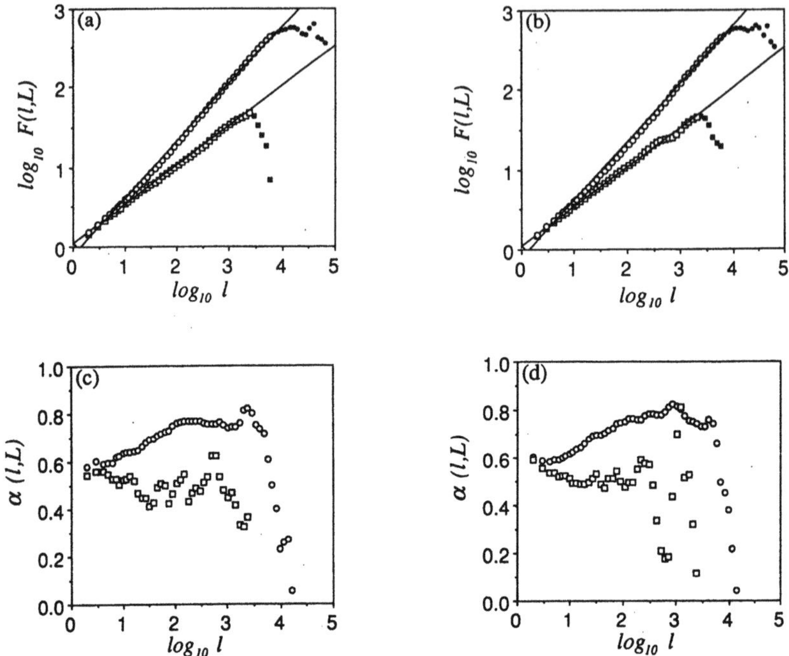

Figure 3: The actual DNA sequences are presented in (a) and (c): the entire HUMHBB sequence (○) and LINE1c sequence (□). The slopes for the linear fits are 0.72 and 0.49 respectively. The Lévy model sequences are presented in (b) and (d): the entire Lévy walk sequence of Fig. 1c (○), a segment of this walk of Fig. 1d (□). The slopes for the linear fits are 0.73 and 0.49 respectively.

where $\sum_{i=1}^{N} l_i = L$, N is the number of sub-strings and L is the total number of steps that the random walker takes.

We consider a generalization of the Lévy walk [42] to interpret recent findings of long-range correlation in non-coding DNA sequences described above. Instead of taking l_j steps in the *same* direction as occurs in a classic Lévy walk, the walker takes each of l_j steps in *random* directions, with a fixed bias probability

$$p_+ = (1 + \epsilon_j)/2 \tag{4}$$

to go up and

$$p_- = (1 - \epsilon_j)/2 \tag{5}$$

to go down, where ϵ_j gets the values $+\epsilon$ or $-\epsilon$ randomly. Here $0 \leq \epsilon \leq 1$ is a bias parameter (the case $\epsilon = 1$ reduces to the Lévy walk). Fig. 2b shows such a generalized Lévy walk for the same choice of l_j as in Fig. 2a.

As shown in Ref. [30], the generalized Lévy walk—like the pure Lévy walk—gives rise to a landscape with a fluctuation exponent α that depends upon the

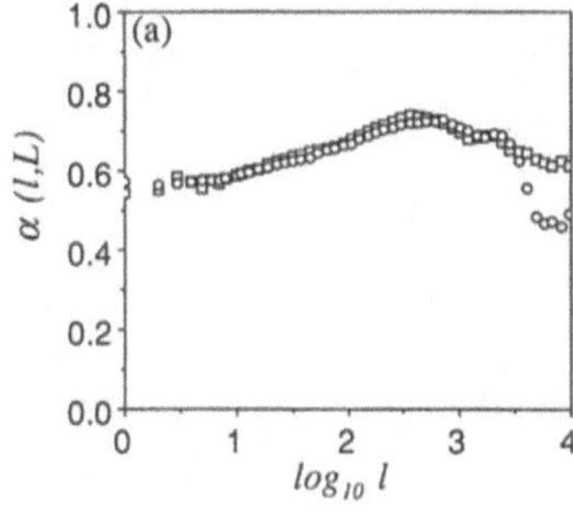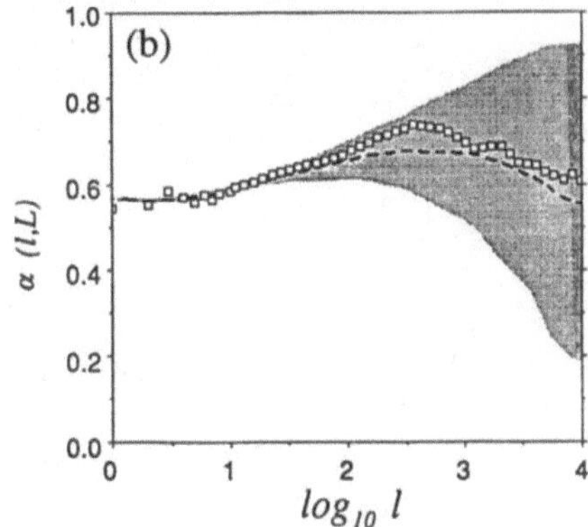

Figure 4: Comparison of successive slopes of the scaling exponent α for yeast chromosome III ($\square$) and (a) successive slopes of a realization of the generalized Lévy walk with parameters $L = 315,000$, $\mu = 2.5$, $l_c = 5$, $\alpha_o = 0.55$, $\epsilon = 0.16$ ($\bigcirc$); (b) average successive slopes over 15 different realization of Lévy walks with the same parameters (dashed line). The shaded area corresponds to two standard deviations of successive slopes of the model, calculated for 15 random realizations. The parameters for Markov process, α_o and ϵ, used in the model are calculated from real DNA sequence of yeast chromosome III.

Lévy walk parameter μ [42, 46],

$$\alpha = \begin{cases} 1 & \mu \leq 2 \\ 2 - \mu/2 & 2 < \mu < 3 \\ 1/2 & \mu \geq 3. \end{cases} \tag{6}$$

i.e., non-trivial behavior of α corresponds to the case $2 < \mu < 3$ where the first moment of $P(l_j)$ converges while the second moment diverges. The long-range correlation property for the Lévy walk, in this case, is a consequence of the broad distribution of Eq. (3) that lacks of a characteristic length scale. However, for $\mu \geq 3$, the distribution of $P(l_j)$ decays fast enough that an effective characteristic length scale appears. Therefore, the resulting Lévy walk behaves like a normal random walk for $\mu \geq 3$.

To be precise, we define our generalized L-step Lévy walk model as follows:

(1) Choose a random number u which is uniformly distributed between 0 and 1, and define $l_j \equiv l_c u^{\mu-1}$ where l_c is some lower cutoff characteristic length. The number (l_j) thus generated will obey the distribution of Eq. (3).

(2) Produce a biased random walk of length l_j (see Ref. [30]) with p_+ and p_- given by Eqs. (4) and (5), where ϵ_j takes on the value $+\epsilon$ or $-\epsilon$ randomly and ϵ is a fixed value close to 0.2 (corresponding to the percentage of purines vs. pyrimidines in real DNA sequences)

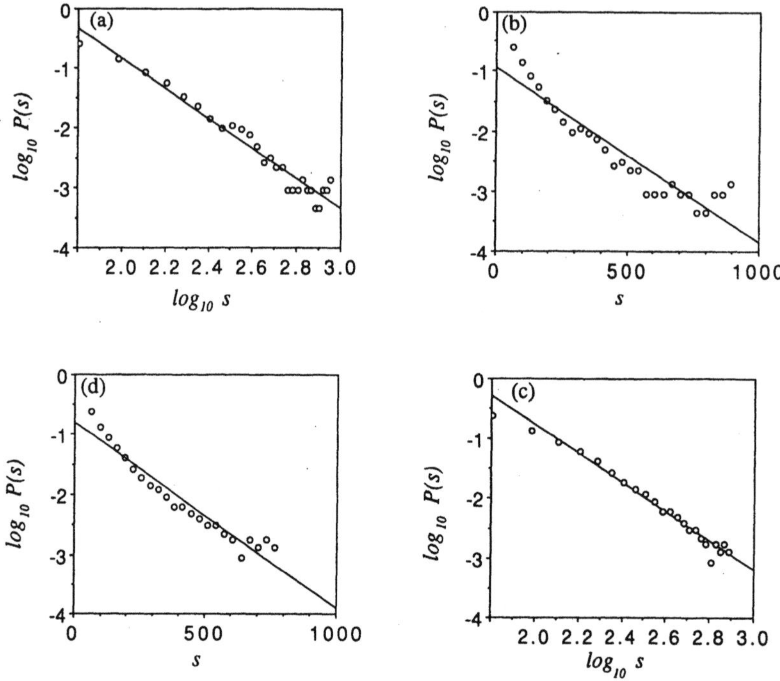

Figure 5: The probability distribution to find a run of certain size s of purines or pyrimidines in the coarse-grained sequence calculated using coarse-grained window size equal to 32. (a) Actual sequence of HUMHBB on log-log plot. (b) Actual sequence of HUMHBB on semi-log plot. (c) Log-log plot for the model sequence, shown in Fig. 1c. (d) Semi-log plot for the model.

(3) Iterate the process, attaching together biased random walk until the total length of the sequence reaches a given value L.

6 Comparison with DNA Data

To test the generalized Lévy walk model, we have adjusted the two parameters, μ and l_c, described in the previous section to best approximate features of an actual DNA sequence the human beta-globin DNA sequence shown in Fig. 1(a). The resulting landscape for the generalized Lévy walk model is presented in Fig. 1(c). The comparison of $F(\ell)$ for the model and DNA sequences is shown in Figs. 3a and 3b.

A more detailed scaling analysis (Figs. 3c, 3d), considers the "local slopes" of successive points in the graphs of Figs. 3a, 3b

$$\alpha(\ell_i, L) \equiv \frac{\log F(\ell_{i+1}, L) - \log F(\ell_i, L)}{\log \ell_{i+1} - \log \ell_i}, \tag{7}$$

where ℓ_{i+1} and ℓ_i are values of two subsequent data points. The local slope changes from $\alpha = 0.6$ for $\ell_1 = 1$ to $\alpha = 0.75$ for $\ell_i = 128$, and stays at this value for about two decades. It eventually drops down when ℓ_i becomes too close to L, since $F(L, L) \equiv 0$ according to Eq. (2.1). This kind of scaling behavior is general for all kinds of DNA sequences that contain non-coding material. The initial monotonic increase in α, however, does not mean that long-range correlations do not exist. Indeed, as seen in Fig. 3d, a similar type of behavior exists in the generalized Lévy walk model. Equation (6) is valid asymptotically for very large ℓ and L and the local value of $\alpha(\ell, L)$ for finite values of ℓ and L may differ considerably from its asymptotic value. The comparison of $\alpha(\ell, L)$ plots for human beta-globin chromosomal region ($L = 73326$) and a Lévy walk model of the same size is made for one of the largest available ($L = 315357$) DNA sequences [20, 49], that of Yeast chromosome III (see Fig. 4a).

For any given size L, it is possible to calculate the average value and standard deviation of $\alpha(\ell, L)$ for the Lévy walk model by calculating $\alpha(\ell, L)$ for a large number k of statistically independent realizations of the model sequence of the size L. The data for yeast chromosome III is well within a 2 standard deviation interval ($k = 15$) for the generalized Lévy walk model with $\mu = 2.5$, which corresponds to observed value of $\alpha = 0.75$ (Fig. 4b).

An alternative test of Lévy walk structure can be made if one analyzes a "coarse-grained" version of the original DNA sequence. To this end, we (i) divide the entire sequence into L/w sub-sequences of equal length w, (ii) replace each sub-sequence by 1 if there is an excess of purines or by 0 if there is an excess of pyrimidines and, (iii) calculate the distribution $P(s)$ of sizes s of long runs of ones and zeroes. These calculations for human beta-globin chromosomal region show that $P(s)$ has a scaling region of roughly one decade, where $P(s) \sim s^{-\mu}$ with $\mu \approx 2.5$. Our results are in good agreement with the value of the exponent $\alpha = 0.75$ (see Fig. 5). Unfortunately, the coarse-graining process requires a long sequence ($> 10^5$ nucleotides) in order that the statistics for the distribution be meaningful. To date, only a few documented long sequences are available, but as longer sequences become available this renormalization test should prove to be increasingly useful.

7   "Mosaic" Nature of DNA

The key finding of this analysis is that a generalized Lévy walk model can account for two hitherto unexplained features of DNA nucleotides: (i) the long-range power law correlations that extend over thousands of nucleotides in sequences containing non-coding regions (e.g., genes with introns and intergenomic sequences), and (ii) the presence within these correlated sequences of sometimes large sub-regions that correspond to biased random walks. This apparent paradox is resolved by the generalized Lévy walk, a mechanism for generating long-range correlations (no characteristic length scale), that with finite (though rare) probability also generates large regions of uncorrelated strand bias. The uncorrelated sub-regions, therefore, are an anticipated feature of this mechanism for long-range correlations.

From a biological viewpoint, two questions immediately arise: (i) What is the significance of these uncorrelated sub-regions of strand bias? and (ii) What is the molecular basis underlying the power-law statistics of the Lévy walk? With respect to the first question, we note that these long uncorrelated regions at least sometimes correspond to well-described but poorly understood sequences termed "repetitive elements", such as the LINE1 region noted above [44, 50]. There are at least 53 different families of such repetitive elements within the human genome. The lengths of these repetitive elements vary from 10 to 10^4 nucleotides [44]. At least some of the repetitive elements are believed to be remnants of messenger RNA molecules that formerly did code for proteins [50, 51, 52]. Alternatively, these segments may represent retroviral sequences that have inserted themselves into the genome [53]. Our finding that these repetitive elements have the statistical properties of biased random walks (e.g., the same as that of active coding sequences) is consistent with these hypotheses.

Finally, what are the biological implications of this type of analysis? Our findings clearly support the following possible hypothesis concerning the molecular basis for the power-law distributions of elements within DNA chains. In order to be inserted into DNA, a macromolecule should form a loop of certain length l with two ends, separated by l nucleotides along the sequence, coming close to each other in real space. The probability of finding a loop of length l inside a very long linear polymer scales as $l^{-\mu}$ [54, 55]. Theoretical estimates of μ made by different methods [55–58] using a self-avoiding random walk model [54] indicate that the value of μ for three-dimensional model is between 2.16 and 2.42. Our estimate made by the Rosenbluth Monte-Carlo Method [59] gave $\mu = 2.22 \pm 0.05$ which yields according to Eq. (6) $\alpha = 0.89$, a larger value than the effective value of $\alpha(\ell, L)$, observed in DNA of finite length. However, the asymptotic value of the exponent α remains uncertain since the statistics of Lévy walks converge very slowly due to rare events associated with the very long strings of constant bias that may occur in the sequence according to Eq. (3). This results in the very large error bars for $\alpha(\ell, L)$ for large values of ℓ and finite length L (see Fig. 4). Even for the sequences of about 300K base-pairs we cannot estimate the limiting value of α with good accuracy. It is clear, however, that the behavior of DNA sequences cannot be satisfactorily explained in terms of only one characteristic length scale even of about $10^3 - 10^4$ base pairs long. The asymptotic behavior of the scaling exponent α and whether it reaches some universal value for long DNA chains must await further data from the Human Genome Project.

Recently, a report appeared that confirms the existence of long-range correlations in DNA [25]. However, where Ref. [25] might appear to disagree with Ref. [15] is in the interpretation of that finding for coding and non-coding regions. Both figures in [25] apply to the complete genome of the phage λ which does not contain non-coding sequences and consists of only three regions of different strand bias (see Fig. 1c of Ref. [15]. Each such region when analyzed separately by the DNA-walk method gives exponent $\alpha \approx 0.5$, close to that of random walk. The combination of three such regions produces a crossover in the local values of $\alpha(l, L) \approx 0.5$ at small length scales l to $\alpha(l, L) \approx 1$ at large l. Thus, for coding sequences, there is indeed no well-defined scaling exponent α for large length-scales.

In contrast, the monotonically increasing local values of $\alpha(l, L)$ followed by a plateau at large l for non-coding sequences are completely explained by the generalized Lévy walk model presented here in terms of a crossover from an uncorrelated random walk at small length scales to a Lévy walk at large length scales. The latter has well-defined scaling with an exponent α related to the exponent μ characterizing the power law distribution of steps of the Lévy walk. Figure 3d of the present work clearly demonstrates that the generalized Lévy walk model accounts for the upward curvature in the values of $\alpha(l, L)$, followed by a plateau with $\alpha(l, L) \approx 2 - \mu/2$ [17, 27].

8 "Linguistic" Analysis

Long-range correlations have been found recently in human writings [60]. A novel, a piece of music or a computer program can be regarded as a one-dimensional string of symbols. These strings can be mapped to a one-dimensional random walk model similar to the DNA walk allowing calculation of the correlation exponent α. Values of α between 0.6 and 0.9 were found for various texts.

An interesting hierarchical feature of languages was found in 1949 by Zipf [61]. He observed that the frequency of words as a function of the word order decays as a power law (with a power close to -1) for more than four orders of magnitude.

In order to adapt the Zipf analysis to DNA, the concept of word must first be defined. In the case of coding regions, the words are the 64 3-tuples ("triplets") which code for the amino acids, AAA, AAT, ... GGG. However for non-coding regions, the words are not known. Therefore Ref. [36] considers the word length n as a free parameter, and performs analyses not only for $n = 3$ but also for all values of n in the range 3 through 8. The different n-tuples are obtained for the DNA sequence by shifting progressively by 1 base a window of length n; hence, for a DNA sequence containing L base pairs, we obtain $L - n + 1$ different words.

The results of the Zipf analysis for all 40 DNA sequences analyzed are summarized in Ref. [36]. The averages for each category support the observation that ζ is consistently larger for the non-coding sequences, suggesting that the non-coding sequences bear more resemblance to a natural language than the coding sequences.

Related interesting statistical measures of short-range correlations in languages are the entropy and redundancy. The redundancy is a manifestation of the *flexibility* of the underlying code. To quantitatively characterize the redundancy implicit in the DNA sequence, we utilize the approach of Shannon, who provided a mathematically precise definition of redundancy [62, 63]. Shannon's redundancy is defined in terms of the entropy of a text—or, more precisely, the

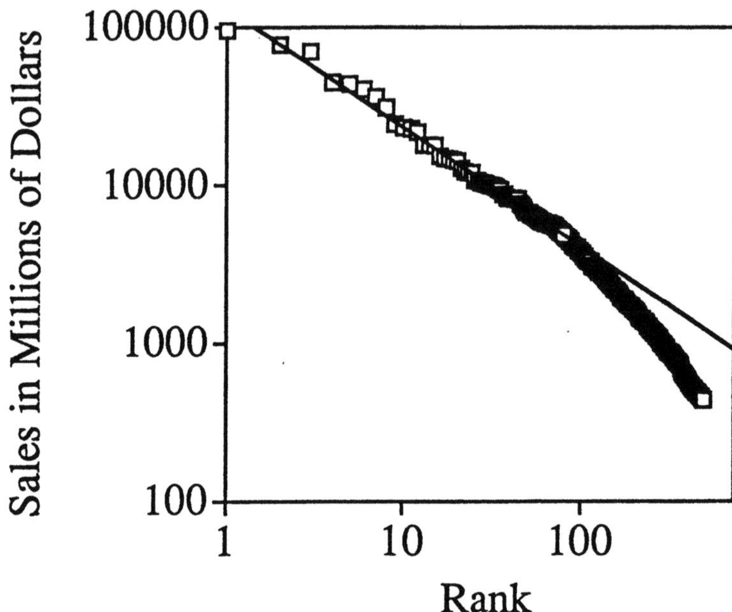

Figure 6: A Zipf plot using the *Fortune 500* for 1993 [68]. On the y-axis of this log-log plot is the sales of a firm in 1985 dollars. On the x-axis is the rank of that firm. The straight line is fit to the first 100 firms. One can see that the first approximately 100 firms are well fit by a straight line, but after approximately firm rank 100, the plot is no longer a straight line. After [66].

"n-entropy"

$$H(n) = -\sum_{i=1}^{4^n} p_i log_2 p_i, \tag{8}$$

which is the entropy when the text is viewed as a collection of n-tuple words. The redundancy is defined through as $R \equiv \lim_{n \to \infty} R(n)$, where

$$R(n) \equiv 1 - H(n)/kn; \tag{9}$$

here $k = \log_2 4 = 2$.

Reference [36] calculates the Shannon n-entropy $H(n)$ for $n = 1, 2, \cdots 6$. The maximum value of n for which it is possible to determine $H(n)$ is $n = 6$—even for very long sequences (e.g., *C. elegans*, 2.2 million nucleotides)—due to the extremely slow convergence to the final value. For shorter sequences, reliable values of $H(n)$ are obtainable only up to a value of n less than 6.

For sufficiently high values of n (for example $n = 4$), we found that the redundancy is consistently larger for the primarily non-coding sequences. In fact, for most of the sequences consisting primarily of coding regions, we find that $R(n)$ is quite close to the value $R(n) = 0$ which we find for a control sequence of random numbers.

In summary, Ref. [36] finds that *non-coding* sequences show two similar statistical properties to those of both natural and artificial languages: (a) Zipf-like scaling behavior, and (b) a non-zero value of Shannon's redundancy function $R(n)$. These results are consistent with the *possible* existence of one (or more than one) structured biological languages present in non-coding DNA sequences.

It appears that linearity of a Zipf plot is generally indicative of hierarchical ordering. For example, it is possible that a wide range of systems result in straight-line behavior when subjected to Zipf analysis [64]. An example that was the subject of some discussion at this meeting is the remarkable linearity of the Zipf plot giving the annual sales of a company as a function of its sales rank. J.P. Bouchaud [65] finds that this plot is linear for European companies, while M.H.R. Stanley [66] finds linearity for American companies (Fig. 6). Furthermore, M.H.R. Stanley et al. [67] find a significant deviation from this apparent linearity at rank ≈ 100, and relate this feature to the log-normal distribution of sales (the "Gibrat law").

9 Summary

There is a mounting body of evidence suggesting that the noncoding regions of DNA are rather special for at least two reasons:

1. They display long-range power-law correlations, as opposed to previously-believed exponentially-decaying correlations.

2. They display features common to hierarchically-structured languages—specifically, a linear Zipf plot and a non-zero redundancy.

These results are consistent with the possibility that the noncoding regions of DNA are not merely "junk" but rather have a purpose. What that purpose could be is the subject of ongoing investigation. In particular, the apparent increase of α with evolution [32] could provide insight.

In the event that the purpose is not profound, our results nonetheless may have important practical value since quantifiable differences between coding and noncoding regions of DNA can be used to help distinguish the coding regions [34].

10 Acknowledgements

We are grateful to many individuals, including M.E. Matsa, S.M. Ossadnik, M.A. Salinger, and F. Sciortino, for major contributions to those results reviewed here that represent collaborative research efforts. We also wish to thank C. Cantor, C. DeLisi, M. Frank-Kamenetskii, A.Yu. Grosberg, G. Huber, I. Labat, L. Liebovitch, G.S. Michaels, P. Munson, R. Nossal, R. Nussinov, R.D.

Rosenberg, J.J. Schwartz, M. Schwartz, E.I. Shakhnovich, M.F. Shlesinger, N. Shworak, and E.N. Trifonov for valuable discussions. Partial support was provided by the National Science Foundation, NIH, the G. Harold and Leila Y. Mathers Charitable Foundation, the National Heart, Lung and Blood Institute, the National Aeronautics and Space Administration, the Israel-USA Binational Science Foundation, and (to C-KP) by an NIH/NIMH Postdoctoral NRSA Fellowship.

References

[1] B.B. Mandelbrot: *The Fractal Geometry of Nature* (W.H. Freeman, San Francisco 1982)

[2] A. Bunde, S. Havlin, eds.: *Fractals and Disordered Systems* (Springer-Verlag, Berlin 1991) A. Bunde, S. Havlin, eds.: *Fractals in Science* (Springer-Verlag, Berlin 1994); T. Vicsek, M. Shlesinger, M. Matsushita, eds.: *Fractals in Natural Sciences* (World Scientific, Singapore, 1994)

[3] J.M. Garcia-Ruiz, E. Louis, P. Meakin, L. Sander, eds.: *Growth Patterns in Physical Sciences and Biology* [Proc. 1991 NATO Advanced Research Workshop, Granada, Spain, October 1991], (Plenum, New York, 1993)

[4] A.Yu. Grosberg, A.R. Khokhlov: *Statistical Physics of Macromolecules*, translated by Y. A. Atanov (AIP Press, New York, 1994)

[5] J.B. Bassingthwaighte, L.S. Liebovitch, B.J. West: *Fractal Physiology* (Oxford University Press, New York, 1994)

[6] A.-L. Barabási, H.E. Stanley: *Fractal Concepts in Surface Growth* (Cambridge University Press, Cambridge, 1995)

[7] B.J. West, A.L. Goldberger: J. Appl. Physiol., **60**, 189 (1986); B.J. West, A.L. Goldberger: Am. Sci., **75**, 354 (1987); A.L. Goldberger, B.J. West: Yale J. Biol. Med. **60**, 421 (1987); A.L. Goldberger, D.R. Rigney, B.J. West: Sci. Am. **262**, 42 (1990); B.J. West, M.F. Shlesinger: Am. Sci. **78**, 40 (1990); B.J. West: *Fractal Physiology and Chaos in Medicine* (World Scientific, Singapore 1990); B.J. West, W. Deering: Phys. Reports **246**, 1 (1994); S.V. Buldyrev, A.L. Goldberger, S. Havlin, C.-K. Peng, H.E. Stanley: in *Fractals in Science*, edited by A. Bunde and S. Havlin (Springer-Verlag, Berlin, 1994), 49–83

[8] T. Vicsek: *Fractal Growth Phenomena, Second Edition* (World Scientific, Singapore 1992)

[9] J. Feder: *Fractals* (Plenum, NY, 1988)

[10] D. Stauffer, H.E. Stanley: *From Newton to Mandelbrot: A Primer in Theoretical Physics* (Springer-Verlag, Heidelberg & N.Y. 1990)

[11] E. Guyon, H.E. Stanley: *Les Formes Fractales* (Palais de la Découverte, Paris 1991); **English translation:** *Fractal Forms* (Elsevier North Holland, Amsterdam 1991)

[12] H.E. Stanley, N. Ostrowsky, eds.: *Random Fluctuations and Pattern Growth: Experiments and Models*, Proceedings 1988 Cargèse NATO ASI (Kluwer Academic Publishers, Dordrecht, 1988)

[13] H.E. Stanley: *Introduction to Phase Transitions and Critical Phenomena* (Oxford University Press, London 1971)

[14] H.E. Stanley, N. Ostrowsky, eds.: *Correlations and Connectivity: Geometric Aspects of Physics, Chemistry and Biology*, Proceedings 1990 Cargèse Nato ASI, Series E: Applied Sciences (Kluwer, Dordrecht 1990)

[15] C.-K. Peng, S.V. Buldyrev, A.L. Goldberger, S. Havlin, F. Sciortino, M. Simons, H.E. Stanley: Nature **356**, 168 (1992)

[16] W. Li, K. Kaneko: Europhys. Lett. **17**, 655 (1992)

[17] S. Nee: Nature **357**, 450 (1992)

[18] R. Voss: Phys. Rev. Lett. **68**, 3805 (1992); R. Voss: Fractals **2**, 1 (1994)

[19] J. Maddox: Nature **358**, 103 (1992)

[20] P.J. Munson, R.C. Taylor, G.S. Michaels: Nature **360**, 636 (1992)

[21] I. Amato: Science **257**, 747 (1992)

[22] V.V. Prabhu, J.-M. Claverie: Nature **357**, 782 (1992)

[23] P. Yam: Sci. Am. **267**[3], 23 (1992)

[24] C.-K. Peng, S.V. Buldyrev, A.L. Goldberger, S. Havlin, F. Sciortino, M. Simons, H.E. Stanley: Physica A **191**, 25 (1992); H.E. Stanley, S.V. Buldyrev, A.L. Goldberger, J.M. Hausdorff, S. Havlin, J. Mietus, C.-K. Peng, F. Sciortino, M. Simons: Physica A **191**, 1 (1992)

[25] C.A. Chatzidimitriou-Dreismann, D. Larhammar: Nature **361**, 212 (1993); D. Larhammar, C.A. Chatzidimitriou-Dreismann: Nucleic Acids Res. **21**, 5167 (1993) C.A. Chatzidimitriou-Dreismann, R.M.F. Streffer, D. Larhammar: Biochim. Biophys. Acta **1217**, 181 (1994); C.A. Chatzidimitriou-Dreismann, R.M.F. Streffer, D. Larhammar: Eur. J. Biochem. **224**, 365 (1994)

[26] A.Yu. Grosberg, Y. Rabin, S. Havlin, A. Neer: Europhys. Lett. **23**, 373 (1993)

[27] S. Karlin, V. Brendel: Science **259**, 677 (1993)

[28] C.-K. Peng, S.V. Buldyrev, A.L. Goldberger, S. Havlin, M. Simons, H.E. Stanley: Phys. Rev. E **47**, 3730 (1993)

[29] N. Shnerb, E. Eisenberg: Phys. Rev. E **49**, R1005 (1994)

[30] S.V. Buldyrev, A.L. Goldberger, S. Havlin, C.-K. Peng, M. Simons, H.E. Stanley: Phys. Rev. E **47**, 4514 (1993).

[31] A. S. Borovik, A. Yu. Grosberg and M. D. Frank Kamenezki, J. Biomolec. Structure and Dynamics **xx**, xxx (1994)

[32] S.V. Buldyrev, A.L. Goldberger, S. Havlin, C.-K. Peng, H.E. Stanley, M.H.R. Stanley, M. Simons: Biophys. J. **65**, 2673 (1993)

[33] C.-K. Peng, S.V. Buldyrev, S. Havlin, M. Simons, H.E. Stanley, A.L. Goldberger: Phys. Rev. E **49**, 1685 (1994)

[34] S.M. Ossadnik, S.V. Buldyrev, A.L. Goldberger, S. Havlin, R.N. Mantegna, C.-K. Peng, M. Simons, H.E. Stanley: Biophys. J. **67**, 64 (1994); H.E. Stanley, S.V. Buldyrev, A.L. Goldberger, S. Havlin, C.-K.Peng, M. Simons: [Proceedings of Internat'l Conf. on Condensed Matter Physics, Bar-Ilan], Physica A **200**, 4 (1993); H.E. Stanley, S.V. Buldyrev, A.L. Goldberger, S. Havlin, S.M. Ossadnik, C.-K. Peng, M. Simons: Fractals **1**, 283-301 (1993)

[35] S.V. Buldyrev, A.L. Goldberger, S. Havlin, R.N. Mantegna, M.E. Matsa, C.-K. Peng, M. Simons, and H.E. Stanley, "Long-Range Correlation Properties of Coding and Noncoding DNA Sequences," Phys. Rev. E (submitted).

[36] R.N. Mantegna, S.V. Buldyrev, A.L. Goldberger, S. Havlin, C.-K. Peng, M. Simons, H.E. Stanley: Phys. Rev. Lett. **73**, 3169-3172 (1994); F. Flam: Science **266**, 1320 (1994); E. Pennisi: Science News **146**, 391 (1994)

[37] S. Tavaré, B.W. Giddings, in: *Mathematical Methods for DNA Sequences*, Eds. M.S. Waterman (CRC Press, Boca Raton 1989), pp. 117-132; J.D. Watson, M. Gilman, J. Witkowski, M. Zoller: *Recombinant DNA* (Scientific American Books, New York 1992).

[38] E.W. Montroll, M.F. Shlesinger: "The Wonderful World of Random Walks" in: *Nonequilibrium Phenomena II. From Stochastics to Hydrodynamics*, ed. by J.L. Lebowitz, E.W. Montroll (North-Holland, Amsterdam 1984), pp. 1–121

[39] G.H. Weiss: *Random Walks* (North-Holland, Amsterdam 1994)

[40] S. Havlin, R. Selinger, M. Schwartz, H.E. Stanley, A. Bunde: Phys. Rev. Lett. **61**, 1438 (1988); S. Havlin, M. Schwartz, R. Blumberg Selinger, A. Bunde, H.E. Stanley: Phys. Rev. A **40**, 1717 (1989); R.B. Selinger, S. Havlin, F. Leyvraz, M. Schwartz, H.E. Stanley: Phys. Rev. A **40**, 6755 (1989)

[41] C.-K. Peng, S. Havlin, M. Schwartz, H.E. Stanley, G.H. Weiss: Physica A **178**, 401 (1991); C.-K. Peng, S. Havlin, M. Schwartz, H.E. Stanley: Phys. Rev. A **44**, 2239 (1991)

[42] M. Araujo, S. Havlin, G.H. Weiss, H.E. Stanley: Phys. Rev. A **43**, 5207 (1991); S. Havlin, S.V. Buldyrev, H.E. Stanley, G.H. Weiss: J. Phys. A **24**, L925 (1991); S. Prakash, S. Havlin, M. Schwartz, H.E. Stanley: Phys. Rev. A **46**, R1724 (1992)

[43] C.L. Berthelsen, J.A. Glazier, M.H. Skolnick: Phys. Rev. A **45**, 8902 (1992)

[44] J. Jurka, T. Walichiewicz, A. Milosavljevic: J. Mol. Evol. **35**, 286 (1992)

[45] M.F. Shlesinger, J. Klafter: in *On Growth and Form: Fractal and Non-Fractal Patterns in Physics*, edited by H.E. Stanley and N. Ostrowsky (Martinus Nijhoff, Dordrecht, 1986), p. 279ff

[46] M.F. Shlesinger, J. Klafter, Y.M. Wong: J. Stat. Phys. **27**, 499 (1982)

[47] M.F. Shlesinger, J. Klafter: Phys. Rev. Lett. **54**, 2551 (1985)

[48] R.N. Mantegna: Physica A **179**, 232 (1991)

[49] The long-range correlations were found to extend over the entire yeast chromosome III region (315,357 nucleotides)—see Ref. [20]. The yeast chromosome III sequence was published by S.G. Oliver et al.: Nature **357**, 38 (1992)

[50] J. Jurka: J. Mol. Evol. **29**, 496 (1989)

[51] R.H. Hwu, J.W. Roberts, E.H. Davidson, R.J. Britten: Proc. Natl. Acad. Sci. USA. **83**, 3875 (1986)

[52] E. Zuckerkandl, G. Latter, J. Jurka: J. Mol. Evol. **29**, 504 (1989)

[53] B. Levin: *Genes IV* (Oxford University Press, Oxford, 1990)

[54] P.-G. de Gennes: *Scaling Concepts in Polymer Physics* (Cornell University Press, Ithaca NY, 1979)

[55] J. de Cloiseaux: J. Physique (Paris) **41**, 223 (1980), p. 223

[56] S. Redner: J. Phys. A **13**, 3525 (1980)

[57] A. Baumgartner: Z. Phys. B **42**, 265 (1981)

[58] H.S. Chan, K.A. Dill: J. Chem. Phys. **92**, 3118 (1990)

[59] T. M. Birshtein, S. V. Buldyrev: Polymer **32**, 3387 (1991)

[60] A. Schenkel, J. Zhang, Y-C. Zhang: Fractals **1**, 47 (1993); M. Amit, Y. Shmerler, E. Eisenberg, M. Abraham, N. Shnerb: Fractals **2**, 7 (1994)

[61] G.K. Zipf: *Human Behavior and the Principle of "Least Effort"* (Addison-Wesley, New York 1949)

[62] L. Brillouin: *Science and Information Theory* (Academic Press, New York 1956)

[63] C.E. Shannon: Bell Systems Tech. J. **80**, 50 (1951)

[64] A. Czirok, R. Mantegna, S. Havlin, H.E. Stanley: Phys. Rev. E (submitted)

[65] J.-P. Bouchaud: "More Lévy distributions in physics", These Proceedings

[66] M.H.R. Stanley: 1994 Westinghouse Report (unpublished)

[67] M.H.R. Stanley, S.V. Buldyrev, S. Havlin, R. Mantegna, M.A. Salinger, H.E. Stanley: Eco. Lett. (submitted)

[68] J. Pivinski, R. Tucksmith, A. Such, C. Haight: Fortune (18 April 1994), p. 224

Lecture Notes in Physics

For information about Vols. 1–415
please contact your bookseller or Springer-Verlag

Vol. 416: B. Baschek, G. Klare, J. Lequeux (Eds.), New Aspects of Magellanic Cloud Research. Proceedings, 1992. XIII, 494 pages. 1993.

Vol. 417: K. Goeke P. Kroll, H.-R. Petry (Eds.), Quark Cluster Dynamics. Proceedings, 1992. XI, 297 pages. 1993.

Vol. 418: J. van Paradijs, H. M. Maitzen (Eds.), Galactic High-Energy Astrophysics. XIII, 293 pages. 1993.

Vol. 419: K. H. Ploog, L. Tapfer (Eds.), Physics and Technology of Semiconductor Quantum Devices. Proceedings, 1992. VIII, 212 pages. 1993.

Vol. 420: F. Ehlotzky (Ed.), Fundamentals of Quantum Optics III. Proceedings, 1993. XII, 346 pages. 1993.

Vol. 421: H.-J. Röser, K. Meisenheimer (Eds.), Jets in Extragalactic Radio Sources. XX, 301 pages. 1993.

Vol. 422: L. Päivärinta, E. Somersalo (Eds.), Inverse Problems in Mathematical Physics. Proceedings, 1992. XVIII, 256 pages. 1993.

Vol. 423: F. J. Chinea, L. M. González-Romero (Eds.), Rotating Objects and Relativistic Physics. Proceedings, 1992. XII, 304 pages. 1993.

Vol. 424: G. F. Helminck (Ed.), Geometric and Quantum Aspects of Integrable Systems. Proceedings, 1992. IX, 224 pages. 1993.

Vol. 425: M. Dienes, M. Month, B. Strasser, S. Turner (Eds.), Frontiers of Particle Beams: Factories with e⁺ e⁻ Rings. Proceedings, 1992. IX, 414 pages. 1994.

Vol. 426: L. Mathelitsch, W. Plessas (Eds.), Substructures of Matter as Revealed with Electroweak Probes. Proceedings, 1993. XIV, 441 pages. 1994

Vol. 427: H. V. von Geramb (Ed.), Quantum Inversion Theory and Applications. Proceedings, 1993. VIII, 481 pages. 1994.

Vol. 428: U. G. Jørgensen (Ed.), Molecules in the Stellar Environment. Proceedings, 1993. VIII, 440 pages. 1994.

Vol. 429: J. L. Sanz, E. Martínez-González, L. Cayón (Eds.), Present and Future of the Cosmic Microwave Background. Proceedings, 1993. VIII, 233 pages. 1994.

Vol. 430: V. G. Gurzadyan, D. Pfenniger (Eds.), Ergodic Concepts in Stellar Dynamics. Proceedings, 1993. XVI, 302 pages. 1994.

Vol. 431: T. P. Ray, S. Beckwith (Eds.), Star Formation and Techniques in Infrared and mm-Wave Astronomy. Proceedings, 1992. XIV, 314 pages. 1994.

Vol. 432: G. Belvedere, M. Rodonò, G. M. Simnett (Eds.), Advances in Solar Physics. Proceedings, 1993. XVII, 335 pages. 1994.

Vol. 433: G. Contopoulos, N. Spyrou, L. Vlahos (Eds.), Galactic Dynamics and N-Body Simulations. Proceedings, 1993. XIV, 417 pages. 1994.

Vol. 434: J. Ehlers, H. Friedrich (Eds.), Canonical Gravity: From Classical to Quantum. Proceedings, 1993. X, 267 pages. 1994.

Vol. 435: E. Maruyama, H. Watanabe (Eds.), Physics and Industry. Proceedings, 1993. VII, 108 pages. 1994.

Vol. 436: A. Alekseev, A. Hietamäki, K. Huitu, A. Morozov, A. Niemi (Eds.), Integrable Models and Strings. Proceedings, 1993. VII, 280 pages. 1994.

Vol. 437: K. K. Bardhan, B. K. Chakrabarti, A. Hansen (Eds.), Non-Linearity and Breakdown in Soft Condensed Matter. Proceedings, 1993. XI, 340 pages. 1994.

Vol. 438: A. Pȩkalski (Ed.), Diffusion Processes: Experiment, Theory, Simulations. Proceedings, 1994. VIII, 312 pages. 1994.

Vol. 439: T. L. Wilson, K. J. Johnston (Eds.), The Structure and Content of Molecular Clouds. 25 Years of Molecular Radioastronomy. Proceedings, 1993. XIII, 308 pages. 1994.

Vol. 440: H. Latal, W. Schweiger (Eds.), Matter Under Extreme Conditions. Proceedings, 1994. IX, 243 pages. 1994.

Vol. 441: J. M. Arias, M. I. Gallardo, M. Lozano (Eds.), Response of the Nuclear System to External Forces. Proceedings, 1994, VIII. 293 pages. 1995.

Vol. 442: P. A. Bois, E. Dériat, R. Gatignol, A. Rigolot (Eds.), Asymptotic Modelling in Fluid Mechanics. Proceedings, 1994. XII, 307 pages. 1995.

Vol. 443: D. Koester, K. Werner (Eds.), White Dwarfs. Proceedings, 1994. XII, 348 pages. 1995.

Vol. 444: A. O. Benz, A. Krüger (Eds.), Coronal Magnetic Energy Releases. Proceedings, 1994. X, 293 pages. 1995.

Vol. 445: J. Brey, J. Marro, J. M. Rubí, M. San Miguel (Eds.), 25 Years of Non-Equilibrium Statistical Mechanics. Proceedings, 1994.

Vol. 446: V. Rivasseau (Ed.), Constructive Physics. Results in Field Theory, Statistical Mechanics and Condensed Matter Physics. Proceedings, 1994. X, 337 pages. 1995.

Vol. 447: G. Aktaş, C. Saçlıoğlu, M. Serdaroğlu (Eds.), Strings and Symmetries. Proceedings, 1994. XIV, 389 pages. 1995.

Vol. 448: P. L. Garrido, J. Marro (Eds.), Third Granada Lectures in Computational Physics. Proceedings, 1994. XIV, 346 pages. 1995.

Vol. 449: J. Buckmaster, T. Takeno (Eds.), Modeling in Combustion Science. Proceedings, 1994. X, 369 pages. 1995.

Vol. 450: M. F. Shlesinger, G. M. Zaslavsky, U. Frisch (Eds.), Lévy Flights and Related Topics in Physics. Proceedigs, 1994. XIV, 347 pages. 1995.

Vol. 452: A. M. Bernstein, B. R. Holstein (Eds.), Chiral Dynamics: Theory and Experiment. Proceedings, 1994. VIII, 351 pages. 1995.

Vol. 453: S. M. Deshpande, S. S. Desai, R. Narasimha (Eds.), Fourteenth International Conference on Numerical Methods in Fluid Dynamics. Proceedings, 1994. XIII, 588 pages. 1995.

Vol. 454: J. Greiner, H. W. Duerbeck, R. E. Gershberg (Eds.), Flares and Flashes, Germany 1994. XXII, 477 pages. 1995.

New Series m: Monographs